Advances in Environmental Science

D.C. Adriano and W. Salomons, Editors

Acidic Precipitation

Volume 3

Sources, Deposition, and Canopy Interactions

Edited by S.E. Lindberg, A.L. Page, and S.A. Norton

Springer-Verlag
New York Berlin Heidelberg
London Paris Tokyo Hong Kong

Volume Editors:

S.E. Lindberg
Environmental Sciences Division
Oak Ridge National Laboratory
P.O. Box 2008
Oak Ridge, TN 37831
USA

A.L. Page
Department of Soil and
Environmental Sciences
University of California
Riverside, CA 92521
USA

S.A. Norton
Department of Geological Sciences
University of Maine
Orono, ME 04469
USA

Library of Congress Cataloging-in-Publication Data

Acidic precipitation.
(Advances in environmental science)
Bibliography: v. 1, p.
Includes index.
Contents: v. 1. Case studies/volume editors,
D.C. Adriano and M. Havas—v. 2. Biological and
ecological effects/D.C. Adriano and A.H. Johnson,
editors—v. 3. Sources, deposition, and canopy interactions/S.E. Lindberg, A.L. Page, and S.A. Norton, editors.
I. Series.
TD19.5.42.A25 1989 363.7'386 88-37418

Printed on acid-free paper.

Typeset by McFarland Graphics and Design, Dillsburg, Pennsylvania.

9 8 7 6 5 4 3 2 1

ISBN 978-1-4612-8781-0 ISBN 978-1-4612-4454-7 (eBook)
DOI 10.1007/978-1-4612-4454-7

Preface to the Series

In 1986, my colleague Prof. Dr. W. Salomons of the Institute for Soil Fertility of the Netherlands and I launched the new *Advances in Environmental Science* with Springer-Verlag New York, Inc. Immediately, we were faced with a task of what topics to cover. Our strategy was to adopt a thematic approach to address hotly debated contemporary environmental issues. After consulting with numerous colleagues from Western Europe and North America, we decided to address *Acidic Precipitation,* which we view as one of the most controversial issues today.

This is the subject of the first five volumes of the new series, which cover relationships among emissions, deposition, and biological and ecological effects of acidic constituents. International experts from Canada, the United States, Western Europe, as well as from several industrialized countries in other regions, have generously contributed to this subseries, which is grouped into the following five volumes:

Volume 1 *Case Studies*
(D.C. Adriano and M. Havas, editors)

Volume 2 *Biological and Ecological Effects*
(D.C. Adriano and A.H. Johnson, editors)

Volume 3 *Sources, Deposition, and Canopy Interactions*
(S.E. Lindberg, A.L. Page, and S.A. Norton, editors)

Volume 4 *Soils, Aquatic Processes, and Lake Acidification*
(S.A. Norton, S.E. Lindberg, and A.L. Page, editors)

Volume 5 *International Overview and Assessment*
(T. Bresser and W. Salomons, editors)

From the vast amount of consequential information discussed in this series, it will become apparent that acidic deposition should be seriously addressed by many countries of the world, in as much as severe damages have already been inflicted on numerous ecosystems. Furthermore, acidic constituents have also been shown to affect the integrity of structures of great historical values in various places of the world. Thus, it is hoped that this up-to-date subseries would increase the "awareness" of the world's citizens and encourage governments to devote more attention and resources to address this issue.

The series editors thank the international panel of contributors for bringing this timely series into completion. We also wish to acknowledge the very insightful input of the following colleagues: Prof. A.L. Page of the University of California, Prof. T.C. Hutchinson of the University of Toronto, and Dr. Steve Lindberg of the Oak Ridge National Laboratory.

We also wish to thank the superb effort and cooperation of the volume editors in handling their respective volumes. The constructive criticisms of chapter reviewers also deserve much appreciation. Finally, we wish to convey our appreciation to my secretary, Ms. Brenda Rosier, and my technician, Ms. Claire Carlson, for their very able assistance in various aspects of this series.

Aiken, South Carolina

Domy C. Adriano
Coordinating Editor

Preface to *Acidic Precipitation*, Volume 3 (*Advances in Environmental Science*)

As a result of pioneering research in the 1960s and because of the real and perceived environmental effects described during the ensuing years, the term *acidic precipitation* has become commonplace in scientific and popular literature. In the last decade, governments throughout the world have responded to public pressure and to the concerns of the scientific community by establishing research programs on national and international scales. These programs have been designed to improve our understanding of the important links between atmospheric emissions and their potential environmental effects in both industrialized and developing nations. Acidic precipitation was studied initially because of its effects on aquatic systems. However, because reports from Western Europe in the early 1980s suggested a link with forest decline, acidic precipitation is now considered a potential environmental stress in terrestrial systems as well as in aquatic systems.

As has been the case with many environmental issues of the twentieth century, acidic precipitation has its origin in emissions to the atmosphere of numerous compounds from both natural and man-made sources. This volume, the third in the series *Advances in Environmental Science: Acidic Precipitation*, emphasizes the atmospheric aspects of acidic precipitation and all that this term has come to include (e.g., toxic gases such as ozone, trace metals, aluminum, and oxides of nitrogen). It progresses from emissions of the precursors of acidic precipitation to their eventual deposition on environmental surfaces.

The chapters in this volume describe the sources of acidic and basic airborne substances, their interactions in the atmosphere and with rain droplets, and their reactions with other airborne constituents such as aluminum and other metals. Also discussed are the use of metals as tracers of sources of the precursors of acidic precipitation and as tracers of historical deposition rates, the processes controlling the removal of airborne material as dry deposition and deposition interactions with the forest canopy, and past and future trends in atmospheric emissions and options for their abatement. With the National Acidic Precipitation Assessment Program of the United States nearing the 1990 completion date, and with the programs of Canada and many European countries accelerating to reach a consensus on the role that atmospheric emissions and acidic precipitation play in the environment, publication of this series is timely.

The editors thank the contributors to this volume for their efforts in describing a wide array of atmospheric topics, all of which are important to an understanding of the acidic precipitation issue.

Oak Ridge, Tennessee *Steven E. Lindberg*

Riverside, California *Albert L. Page*

Orono, Maine *Stephen A. Norton*

Contents

Contributors

Cliff I. Davidson, Departments of Civil Engineering and Engineering & Public Policy, Carnegie Mellon University, Pittsburgh, PA 15213, USA

Peter J. Dillon, Ontario Ministry of the Environment, Dorset Research Centre, Dorset, Ontario P0A 1E0, Canada

E.C. Ellis, Southern California Edison Company, System Planning & Research, P.O. Box 800, Rosemead, CA 91770, USA

R.E. Erbes, Geoscience Consultants, Ltd., 500 Copper Avenue NW, Suite 200, Albuquerque, NM 87102, USA

R. Douglas Evans, Trent Aquatic Research Centre, Trent University, Peterborough, Ontario K9J 7B8, Canada

J.K. Grott, Southern Company Services, P.O. Box 2625, Birmingham, AL 35202, USA

Jeffrey S. Kahl, Department of Geological Sciences, University of Maine, Orono, ME 04469, USA

Gregory Mierle, Ontario Ministry of the Evnironment, Dorset Research Centre, Dorset, Ontario P0A 1E0, Canada

C.E. Murphy, Jr., Westinghouse Savannah River Co., Savannah River Laboratory, Aiken, SC 29802, USA

Stephen A. Norton, Department of Geological Sciences, University of Maine, Orono, ME 04469, USA

Jozef M. Pacyna, Norwegian Institute for Air Research, P.O. Box 64, 2001 Lillestrøm, Norway

William A. Reiners, Department of Botany, University of Wyoming, Laramie, WY 82071, USA

Douglas A. Schaefer, Syracuse University, Department of Civil Engineering, 220 Hinds Hall, Syracuse, NY 13244, USA.

J.T. Sigmon, Department of Environmental Science, University of Virginia, Charlottesville, VA 22903, USA

Roger L. Tanner, Environmental Chemistry Division, Department of Applied Science, Brookhaven National Laboratory, Upton, NY 11973, USA

John W. Winchester, Department of Oceanography, Florida State University, Tallahassee, FL 32306, USA

Yee-Lin Wu, Departments of Civil Engineering and Engineering & Public Policy, Carnegie Mellon University, Pittsburgh, PA 15213, USA

Sources of Acids, Bases, and Their Precursors in the Atmosphere

Roger L. Tanner*

Abstract

A summary of the sources of acidic and basic substances in the atmosphere is given. The primary combustion sources and secondary homogenous and heterogenous pathways by which acidic sulfate aerosols and gaseous nitric and hydrochloric acids enter or are formed in the atmosphere are delineated, and emission sources and fates of the principal atmospheric base, ammonia, are discussed. Homogenous reactions of SO_2 and NO_2 with OH radicals are important for sulfuric and nitric acid formation, respectively, but aqueous-phase oxidation reactions also convert dissolved S(IV) to sulfuric acid within typical cloud and rain droplet lifetimes. Evidence for rapid reaction of ammonia emissions with the principal atmospheric acids (H_2SO_4 and HNO_3) is reviewed. Discussion then follows concerning how these reactions combine with boundary layer mixing processes to establish vertical gradients in the mixing ratios of acidic aerosols and gases.

I. Introduction and Definitions

This chapter summarizes the currently known facts concerning the sources of acids and acidic precursors in the atmosphere. This topic has been the subject of extensive research and monitoring efforts in the past decade, as concerns have escalated about the effects of acidic materials on marine and terrestrial ecosystems and on man-made materials. Indeed, the National Acid Precipitation Assessment Program (NAPAP) in the United Stated was mandated (among other things) to investigate primary atmospheric sources of acids and how these acids are formed in the atmosphere from emitted precursors. This chapter is thus intended to be an introduction to the research literature so that interested readers may obtain more details on sources and occurrences of atmospheric acids and bases.

---

*Environmental Chemistry Division, Department of Applied Science, Brookhaven National Laboratory, Upton, NY 11973, USA.

A. Acids and Bases

Acids and bases are defined by chemists and other measurement scientists in accordance with Bronsted–Lowry theory in terms of their propensity to donate or accept hydrated protons in aqueous solution. This method of definition is readily extended to donation or acceptance of solvated protons in nonaqueous solvents. An *acid* is thereby a substance with a tendency to dissociate into a hydrated proton and its conjugate base in solution. As the concentrations of hydrated protons (abbreviated H^+) in solution may vary over many orders of magnitude, chemists usually use a logarithmic scale to express H^+ as pH, where $pH = -\log[H^+]$.

A strong acid is one in which the equilibrium of the dissociation Reaction 1 (shown for water as the solvent) lies far to the right.

$$HA + H_2O \longleftrightarrow H_3O^+ + A^- \tag{1}$$

This means that the strong acid, HA, is essentially all dissociated in atmospheric water, and the free acid (i.e., hydrated proton) concentration in solution equals the added HA concentration. Weak acids are only partially dissociated in atmospheric liquid water, and the degree of dissociation is quantitatively expressed in terms of the equilibrium constant, K_a, for reaction 1 above. For example, if the pH of a water sample equals the pK_a ($-\log K_a$), 50% of the weak acid is dissociated.

A base in aqueous solution is defined, analogously, in terms of its likelihood to accept a proton relative to H_2O (Reaction 2).

$$ROH + H_2O \longleftrightarrow R^+ + OH^- \ (+ H_2O) \tag{2}$$

A strong base would then be nearly completely dissociated in aqueous solution to form free hydroxide ions. Dissociation of water into H^+ and OH^- (with a dissociation constant of $\sim 10^{-14}$) is the process determining coexisting H^+ and OH^- concentrations in aqueous solutions that are acidic or basic, that is, in which $H^+ > OH^-$ or vice versa.

In addition to pH, other terms have been used to describe acidic levels in environmental samples—*titratable acidity, titratable strong acidity,* and *total acidity*. Titratable acidity is defined as the acidic content in aqueous solution that is neutralizable by addition of strong base; this term has meaning only if the pH at the end point of the titration is defined. If a low end point pH is chosen (e.g., pH = 2), this quantity becomes total strong acidity. If, however, a higher pH is chosen (e.g., pH = 7), this quantity becomes total strong and weak acidity; an even higher pH (pH = 10) becomes total acidity, as now such conjugate acidic species as ammonium are titrated. The degree of neutralization at a specified pH depends on concentration of the acidic species in solution as well as the equilibrium constant. In this chapter, I use pH to describe free acidity in solution and report the titratable acidity of a sample when the end point pH and total acidic species concentration can be specified.

B. Acidic Precursors

Acidic precursors are those species that are known to produce strong or weak acids during secondary reactions in the atmosphere. The principal examples are SO_2,

NO, and NO_2; the sources of these acid precursors have previously been extensively evaluated (see, for example, NAPAP, 1987) and will be be discussed herein. Their acid-forming reactions include the following: Sulfur dioxide reacts with OH radicals in the gas phase, with air and/or oxygen with or without metal catalysis in the aerosol phase, and with H_2O_2 or O_3 in cloud or rain liquid water phases to form sulfuric acid (Middleton et al., 1980). Gas-phase homogeneous oxidation by OH radicals and oxidation by peroxides in liquid water phases are thought to be the most important mechanisms for producing H_2SO_4 in the atmosphere (Calvert et al., 1985). If formed in the gas phase, this strong acid rapidly condenses on the surface of existing aerosol particles and is partially neutralized by any existing bases in the aerosol particles and, more likely, by ammonia gas absorbed from the gas phase. Nitric acid is formed principally from the reaction of NO_2 with hydroxyl radicals in the daytime (Rodhe et al., 1981) and possibly from the nighttime chemistry of NO_3 (formed by reaction of O_3 with NO_2) (Calvert and Stockwell, 1983). Nitrogen dioxide is formed by the rapid reaction of the principal primary source of nitrogen oxides, NO, with ambient ozone. More details on these processes will be given below. Precursors of weak acids, particularly organic acids, and the conversion processes by which they are formed are not as well established, although oxidation of aldehydes is probably involved.

II. Sources of Acids

A. Sources of Aerosol Strong Acids

In this section I establish that acidic sulfate species are the principal constituents of aerosol strong acids, summarize the extent of primary emissions of acidic sulfate aerosol species, and discuss the major secondary paths of atmospheric sulfate formation involving oxidation of gaseous SO_2.

1. Strong Acidic Species

The presence of strong acids in aerosol particles has been known for many years and quantitated since the mid-1960s (Commins, 1963; Scarengelli and Rehme, 1969; Junge and Scheich, 1971). An extensive data set was reported by Brosset and coworkers (1975, 1978) by the mid-1970's, followed by aerosol strong acidic data obtained by several groups in the United States (Tanner et al., 1977, and Stevens et al., 1978, among others). Acquisition of quantitative data was dependent on the development and application of artifact-free, neutral filter media and on the use of titration procedures that were able to distinguish between strong and weak acids and obtain total strong acidic content (as defined above) of aerosol samples. It was suspected (Junge and Scheich, 1971) that aerosol strong acidity was due to the presence of sulfuric acid and its ammonia-neutralization products because SO_2 oxidation should take place in the atmosphere in the presence of varying amounts of NH_3 to form sulfuric acid and ammonium sulfate products of low volatility, which would be readily condensible onto aerosol particles. Direct observation of H_2SO_4 was reported in a few cases.

Nitrate ion was also reported to be an aerosol constituent, although early reported data are strongly suspect due to measurement difficulties (Spicer and Schumacher, 1977). Evidence now strongly suggests that nitrate is present in the fine aerosol phase due to adsorption into sufficiently neutralized aerosol particles of nitric acid vapor, the latter formed by homogeneous gas-phase oxidation of NO_2 as described below in section II,B. Phase equilibrium considerations dictate that, contrary to sulfuric acid vapor, which exists with $99^{+}\%$ abundance in the particle phase under all lower tropospheric conditions, nitric acid is distributed between gaseous and fine particulate phases. The fraction present as nitric acid depends on the temperature, the ammonia concentration, and the degree of neutralization of the aerosol particle or droplet. Other strong acids ($pK_a < 2$) that could possibly be present in ambient atmospheric aerosols under some conditions include hydrochloric acid, oxalic acid, and methanesulfonic acid. Their presence has been reported (e.g., Matusca et al., 1984; Norton et al., 1983; Ayers et al., 1986), but data indicating widespread distribution at significant levels are lacking to date.

2. Primary Sulfuric Acid/Sulfate Emissions

Primary acidic sulfates are emitted from fossil fuel combustion and a variety of industrial processes including petroleum refining, nonferrous smelting, pulp milling, and the manufacture of sulfuric acid. The techniques for measuring primary sulfates in stack gas emissions and other source in the presence of large excesses of SO_2 have been greatly improved in the past decade and, with the standardized procedure and new data reported by Homolya (1985), the estimates in the NAPAP *Interim Assessment* (1987) are considered to be reasonably accurate. Goklany and others (1984) suggest that the estimates of total primary sulfate emissions may be systematically high, but by how much is unclear. More importantly for this discussion, these authors report the acidic sulfate ratio to total sulfate emissions for various source types. Combined with the fractional primary sulfate emissions by source reported in the NAPAP document, the emissions of acidic sulfate may be estimated (see Table 1–1). *Acidic sulfate* is defined as SO_3 +

Table 1-1. Estimate of acidic sulfate emissions.

Emission source	Emissions[a]	Acidic sulfate ratio[b]	Acidic emissions[a]
Coal-fired power plants	0.23	20–86	0.05–0.20
Coal-fired industrial	0.04	81	0.03
Oil-fired power plants	0.03	~70 (>1% S)	0.02
		~10 (<1% S)	0.00
Other oil combustion	0.07	76–82	0.05–0.06
Transportation	0.02	~100[c]	0.02
Industrial processes	0.09	76–81	0.07
Other	0.12	~80	0.10

[a] Emissions in millions of metric tons of SO_2 equivalent per year.

[b] Ratio of emissions (moles) of acidic sulfate to total sulfate, expressed as %.

[c] Estimate based on Stevens et al. (1976).

H_2SO_4 for the purposes of this discussion. The variation in the acidic emissions column reflects the range of values for that source type depending on control method used. The total emissions of primary sulfate as estimated here for 1980 are 0.60, in units of millions of metric tons of SO_2 equivalent per year, with a range of acidic emissions of 0.34 to 0.50 (57% to 83%), depending on emission controls. Estimates of direct SO_2 emissions for 1980 are 24.5 ($\times 10^6$) metric tons per year (NAPAP, 1987); hence, direct emissions of acidic sulfates constitute at most 2% of emitted sulfur from anthropogenic emissions. Because it can be estimated from atmospheric lifetimes and concentrations that on the order of 20% of SO_2 emitted from the United States is converted to sulfate during its atmospheric lifetime, secondary processes are likely the dominant source of acidic sulfates in the atmosphere. This is true independent of the extent of neutralization of acidic sulfate by ammonia (or other bases, if present) after emission or formation in the atmosphere. Nevertheless, it is important to be able to measure accurately and account properly for these primary acidic emissions for purposes of sulfur budgets and exposure calculations.

3. Secondary Acidic Sulfate Formation

The formation of sulfuric acid in the atmosphere, principally by oxidation of anthropogenic SO_2, is the major source of acidity in aerosols. Further, in the absence of complete neutralization, these acidic aerosols constitute a major source of strong acid in precipitation. Other processes may produce H_2SO_4 in the atmosphere, for example, oxidation of natural sources of reduced sulfur compounds such as dimethyl sulfide from marine emission sources and H_2S from certain marshes. However, I will focus on the two major H_2SO_4-formation processes: homogenous gas-phase oxidation of SO_2 and aqueous phase oxidation of dissolved S(IV).

There is now abundant evidence that oxidation of SO_2 by the hydroxyl radical (Reaction 3) is the principal gas-phase route for H_2SO_4 formation:

$$SO_2 + OH \xrightarrow{(M)} HSO_3 \tag{3}$$

Greater than 98% of gas-phase oxidation proceeds thereby, according to the estimate of Calvert and others (1985). The reactions with other common radical and odd oxygen species—in particular, methylhydroperoxyl and $O(^3P)$—are much slower and not significant, given typical atmospheric residence times for SO_2. Conversion of HSO_3 to sulfuric acid is believed to proceed by Reactions 4 and 5:

$$HSO_3 + O_2 \longrightarrow SO_3 + HO_2 \tag{4}$$

$$SO_3 + H_2O \longrightarrow H_2SO_4 \tag{5}$$

Indeed, the kinetics of Reaction 4 have recently been reported (Gleason and Howard, 1987). The significance of this finding is that gas-phase SO_2 oxidation is therefore not a sink of radical species (including OH), and the amount of H_2SO_4 formed by this process is expected to be linear with $SO_2(g)$ over the ambient range of concentrations. The rate of oxidation of SO_2 is dependent on the OH, but with typical summer conditions, the rate is about 0.5% per hour on a diurnally averaged basis. Winter oxidation rates are about an order of magnitude lower under midlatitude tropospheric conditions.

The second type of process by which secondary acidic sulfate may be formed in the atmosphere is aqueous-phase oxidation in cloud liquid water or precipitation. This process is important from two aspects—the acidic sulfate formed by aqueous-phase oxidation may be directly deposited in precipitation or, as occurs in about 90% of cases (Pruppacher and Klett, 1978), the cloud evaporates, acidic sulfate formed by aqueous-phase oxidation is restored to the aerosol phase, and it is available for transport and subsequent deposition at another location.

Several authors have reviewed the kinetics of various aqueous oxidation mechanisms that could be important in converting SO_2 to acidic sulfate under atmospheric conditions (Penkett et al., 1979, Martin, 1984; Schwartz, 1984). For example, the review of Calvert and others (1985) identified several potential oxidants on the basis of theoretical instantaneous rates of dissolved S(IV) in equilibrium with 1 ppbv of SO_2(g): O_3; H_2O_2 and its organic analogues, peroxyacetic acid and methyl hydroperoxide; O_2 catalyzed by Fe(III) or Mn(II); graphitic carbon; OH and HO_2 radicals; and peroxyacyl nitrates.

Essentially, the results of this analysis are that H_2O_2-S(IV) reaction rates are high enough at all pHs to cause a major portion of S(IV) to be converted to acidic sulfate during the lifetime of most clouds. Oxidation by ozone is significant above about pH 5, a result of the strong positive pH dependence of the S(IV)-O_3(aq) reaction rate. No other reaction is fast enough under rural continental boundary layer conditions to convert significant amounts of S(IV), although some contributions are probably made by organic peroxides (Lind and Kok, 1986), by Fe(III)/Mn(II), and/or by OH and HO_2 (if the mass-accommodation coefficients for these species are > ca. 10^{-3}).

As many reports have shown, the pH dependence of Reaction 6 is small because

$$H_2O_2 + HSO_3 \rightarrow H_2O + H^+ + SO_4^{2-} \tag{6}$$

the H^+ catalysis is offset by the acid-base dissociation of $SO_2 \cdot H_2O$ in solution. The rates of oxidation of S(IV) by O_3 (30 ppbv) and H_2O_2 (1 ppbv) are shown by, for example, Schwartz (1984) to be equal to each other and equal to several hundred percent per hr at about pH 5.6.

This rate analysis has been evaluated against field-derived data in some cases. Lee and others (1986) has shown that the rate of Reaction 6 is similar (about 30% lower) in authentic rain samples compared to laboratory values in "purest possible" water. Thus, conclusions concerning extent of S(IV) oxidation by H_2O_2 based on laboratory studies are expected to be valid for atmospheric water samples. Second, indirect evidence for the importance of Reaction 6 was obtained by Daum and others (1984), that is, the noncoexistence of interstitial SO_2(g)—which is assumed to be in equilibrium with S(IV)—and aqueous H_2O_2 in nonprecipitating clouds. It is inferred that Reaction 6 in the clouds sampled (mostly stratus and stratocumulus) was sufficiently fast to exhaust the limiting reagent within the lifetime of the clouds.

A third source of acidic sulfate should be considered: sorption of SO_2 into aerosol-sized droplets and oxidation therein to acidic sulfate. Direct experimental evidence for the process has been very difficult to obtain. The work of McMurray and Wilson (1982) on accumulation-mode aerosol growth certainly suggests that oxidation of SO_2 in aerosol particles occurs in the atmosphere (at least under polluted summertime conditions), but the relative importance of this acid-formation process is still not well established.

B. Sources of Gaseous Strong Acids

The principal gaseous strong acids in the atmosphere are HCl and HNO_3. Homogenous gas-phase oxidation to form nitric acid vapor has been identified as the major route by which acidic nitrate is available for deposition in precipitation (Calvert et al., 1985), but also because most inorganic nitrate is found in the gas phase as HNO_3. The mechanism for formation of atmospheric HCl(g) is not as well established as that of HNO_3.

1. Nitric Acid Formation

The oxidation of NO_2 by OH through Reaction 7 is about ten times more rapid than the corresponding OH reaction with SO_2 (Reaction 3), and is thus the major route of formation of nitric acid in the boundary layer troposphere during daylight hours.

$$OH + NO_2 \xrightarrow{(M)} HNO_3 \qquad (7)$$

Examination of the seasonal dependence of nitrate deposition demonstrates that this process dominates in winter as in summer (Kelly, 1985); that is, the lower formation rate in winter is compensated for by lower dry deposition rates or perhaps by enhanced below-cloud scavenging to precipitation in the form of snow (Raynor and Hayes, 1982).

A second mechanism for nitric acid formation is now thought to be important at night. This involves oxidation of NO_2, the predominant form of NO_y in the absence of NO_2 photolysis, by ozone (Reaction 8).

$$O_3 + NO_2 \longrightarrow NO_3 + O_2 \qquad (8)$$

The nitrate radical, NO_3, is not stable during daylight hours due to photolysis and/or reaction in the presence of NO. These decomposition paths become slow at night, however, and other NO_3 chemistry can become important, namely, reaction with NO_2 and H_2O (Reactions 9 and 10) or with gaseous aldehydes (Reaction 11), forming nitric acid in both cases.

$$NO_3 + NO_2 \xrightarrow{(M)} N_2O_5 \qquad (9)$$

$$N_2O_5 + H_2O \longrightarrow 2HNO_3 \qquad (10)$$

$$NO_3 + RCHO \longrightarrow HNO_3 + RCO \qquad (11)$$

Reaction 10 is reported to be slow in the absence of condensed phase water; hence, nitric acid formation may occur by the NO_3 route only in the presence of clouds or fog.

The only other significant gas-phase strong acid in the atmosphere is likely to be HCl. Below I discuss the phase equilibrium between ammonium salts of HCl, HNO_3, and H_2SO_4 and the gas-phase acids and ammonia. For the consideration of sources, it is sufficient to note that NH_4Cl is more volatile than NH_4NO_3 to the extent that with ambient levels of aerosol strong acid in fine particles, chloride originally in fine particles is found almost entirely in the gas phase as HCl (Hitchcock et al., 1980). Indeed, the presence of HCl(g) may be predominantly due to the disproportionation of Cl^- from fine particles as they are acidified by incorporation of secondary H_2SO_4 or HNO_3 (e.g., Reaction 12).

$$Cl^- + HSO_4^- \longrightarrow HCl(g) + SO_4^{2-} \tag{12}$$

The presence of oxalic acid has been reported by Norton and others (1983); methanesulfonic acid (CH_3SO_3H) has been reported in marine aerosols by Ayers and others (1986) and Saltzman and others (1983), and both dimethyl sulfate and methyl hydrogen sulfate have been reported in near-source aerosols by Eatough and co-workers (1986). On the basis of the definition in section I above, dissociation of oxalic acid into bioxalate and of CH_3SO_3H into methanesulfonate is essentially complete at pH 4 (a typical pH for rain and cloud water in eastern North America); hence, both qualify as strong acids. The occurrence of oxalic acid at significant levels in the troposphere has not been well documented, although it could exist, on the basis of its volatility and that of its ammonium salt, in both the particulate and gas phases under typical conditions. Methanesulfonic acid is expected to be found only in the particulate phase. Dimethyl sulfate and CH_3OSO_3H are acute toxics that hydrolyze rapidly under most atmospheric conditions.

2. Sources of Atmospheric Weak Acids

The principal weak acids of concern in the troposphere are the carboxylic acids, the low-molecular-weight homologues of which are found in the gas phase and higher homologues in the aerosol phase under typical conditions. Other weak acids include nitrous acid, sulfurous acid, $FeOH^{+2}$, and possibly hydroxymethanesulfonic acid and pernitric acids; however, these are unlikely to be present in amounts comparable to nitric and sulfuric acids and/or sufficient to affect the pH of wet-deposited hydrometeors.

In contrast, measurements of formic and acetic acids show significant quantities even in relatively remote areas (Keene et al., 1983). The terrestrial and aquatic effects of these weak acids are not large because they are subject to rapid microbial decomposition, but they certainly affect the pH of clouds and precipitation and thus may limit the extent of pH-dependent processes such as the oxidation of dissolved S(IV) by ozone. The photochemical processes by which organic acids are formed are complex. It is not at all clear whether the major pathways are gas-phase oxidation of precursors, for example, the corresponding aldehydes, followed by scavenging into cloud or rain droplets, or dissolution of precursors followed by aqueous-phase-oxidation. As described, for instance, by Pierson and Brachaczek (1989), Henry's Law solubilities of the pertinent species are sufficient for either process to be significant at the pHs observed in atmospheric liquid H_2O in northeastern North America. The sources of the precursors of organic acids are not well established, although they no doubt include both natural and anthropogenic hydrocarbons.

III. Sources of Acid-Neutralizing Substances (Bases)

A. Sources of Ammonia

Ammonia is the most abundant basic gas in the atmosphere, and its reactions, together with those of basic coarse particulate matter, provide the principal means by which acids formed in the atmosphere are neutralized prior to deposition. Despite the importance of ammonia in atmospheric chemistry, until recently there was no comprehensive emissions inventory for NH_3. A modest amount of surface concentration data suggested that ambient levels were of the order of 0.1 to 10 ppbv and that emissions were taking place mostly from the earth's surface. Data that have become available are consistent in identifying animal wastes as the largest single source of ammonia in continental areas, but considerable variation in the emission strengths of other ammonia sources is evident from the literature. Table 1–2 summarizes the fractional contributions to NH_3 emissions for various sources in the United States from the data of Harriss and Michaels (1982) and from Misenheimer and others (1987), as quoted in the NAPAP *Interim Assessment* (1987), along with the European estimates of Buijsman and others (1987). Despite the considerable difficulty in reconciling the differences between various estimates of ammonia emissions, it does appear that emissions from animal wastes at the earth's surface are the dominant source. This leads to a strong vertical gradient in $(NH_3)_g$ in the first 100 to 200 m of the atmosphere, and as a result, the amount of strong acid available for scavenging into clouds and precipitation depends

Table 1–2. Anthropogenic emission source estimates for ammonia in the USA and Europe.

<table>
<tr><th>Emission source</th><th>USA = I (%)[a]</th><th>USA = II (%)[b]</th><th>Europe (%)[c]</th></tr>
<tr><td>Livestock waste</td><td>65</td><td>62</td><td>81</td></tr>
<tr><td>Ammonia application</td><td>6</td><td>12</td><td>17</td></tr>
<tr><td>Fossil fuel combustion</td><td>2</td><td>20</td><td><1</td></tr>
<tr><td>Fertilizer production</td><td>13</td><td rowspan="2">6</td><td rowspan="2">2</td></tr>
<tr><td>Other industrial processes</td><td>14</td></tr>
<tr><td>Total (10^5 metric tons/yr)</td><td>8.3</td><td>34</td><td>64</td></tr>
</table>

[a] From Harriss and Michaels (1982).

[b] From Misenheimer et al. (1987), as quoted in NAPAP (1987).

[c] From Buijsman et al. (1987).

strongly on how efficiently this NH_3-rich surface layer gets mixed into the remainder of the boundary layer. One large uncertainty remains: Is there a soil source of ammonia, and is it possible that soil emissions and vegetative absorption form a "closed loop" with respect to ammonia availability in the atmosphere, as has been suggested by Stedman and Shetter (1983)? If soil sources are important, their magnitude is certainly dependent on both soil pH and temperature, and they should be included in inventories, especially for areas like parts of the western United States that have basic soils and wide temperature ranges.

B. Sources of Basic Aerosols

Alkaline materials in aerosols form can strongly influence the acidity of precipitation, as first noted by Junge and Werby (1958). Such materials are thought to be principally of crustal origin and provide the major route by which alkaline earth cations (e.g., Cal and Mg) are incorporated into rain. The continental sources of alkaline dust have been linked to soil type (Gatz et al., 1986). More than 90% of the emissions are attributed to "open sources," with the remainder due to industrial and miscellaneous sources (NAPAP, 1987). Of the open sources, traffic on unpaved roads is the largest (67%), followed by wind erosion (28%), and agricultural tilling (5%) (NAPAP, 1987).

An important property of these mechanically generated basic aerosols is the fact that they are principally found in the coarse particle size range (>2.5 μm); hence, although there are large emission fluxes, most of the dust settles close to the emission site; only a small portion is borne aloft for significant distances (Stensland et al., 1985). It should be recalled that most acidic sulfate aerosol particles are present in the fine particle range and that nitric acid (the other major source of H^+) and ammonia (the other significant atmospheric base) are both gases. As a result, it can be shown that acidic fine particles may coexist in the atmosphere with basic coarse particles, at least until they are co-collected on a filter. Indeed, previously reported data (Tanner, 1980) suggest that total aerosol samples from urban sites may be on balance basic, even though the fine fraction contains acidic particles, because of the presence of alkaline, resuspended dust.

IV. Distribution of Atmospheric Acids and Bases*

A. Aerosol Acidity

Aerosol characterization studies have been conducted in enough locations, using sufficiently sensitive and specific techniques, that certain conclusions can now be made concerning the spatial and temporal distribution of aerosol strong acid

*The contents of this section are taken, in large part, from Tanner, R. L. 1989. "State of the Art Review." *In* J. P. Lodge, Jr., ed. *Methods of Air Sampling and Analysis,* 3d ed. Lewis Publishers, Chelsea, Mich.

concentrations in the lower troposphere. The available data are most complete for the northeastern United States, adjoining portions of Canada, the Los Angeles area, and selected portions of western Europe. They are adequate for the southeastern and midwestern United States, Japan, and remainder of southern Canada, and western Europe but are relatively sparse elsewhere. A selected summary of U.S. and Canadian surface data is shown in Table 1–3 reprinted from Tanner (1989). A more extensive summary has also been prepared by Lioy and others (1987), extending the report of Lioy and Lippman (1986), but the principal findings are largely the same.

In general, for data from sampling in the northeastern United States, the molar ratio of NH_4^+/SO_4^{2-} varies from 1 to 2, whereas the H^+/SO_4^{2-} ratio is usually <0.5 and averages about 0.25. Acid-to-sulfate ratios are higher on average in summer than in other seasons, with documented episodes during which H^+/SO_4^{2-} exceeded 1. Correlation of sulfate levels with H^+ and NH_4^+ and especially with their sum (in molar or equivalence units) is highly significant under nearly all conditions, but ammonium-to-sulfate ratios may exceed 2 when significant levels of aerosol nitrate are present.

Summaries of aerosol composition studies in Europe (Brosset, 1976; Harrison and Pio, 1983; Clarke et al., 1984) and in the midwestern and western United States (Countess et al., 1980; Appel et al., 1980) show that the sulfate/nitrate content of aerosols is usually more completely neutralized by ambient $NH_3(g)$ in these locations; this observation is consistent with higher measured ammonia concentrations in Europe than those obtained in northeastern North America. This is consistent with the hypothesis that observed acidity in ambient aerosols is that remaining after nitric and sulfuric acids are partially neutralized by available ammonia. Steady-state aerosol composition may not be attained because the boundary layer mixing processes that produce the steady state may have characteristic times that are comparable to the rate of sulfuric acid or HNO_3 formation from precursor gases. This leads to temporal and spatial gradients in the acidity available for incorporation into precipitation.

Several studies have demonstrated, using both filter-integrated and continuous sampling techniques, that H_2SO_4 is occasionally present in the atmosphere, usually in slow-moving anticyclonic air masses with high concentrations of aerosol sulfate (Charlson et al., 1974; Tanner et al., 1977; Stevens et al., 1980; Appel et al., 1980; Pierson et al., 1980; Morandi et al., 1983; Slanina et al., 1985; Spengler et al., 1986). Acid-to-sulfate ratios are usually higher in rural areas than at nearby urban sites, probably due to additional sources of ammonia and alkaline dust in the latter (Tanner et al., 1981). However, there are reported cases of free H_2SO_4 measured at urban sites (Cobourn and Husar, 1982). Indications from case studies to date are that average H_2SO_4 levels are $<1\ \mu g/m^3$, even in the summer season. Levels of sulfuric acid are lower in the wintertime, but the relative acidity (equivalent ratio, $H^+/[2\ SO_4^{2-} + NO_3^-]$) may actually be higher in winter, at least in more northern locations. (Daum et al., 1988).

In summary, free sulfuric acid is occasionally found in stagnant air masses in northeastern North America but has only rarely been reported in measurements

Table 1-3. Strong acid- and ammonium-to-sulfate molar ratios in ambient aerosols (from Tanner, 1980).

Location	Date	Sampling period, hr	*N* of samples	Mean sulfate ($\mu g/m^3$)	Molar $H^+/SO_4^{2-} \pm$ SD	Molar $NH_4^+/SO_4^{2-} \pm$ SD
Brookhaven HiVol						
New York, NY	8/76	12	37	12.2	0.24 ± 0.20	1.37
Brookhaven (LI), NY	10–11/76	3,6	19	9.8	0.20 ± 0.24	1.61
New York, NY	2/77	6	46	10.8	0.26	1.61
Brookhaven (LI), NY	8/77	6	22	16.7	0.13 ± 0.15	1.15
High Point, NJ	8/77	12	55	11.7	0.57 ± 0.40	1.02
New Haven, CT	8/77	6	56	11.7	0.08 ± 0.06	1.90
New York, NY	8/77	6	35	14.1	NA	1.34
Brookhaven (LI), NY	1/78	6	11	5.5	0.25	1.01
Whiteface Mountain, NY	7/82	12	43	1.46	0.48	1.77
Argonne Impactor						
State College, PA	1/77–2/80	4	1,369	1.33	0.31	2.7
Charlottesville, VA	5/77–2/80	4	988	1.48	0.24	2.7
Rockport, IN	7/77–12/79	4	819	2.17	0.07	3.2
Brookhaven (LI), NY	12/77–1/80	4	843	1.26	0.21	2.7
Raquette Lake, NY	10/78	4	442	0.78	0.27	2.7

EPA HiVol, LoVol						
New York, NY	8/77	4	48	NA	NA	1.52 ± 0.30
Research Triangle Park, NC	8/77	4	16	17	0.6	1.4
Cedar Island, NC	8/78	24	4	NA	0.22 ± 0.06	1.18 ± 0.58
Great Smokey Mountains	9/78	12	14	11.2	0.86 ± 0.18	1.10 ± 0.26
Shenandoah Valley, VA	7–8/80	12	28	14.3	0.50 ± 0.34	1.68 ± 0.48
Houston, TX	9/80	12	15	13.7	0.38 ± 0.12	1.38 ± 0.12
Abastumani, USSR	7/79	24	17	4.7	NA	1.9
Other data, locations						
Allegheny Mountain, PA	7–8/77	12	35	14.1	1.06	0.88
Greater Los Angeles	2/79	4	18	~2.5	0.01–1.0	1–4
Shenandoah Valley, VA	7–8/80	5 min	$>10^3$	Day 14.6 Night 13.4	NA	1.44 1.71
Houston, TX	8/80	5 min	$>10^3$	Day 13.9 Night 11.1	NA	1.25 1.46

NA = data not available.

made at other locations. The principal source of strong acid in aerosols is the bisulfate ion, usually associated with ammonium in water-soluble fine particles.

B. Ammonia and Alkaline Dust

Ammonia gas is emitted predominantly at the earth's surface with high temporal and spatial variability. Adequate monitoring data for ammonia are not available in the United States, but limited data suggest that surface concentrations are in the range of 0.1 to 10 ppbv. This is consistent with emission data for NH_3 sources reported above and with the emission density data of Husar and Holloway (1983), even if equivalent additional soil emissions of NH_3 were present. More extensive data sets are available in Europe, indicating that ammonia levels are somewhat higher (mostly 1–10 ppbv) there than in the United States and Canada. The latest European data of Erisman and others (1988) indicate that sharply decreasing gradients in $(NH_3)_g$ in the first approximately 200 m of the boundary layer are both expected and observed (Netherlands data) because of the dominance of surface sources and, as discussed in the next section, the reaction of ammonia with acidic aerosols (and likely HNO_3 as well). The profile data over North America, especially in the lowest portion of the boundary layer, are inadequate to confirm the European results.

Alkaline dust, being present mostly in coarse aerosol particles (see above), is limited in its effects on aerosol acidity to the vicinity of its sources. Atmospheric levels of alkaline dust are highly variable in space and time, and its effects (excepting reducing the apparent acidity of co-collected fine particles) may be limited to below-cloud scavenging into rain droplets or onto snow.

V. Gas-Aerosol Equilibria and Boundary Layer Mixing

Two important and interrelated considerations in assessing the availability of acidic aerosols and gases for incorporation into precipitation involve the size-dependent composition of atmospheric aerosols and whether the existence of gas-aerosol equilibria cause vertically dependent aerosol and gas concentrations in imperfectly mixed atmospheric boundary layers. Current descriptions of atmospheric aerosols place most aerosol mass in a fine, internally mixed mode (0.1 to about 2 μm diameter) and a mechanically generated, probably externally mixed coarse mode (>2 μm).

The fine mode contains mostly sulfate, nitrate, ammonium, carbon/organics, water, and some trace elements. Substantial evidence suggests that the fine-mode aerosol mixture is in equilibrium with gaseous ammonia and nitric acid (Reaction 13),

$$NH_4NO_3(\text{aerosol}) \longleftrightarrow NH_3(g) + HNO_3(g) \qquad (13)$$

when these gases are present in large enough amounts to equal or exceed their equilibrium partial pressure product with aerosol ammonium nitrate under the existing temperature, relative humidity, and aerosol chemical composition. There is more limited evidence of interaction of pollutant gases with coarse aerosol particles, with firm evidence only for sorption of HNO_3(g) onto sea salt particles to form a surface layer of sodium nitrate. The evidence dictates equilibrium via Reaction 12 whenever fine aerosol nitrate is present. Calculated ammonia vapor pressures over ammonium acid sulfates do not directly pertain; indeed, observed ammonia levels greatly exceed calculated NH_3 partial pressures over ammonium acid sulfates due to the presence of mixed nitrate-sulfate salts in ambient aerosol particles (Bassett and Seinfeld, 1983). The presence of aerosol strong acids does strongly affect the nitrate–nitric acid distribution between phases.

Experimental evidence provided by Georgii (1978) and by Tanner and others (1984) demonstrate that aerosol strong acid-to-sulfate ratios generally increase with altitude (NH_4^+/SO_4^{2-} decrease), occasionally remain constant within certain layers, and almost never decrease, at least up to about 3 km. In contrast, NH_3 decreases sharply, and the fraction of inorganic nitrate present as HNO_3 increases with altitude. These vertical distributions apparently result from the combination of surface emissions of ammonia, atmospheric mixing at rates comparable to acid formation processes, and a dynamic equilibrium as indicated in Reaction 13.

A particularly useful set of vertical profiles has been recently reported by Erisman and others (1988) for the first 200 m of the boundary layer in a high-NH_3 emissions area of the Netherlands, a summary of which is shown in Figure 1–1. Of special note is the difference in profiles between daylight and nighttime hours. It is clear that the rate at which boundary layer mixing processes produce microscale aerosol equilibrium with gaseous NH_3 and HNO_3 is critical in determining the availability of acidic aerosols and gases for incorporation into clouds and precipitation. In the absence of detailed boundary layer mixing information, surface concentration data cannot be used in models of acidic wet-deposition processes. Future studies, including model-validation exercises, will likely need to include aircraft and/or tower measurements of acidic species and their precursors in order to test the validity of model estimates of dry and wet deposition of acidic substances.

VI. Summary of Significant Acid-Formation Pathways

The significant pathways by which strong acidic sulfur and nitrogen compounds appear in the troposphere are summarized as follows:

1. Direct emission of acidic sulfate aerosols from power plants and other industrial sources
2. Formation of sulfuric acid from sulfur dioxide via a homogenous, gas-phase oxidation reaction with OH radicals
3. Formation of sulfuric acid from sulfur dioxide by an aqueous-phase reaction with hydrogen peroxide or ozone
4. Formation of nitric acid from nitrogen dioxide via a homogenous, gas-phase oxidation reaction with OH radicals

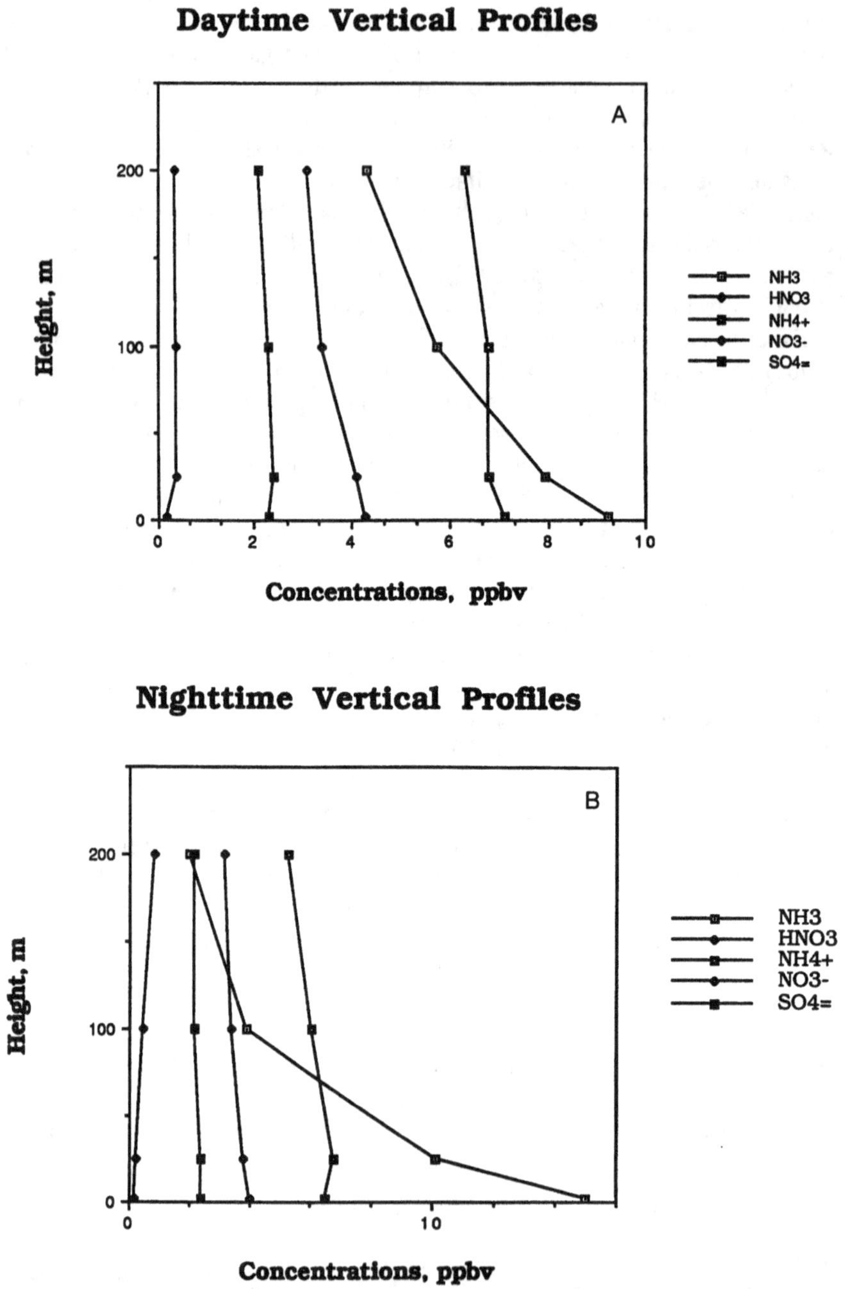

Figure 1–1. Average vertical profiles of aerosols and gases obtained at Cabauw, the Netherlands, condensed from data reported by Erisman et al. (1988). (A) Average profiles for daytime hours; (B) average profiles for nighttime hours. Aerosol concentrations converted to mixing ratios (ppbv at 20°C).

5. Formation of nitric acid (at night) by a combination of gas-phase reaction of nitrogen dioxide with ozone, reaction of NO_3 with NO_2, and aqueous-phase conversion of N_2O_5 to nitric acid

These strong acids are partially neutralized in the atmosphere prior to deposition by reaction with ammonia and, to a lesser extent, with alkaline dust. The resulting, partially neutralized aerosols and nitric acid vapor are available for scavenging into clouds and rain.

Acknowledgments

The author acknowledges many informative discussions with his colleagues at the Brookhaven National Laboratory, especially Peter Daum, Thomas Kelly, Yin-Nan Lee, and Stephen Schwartz. Stimulating discussions with Jeffrey Gaffney (Los Alamos National Laboratory) are also acknowledged. This work was conducted as part of the PRECP program within the National Acid Precipitation Assessment Program, and was performed under the auspices of the U.S. Department of Energy, under Contract No. DE-AC02-76CH00016.

References

Appel, B. R., S. M. Wall, M. Haik, E. L. Kothy, and Y. Tokiwa. 1980. Atmos Environ 14:559–563.

Ayers, G. P., J. P. Ivey, and H. S. Goodman. 1986. J Atmos Chem 4: 173–185.

Bassett, M., and J. H. Seinfeld. 1983. Atmos Environ 17:2237–2252.

Brosset, C. 1976. Ambio 5:157–163.

Brosset, C., K. Andreasson, and M. Ferm. 1975. Atmos Environ 9:631–642.

Brosset, C., and M. Ferm. 1978. Atmos Environ 12:909–916.

Buijsman, E., J. F. M. Maas, and W. A. H. Asman. 1987. Atmos Environ 21:1009–1022.

Calvert, J. G., A. Lazrus, G. L. Kok, B. B. Heikes, J. G. Walega, J. Lind, and C. A. Cantrell. 1985. Nature 317:27–35.

Calvert, J. G., and W. R. Stockwell. 1983. Environ Sci Technol 17:428A–443A.

Charlson, R. J., A. H. Vanderpohl, D. S. Covert, A. P. Waggoner, and N. C. Ahlquist. 1974. Atmos Environ 8:1257–1268.

Clarke, A. G., M. J. Willison, and E. M. Zeki. 1984. *In* B. Versino and G. Angeletti, eds., *Proc. 3d European symp. on physico-chemical behavior of atmos. pollutants,* 331–338. D. Reidel, Dordrecht, FRG.

Coboum, W. G., and R. B. Husar. 1982. Atmos Environ 16:1441–1450.

Commins, B. T. 1963. Analyst 88:364–367.

Countess, R. J., G. T. Wolfe, and S. H. Cadle. 1980. J Air Pollut Contr Assoc 30:1194–1200.

Daum, P. H., T. J. Kelly, S. E. Schwartz, and L. Newman. 1984. Atmos Environ 18:2671–2684.

Daum, P. H., T. J. Kelly, R. L. Tanner, X. Tang, K. Anlauf, J. Bottenheim, K. A. Brice, and H. A. Wiebe. 1989 Atmos Environ 23:161–173.

Eatough, D. J., V. F. White, L. D. Hansen, N. L. Eatough, and J. L. Cheney. 1986. Environ Sci Technol 20:867–872.

Erisman, J.-W., A. W. M. Vermetten, W. A. H. Asman, A. Waijers-Ijpelaan, and J. Slanina. 1988. Atmos Environ 22:1153–1160.

Gatz, D., W. Barnard, and G. Stenslund. 1986. Water Air Soil Pollut 30:245–251.

Georgii, H. W. 1978. Atmos Environ 12:681–690.

Gleason, J., and C. J. Howard. 1987. J Phys Chem 91:719–724.

Goklany, I. M., G. F. Hoffnagle, and E. A. Brackbill. 1984. J Air Pollut Contr Assoc 31:123–134.

Harrison, R. M., and C. A. Pio. 1983. Tellus 35B:155–159.

Harriss, R. C., and J. T. Michaels. 1982. *In Proc. of the 2d symp. on composition on the nonurban troposphere,* pp 33–35, American Meteorological Society, Boston.

Hitchcock, D. R., L. L. Spiller, and W. E. Wilson. 1980. Atmos Environ 14:165–182.

Homolya, J. G. 1985. *Primary sulfate emission factors for the NAPAP emissions inventory,* EPA-600/7-85-037, U.S. Environmntal Protection Agency, Washington, D.C.

Husar, R. B., and J. M. Holloway. 1983. *In Ecological effects of acid precipitation,* Report PM 1636, 95–115. National Swedish Environment Protection Board, Stockholm.

Junge, C., and G. Scheich. 1971. Atmos Environ 5:165–175.

Junge, C. E., and R. T. Werby. 1958. J Meteor 15:417–425.

Keene, W. C., J. N. Galloway, and J. D. Holden, Jr. 1983. J Geophys Res 88:5122–5130.

Kelly, T. J. 1985. *Trace gas and aerosol measurements at Whiteface Mountain, New York.* ESEERCO Report EP83-13, Empire State Electric Energy Research Corp., New York.

Lee, Y.-N., J. Shen, P. J. Klotz, S. E. Schwartz, and L. Newman. 1986. J Geophys Res 91:13264–13274.

Lind, J. A., and G. L. Kok. 1986. J. Geophys Res 91:7889–7895.

Lioy, P. J., T. Buckley, R. L. Tanner, and K. Ito. 1988. *In Acid Aerosols Issue Paper,* Chapter 2, Report EPA-600/8-88-005A. U.S. Environmental Protection Agency, Washington, D.C.

Lioy, P. J., and M. Lippmann. 1986. *In* S. D. Lee, T. Schneider, L. D. Grant, and P. J. Verkerk, eds. *Aerosols: Research, risk assessment and control strategies,* Lewis Publishing, Chelsea, Mich.

McMurray, P. H., and J. C. Wilson. 1982. Atmos Environ 16:121–134.

Martin, L. R. 1984. *In* J. G. Calvert, ed. *SO_2, NO and NO_2 oxidation mechanisms: Atmospheric considerations*, 63–100. Butterworth, Boston.

Matusca, P., B. Schwarz, and K. Bachmen. 1984. Atmos Environ 18:1667–1674.

Middleton, P., C. S. Kiang, and V. A. Mohnen. 1980. Atmos Environ 14:463–472.

Misenheimer, D., T. W. Warn, S. Zelmanowitz, and D. J. Zimmerman. 1987. *Ammonia Emission Factors for the NAPAP Emission Inventory*. EPA-600/7-87-001, U.S. Environmental Protection Agency, Washington, D.C.

Morandi, M., T. Kneip, J. Cobourn, R. Husar, and P. J. Lioy. 1983. Atmos Environ 17:843–848.

National Acid Precipitation Assessment Program. 1987. C. N. Herrick and J. L. Kulp, eds. *Interim assessment: The causes and effects of acidic deposition,* Vol. II. U.S. Government Printing Office, Washington, D.C.

Norton, R. G., J. M. Roberts, and B. J. Huebert. 1983. Geophys Res Lett 10:517–520.

Penkett, S. A., B. M. R. Jones, K. A. Brice, and A. E. J. Eggleton. 1979. Atmos Environ 13:123–137.

Pierson, W. R., and W. W. Brachaczek. 1989. Aerosol Sci Technol (in press).

Pierson, W. R., W. W. Brachaczek, T. J. Truex, J. W. Butler, and T. J. Korniski. 1980. Ann NY Acad Sci 338:145–173.

Pruppacher, H. R., and J. D. Klett. 1978. *Microphysics of clouds and precipitation*. D. Reidel, Hingham, Mass.

Raynor, G. S., and J. V. Hayes. 1982. Atmos Environ 16:1647–1656.

Rodhe, H., P. Crutzen, and A. Vanderpol. 1981. Tellus 33:132–141.

Saltzman, E. S., D. L. Savoie, R. G. Zika, and J. M. Prospero. 1983. J Geophys Res 88:10897–10902.

Scarengelli, F. P. and K. A. Rehme. 1969. Anal Chem 41:707–713.

Schwartz, S. E. 1984. *In* J. G. Calvert, ed. *SO_2, NO, and NO_2 oxidation mechanisms: Atmospheric considerations*, Butterworth, Boston. 173–208.

Slanina, J., C. A. M. Schoonbeek, D. Klockow, and R. Niessner. 1985. Anal Chem 57:1955–1960.

Spengler, J. D., G. A. Allen, S. Foster, P. Severance, and B. Ferris, Jr. 1986. S. D. Lee, T. Schneider, L. D. Grant, and P. Verkerk, eds. *Aerosols: Research, risk assessment and control strategies*, 107–120. Lewis Publishing, Chelsea, Mich.

Spicer, C. W., and P. M. Schumacher. 1977. Atmos Environ 11:873–876.

Stedman, D. H., and R. E. Shetter. 1983. *In* S. E. Schwartz, ed. *Trace atmospheric constituents*, 411–454. John Wiley, New York.

Stenslund, G., W. Barnard, and D. Gatz. 1985. Unpublished data.

Stevens, R. K., T. G. Dzubay, G. Russworm, and D. Rickel. 1978. Atmos Environ 12:55–68.

Stevens, R. K., P. J. Lamothe, W. E. Wilson, J. L. Durham, and T. G. Dzubay, eds. 1976. *The General Motors/Environmental Protection Agency sulfate dispersion equipment*. Report EPA-600/3-76-035, U.S. Environmental Protection Agency, Research Triangle Park, N.C.

Stevens, R. K., T. G. Dzubay, R. W. Shaw, Jr., W. A. McClenny, C. W. Lewis, and W. E. Wilson. 1980. Environ Sci Technol 14:1491–1498.

Tanner, R. L. 1980. Ann NY Acad Sci 338:39–49.

Tanner, R. L. 1989. *In* Y. P. Lodge, Jr., ed. *Methods of air sampling and analysis*, 3d ed., 703–714. Lewis Publishers, Chelsea, Mich.

Tanner, R. L., R. Cederwall, R. Garber, D. Leahy, W. Marlow, R. Meyers, M. Phillips, and L. Newman. 1977. Atmos Environ 11:955–966.

Tanner, R. L., R. Kumar, and S. Johnson. 1984. J Geophys Res 89:7149–7158.

Tanner, R. L., B. P. Leaderer, and J. D. Spengler. 1981. Environ Sci Technol 15:1150–1153.

Aerosol Sulfur Association with Aluminum in Eastern North America: Evidence for Solubilization of Atmospheric Trace Metals before Deposition

John W. Winchester*

Abstract

Statistical evidence has been found for the association of atmospheric aerosol sulfur with soil mineral elements in the eastern United States. The variability in time sequence aerosol composition, measured by proton-induced X-ray emission (PIXE) analysis of 3,600 two-hourly filter samples at nine sites during October 1977, has been examined using factor analysis and multiple linear regression. At all sites, 84% to 91% of the variance is accounted for by three or four factors, including a mainly sulfur component that also contains Al and other soil mineral metals, as well as one or two dust components and a Pb-Br component attributed to automotive emissions. The first of these is interpreted to represent sulfuric acid taken up by airborne alkaline mineral dust. At ambient relative humidities sulfuric acid aerosol may reach pH 0 or below, and the alkaline soil mineral particles are expected to be rendered soluble by reaction with their strongly acidic surface coatings. The resulting solubilized trace metals, after wet or dry deposition to the surface, may be more biologically available than if the metals were deposited in an insoluble mineral form. At two northeastern U.S. sites, the Al concentrations in the S-rich aerosol component are high enough to account for much of the soluble Al reported to be present in Adirondack lakes. At the midwestern sites, the Al concentrations in the S-rich component are higher, corresponding to the expected higher dust fluxes to the atmosphere by wind erosion of soils. At all sites the atmospheric deposition fluxes of metals, after solubilization by reaction of airborne soil mineral particles with sulfuric acid, may be an important input to aquatic and possibly terrestrial ecosystems.

I. Statement of the Problem

Recent reviews of the effects of acidic rain on soils (Tabatabai, 1985) and freshwater ecosystems (Schindler, 1988) have emphasized research interest in soluble trivalent aluminum, Al(III), because of its potential toxicity, but they

---

*Department of Oceanography, Florida State University, Tallahassee, FL 32306, USA.

report only studies of processes for the mobilization of Al within soils or water after acidification. That these processes can occur is not in dispute (Cronan et al., 1986; Walker et al., 1988). Studies of processes that take place within the atmosphere, for example, those by Lindberg and others (1986, 1988) have focused on the magnitudes of trace metal and nutrient fluxes to the surface more than on atmospheric transformations, such as by acidic attack of suspended mineral dust, that may affect the biological availability of the substances after deposition. The possibility that mineral Al may be rendered soluble in the atmosphere by reaction with acidic air pollutants and then deposited to the surface at a flux sufficient to supply appreciable amounts of soluble Al to the surface appears not to have been addressed. In this chapter evidence for such a possibility is presented, based on analysis of existing aerosol composition data and reference to pertinent literature.

Soil dust is generally alkaline and can neutralize acidic air pollutants, including gaseous sulfur dioxide and nitric acid vapor, molecules that diffuse rapidly to particle surfaces and are retained through chemical acid-base neutralization reactions. This neutralization could mitigate some effects of acidic deposition to the earth surface. However, new products of the reactions of dust with the acidic pollutants are formed, some of which, after their subsequent deposition, could lead to important ecological or human health effects. One objective of research on the role of dust in precipitation chemistry should be to evaluate the extent to which such reactions occur and the magnitude of any resulting effects.

In western North America, the flux of alkaline dust to the atmosphere, from wind erosion of soils, may exceed the fluxes of acidic pollutants, whereas in eastern North America the reverse may be true (Gillette and Hanson, 1989). Unpaved roads are an additional source of dust (Barnard et al., 1986). In either case, soil mineral dust occurs everywhere in the United States, and acid-base neutralization reactions with acidic air pollutants may occur and mineral solubilization result. Some minerals, such as airborne calcium carbonate crystals, can be expected to dissolve rapidly, but more slowly reacting clay mineral particles may also be broken down by adsorbed sulfuric acid. Major elements and accessory heavy metals present in the minerals may thus become solubilized in the atmosphere before their wet or dry deposition to the earth's surface. In a soluble form these metals are likely to be more biologically available than if deposited as sparingly soluble mineral grains.

At ambient relative humidities, the H_2SO_4 concentration in a liquid aerosol phase can be high. Phase equilibrium data (Tang et al., 1978) can be used to calculate concentrations of acidic sulfate solutions in equilibrium with air at any relative humidity. Figure 2–1 shows the result for the H_2SO_4-$(NH_4)_2SO_4$-H_2O system. The presence of free H_2SO_4 in aerosols has been confirmed by direct measurement of H^+, SO_4^{2-}, and NH_4^+ ion concentrations, as shown in Table 2–1, and a calculated average $[H^+] > 1$ molal and $pH < 0$ has been found in both Ohio and Florida (Ferek et al., 1983; Fan, 1985). The low pH has been confirmed directly by aerosol sampling on pH paper, not only in the United States but also in China (Ma et al., 1987). Clays tend to break down if $pH < 2$, so that analytical chemists must take precautions in certain procedures to avoid this. Conditions in

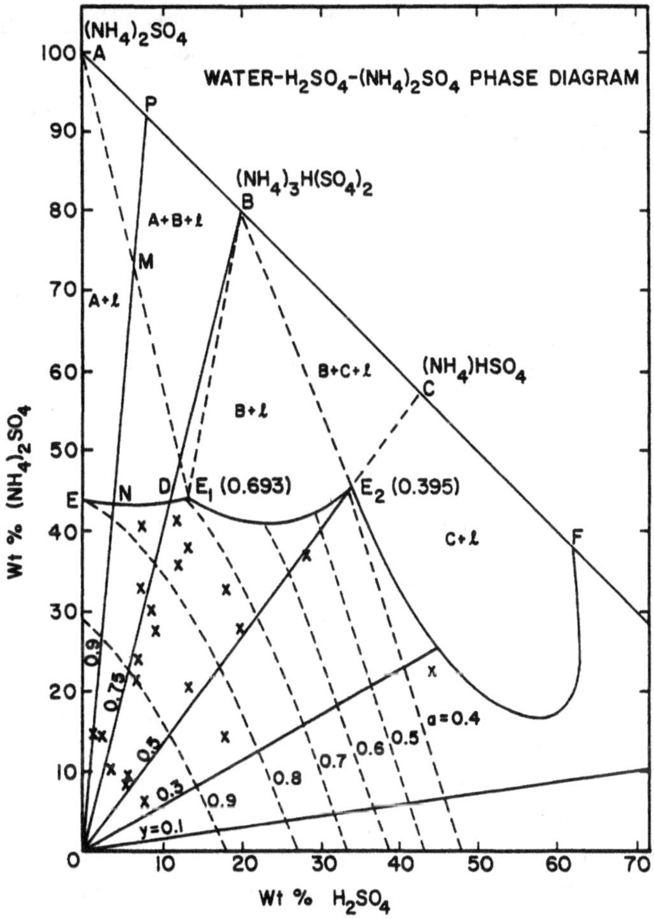

Figure 2–1. Phase diagram for aqueous aerosols containing $(NH_4)_2SO_4$,H_2SO_4, and H_2O. Dashed curves for water activity (a) = 0.9 to 0.4 (relative humidity 90–40%) indicate compositions in liquid phase region where crystalline phases cannot coexist. Dashed lines above the liquid phase region delineate regions where two of three solid phases can coexist with a eutecic aqueous solution phase. Points (x) represent aerosol compositions measured in Tallahassee, Florida, for aerosol pH calculation.

the atmosphere may be considerably more favorable than on the ground within soils for the leaching and eventual dissolution of clay minerals by H_2SO_4.

II. Approach to the Problem

Suspended particulate matter in ambient air is invariably an external mixture of components that have different sources and histories of atmospheric chemical transformation. The present inquiry is into the chemical changes that may have

Table 2-1. Aerosol sulfuric acid concentrations in southeastern United States.[a]

	Ohio River region	Tallahassee, FL
Number of samples	22	22
H_2SO_4 conc., $\mu g\ m^{-3}$, median	2.2	2.8
geometric mean (× SD)	2.6 (2.6)	2.5 (2.5)
H^+ conc., neq m^{-3}, median	45.3	57.0
geometric mean (× SD)	53.7 (2.6)	51.3 (2.5)
Aerosol pH, median	−0.03	−0.17
average ± SD	0.11 ± 0.59	−0.17 ± 0.33
Relative humidity, %, median	79	83
average ± SD	80 ± 11	82 ± 15

[a]In all samples, aerosol particles were aqueous solutions at ambient relative humidity. Ohio River region (filter samples): April 1979 (6), November 1979 (5), July–August 1980 (13). Tallahasseee, FL (fine fraction samples): Sept.-October 1984 (22).

occurred in any of these types of particles based on their internal chemical composition. Individual particle analysis by electron microscopy can be used to identify particles of different types, but it is not well suited to determine their average concentration in air over extended periods of time. Instead, bulk aerosol measurements should be employed; however, to ascertain whether internal mixtures of chemically reacting constituents occur in the particles of one or another component of the bulk aerosol mixture, the compositions of the component must be resolved from the bulk measurement data. The following approach has been used here to identify aerosol components in ambient air and determine their average compositions:

- Collect 400 or more samples in time sequence at ground-level sites under conditions where large variability is expected in the mixing ratios of externally mixed aerosol components.
- Measure concentrations of elements that are considered important to establish relationships between sulfur, soil mineral, and other aerosol constituents, such as S, Al, Si, Fe, K, Mg, Ca, Pb, Br, Cl, and P.
- Apply multivariate statistical factor and multiple regression analysis to resolve aerosol components by a linear mixing model.
- Identify components that contain both S and Al, suggesting an internal mixture from chemical interaction of sulfuric acid with individual soil dust particles.
- Calculate atmospheric concentrations of S, Al, and other elements in this aerosol component.
- Estimate deposition fluxes by multiplying atmospheric concentrations of elements in the component by a suitable deposition velocity.
- Compare these fluxes with dust and pollution sulfur emission fluxes from appropriate model calculations and emissions inventories and with measured surface water composition.

The present study is based on absolute principal factor analysis applied to 3,600 two-hourly samples taken in time sequence during October 1977 at nine Class I SURE network stations and analyzed for 10 elements. These are regionally representative nonurban sites in the U.S. Midwest and Northeast, from Indiana to Massachusetts, sampled when most gasoline contained ethyl fluid lead additive. The locations and some characteristics of the nine sites are presented in Figure 2–2. At each site aerosol particles were collected continuously on a "streaker" sampler (Nelson, 1977) consisting of a 0.4 μm Nuclepore filter strip with a moving sucking orifice that produced a streak sample over 1 week of a single filter. The streak was analyzed for elemental composition in steps corresponding to 2-hour time increments by proton-induced X-ray emission (PIXE) at Florida State University (Johansson et al., 1975).

Absolute principal factor analysis (Thurston and Spengler, 1985) and multiple linear regression were then applied to these 2-hourly concentration measurements of 10 elements at the nine sites, about 400 samples over 5 weeks at each site. These statistical procedures serve to summarize the main numerical relationships in large amounts of measurement data so as to help in recognizing patterns that suggest physical components and processes. The statistical analysis by itself, based on correlations between measured variables, does not prove cause and effect, nor does it prove uniqueness or physical significance of resolved components. However, with additional information, intuition, or imagination, the physical significance of the components may be argued. The connection between perception and reality, of course, is an age-old philosophical problem.

III. Results

The results, which are discussed more extensively in the next section, show these general features: Only three or four factors are needed to account for 84% to 91% of the variance in the data sets of measured elemental concentrations in air at each of the nine sites, consistent with the assumption made in factor analysis of a linear mixing model of physically discrete noninteracting components. In approximately descending order of variance accounted for by each, these are one or two dust components with some S enrichment, a Pb-Br component, and a mainly S component. This last contains 56% to 92% of the average S concentration during the 5-week measurement period and is interpreted to represent sulfuric acid aerosol formed largely from pollution sulfur dioxide. At the five midwestern U.S. sites (where dust may be most prevalent), but not at the more easterly four sites, the mainly S component is negatively correlated with Ca. This hints that Ca-containing dust, readily reactive with acidic gases, plays an active role in SO_2 scavenging before other processes that form H_2SO_4 aerosol can occur. The SO_2 thus removed should lead to some S enrichment in the mainly soil dust component and suppress formation of the mainly S component.

At two midwestern U.S. sites, the S component contains half or more of the average total Al measured, whereas at the remaining seven sites smaller amounts

Figure 2–2. Locations of SURE Class I sites, at which time sequence aerosol compositions were determined by continuous sampling on Nuclepore filters and PIXE analysis of 2-hourly time steps. Site 1, Montague, MA, was on generally flat terrain in a clearing 30 m wide. Site 2, Scranton, PA, alt. 300 m, was near farmland in hilly terrain on a crest overlooking the Susquehanna River 1.5 km from a paper mill. Site 3, Delmarva, Indian River, DE, was in a mainly flat sandy coastal area 120 m from a tree line. Site 4, Duncan Falls, OH, was in a hilly area not screened by trees, near the crest of the highest ridge and 20 m from a road. Site 5, Rockport, IN, was on flat farmland 250 m from a tree line and 8 km from a metal reclamation plant. Site 6, TVA, Giles County, TN, was in hilly terrain on a crest 100 m from a tree line. Site 7, Fort Wayne, IN, was on flat land without tree shadowing and adjacent to an electric substation transformer. Site 8, Research Triangle Park, NC, was halfway up a slope among small hills and valleys, 150 m from trees and 300 m from an interstate highway. Site 9, Lewisburg, WV, was in generally rolling terrain on a hill crest 200 m from an airport.

of Al are present in the S component. The two high-Al sites are within the highest predicted soil erosion dust areas of Indiana and Ohio (Gillette and Hanson, 1989). Thus, wind erosion of soils may be the most important source of clay mineral dust that reaches sufficiently high altitude for reaction with pollution SO_2.

At all of the nine sites, Al is present at statistically significant levels in the S component. Although generally explaining less of measured data variance than the other two or three factors, the S-rich factor can be considered as significant because the sum of squared loadings (correlation coefficients) of its elemental constituents exceeds unity. By multiple regression of Al concentration as a linear combination of its content in each of the three or four factors, the Al content of the S component is always found to be highly significant. Over the nine sites, S in this component falls in the narrow range of 3.5 to 1.3 $\mu g\ m^{-3}$, whereas that of Al is broader, 0.39 to 0.025 $\mu g\ m^{-3}$, with highest Al concentrations at the midwestern U.S. sites. On a chemical equivalent basis, the sum Al + Si + Fe, a measure of alkaline element concentrations, is always less than that of S, indicating an excess of H_2SO_4 over alkaline capacity of these clay mineral elements in the aerosol component and sufficient acidity to render the mineral particles soluble under favorable atmospheric conditions.

A. Factor Analysis of Aerosol Composition Data

Factor, or principal component, analysis is a multivariate statistical procedure that may be viewed as an elaborate way to examine the correlations among variables, the elements with concentrations measured in the samples. Cause and effect are never proved by correlation alone, and a unique solution may not be found, but valuable clues or insights may be gained by summarizing the main features of large data sets more compactly through factor analysis. Essentially, one tries to represent variability in the data as a sum of contributions from a small number of components of fixed multielemental composition, that is, to impose a linear mixing model of components on the data. The procedure, which includes a varimax rotation, computes the compositions of factors that account for data variance most economically. If each has a sum greater than unity of the variances due to its constituent elements, $\Sigma r^2 > 1$, it is a more powerful predictor than any individual element and considered to be a significant factor. If a large percentage of total variance can be accounted for by the sum of factors, the model fits the data quite well. The factor analyses in this study were carried out using IBM PC–compatible Statgraphics Version 2.6 software (STSC, 1987).

Table 2–2 presents some features of the data and factor analysis results. Using concentration measurements of eight elements in about 400 samples at each site, only three or four factors are needed to describe 84% to 91% of the total variance. At most sites factors were retained with minimum eigenvalues less than unity before varimax rotation, but after rotation they were significant with $\Sigma r^2 > 1$. At all sites one of the factors contains mainly S, always significant although explaining less of the data variance than can soil dust or automotive lead aerosol components; the S component tends to be less variable than either dust or lead aerosol.

Table 2–3 shows results obtained by multiple linear regression of each element as a linear combination of three or four factors at the nine sites. The eight most reliably measured elements, S to Br, were used to define the factors, but Cl and P regressions were also carried out. Breakdowns of the contributions of each

Table 2-2. Factor analysis of element concentration data at 9 SURE sites, October 1977.

SURE site	Samples (n + 1)	Number of factors	Cumulative percent variances	Minimum eigenvalue	Σr^2 of S factor
Duncan Falls, OH (4)	414	4	87.9	0.747	1.242
Fort Wayne, IN (7)	412	3	84.0	1.048	1.517
Rockport, IN (5)	413	3	91.0	0.985	1.101
Lewisburg, WV (9)	437	3	84.1	0.793	1.126
Giles County, TN (6)	413	3	88.7	1.027	1.060
Montague, MA (1)	412	4	91.6	0.489	1.177
Scranton, PA (2)	373	4	90.8	0.682	1.103
Indian River, DE (3)	408	4	88.9	0.633[a]	1.035
Research Triangle Park, NC (8)	425	4	90.8	0.611	1.098

NOTE: Aerosol sampling time, 2 hours for 31 to 36 days. Number of samples, $372 < n < 436$. Absolute factor analysis used an additional zero sample. Factors were defined using 8 elements: S, Al, Si, K, Ca, Fe, Pb, Br. Multiple linear regressions used 10 elements, including Cl and P.

[a] Retains factor 4, Ca-rich, $\Sigma r^2 = 0.98$. Factor 3 is S-rich, $\Sigma r^2 = 1.035$. At all other sites, minimum eigenvalue retains the last (3d or 4th) S-rich factor, Σr^2 as indicated.

element to the variances, $\%R^2$, are given, and the eight element averages agree with the cumulative percent variances in Table 2–2. The percentage of element concentration in the constant term of each regression equation is generally quite small, for example, averaging 5.2% ± 1.2% for S and −0.7% ± 3.1% for all eight elements over the nine sites.

Tables 2–4 through 2–7 present the compositions of factors at the nine sites that are denoted as being mainly sulfur, aluminum-rich dust, lead, and second dust. In each the factor number and number of factors are indicated, for example, in Table 2–4 the sulfur factor at Duncan Falls, Ohio, is factor four of four factors, F4/4. Both elemental concentrations, ng m^{-3}, and the percentage of the element contained in the factor are given, for example, in Table 2–4, S = 2,136.16 ng m^{-3} in factor 4, or 59.27% of S in all four factors at Duncan Falls. Also given are the weight ratios Si/Al ± SD. These should be compared with the earth crust average of 3.41, which can be calculated from the crustal composition data of Mason and Moore (1982) in Table 2–8. In the dust factors, Si/Al is generally close to the crustal ratio, but in the mainly sulfur factor, Al is somewhat enriched relative to Si. Table 2–6 for the lead factor also gives the ratio Br/Pb, generally about 0.2, a value expected for automotive exhaust aerosol from leaded gasoline.

The results for Ca deserve special mention. Table 2–4 shows negative contributions of Ca to the sulfur factor at the five western sites but positive at the four eastern sites; in all cases these are highly significant. Table 2–5 shows that the dust factor concentrations of Al, Si, and Fe average about the same at all nine sites, but Ca is much higher at the five western sites than at the four eastern. This difference suggests that rapidly reactive Ca-containing dust at the midwestern

U.S. sites is abundant enough to withdraw SO_2 from the air and substantially inhibit the formation of H_2SO_4 in the mainly sulfur factor. The amounts of SO_2 withdrawn, bound as sulfate in the dust particles, should vary with the supply of both the Ca-containing dust and gaseous SO_2 to a reaction altitude of the atmosphere and probably fluctuate. Perhaps because of the complexity of such a process, no difference between the five western and four eastern sites in the average S content of the dust factor can be discerned.

Next the relative amounts of S and the elements from potentially alkaline minerals in the resolved sulfur and dust components should be examined. For this weight concentrations are converted to a chemical equivalent basis, making use of the ionic charges of principal species and equivalent weights in Table 2–8. Tables 2–9 and 2–10 give the equivalent ratios of the elements to S in the mainly sulfur and Al-rich dust factors for the nine sites, together with the standard deviations of the ratios obtained from multiple linear regression. These ratios are then used to construct Table 2–11, giving the relative amounts of Al + Si + Fe and S in the two factors, a measure of the ratio of slowly reacting alkaline mineral elements to potentially acidic S. (The ratios Al + Si + Fe + Ca to S are a little higher, as can be estimated from data in Tables 2–4 and 2–5.) The dust factor ratio is much greater than unity, indicating excess alkalinity over any sulfuric acid. However, the sulfur factor ratios are always less than unity, indicating an excess amount of acid, assuming S represents H_2SO_4 formed from airborne SO_2. (At the four eastern sites where Ca is positive, S is also greatly in excess of Al + Si + Fe + Ca.) Given appropriate atmospheric conditions, alkaline mineral particles coated with sulfuric acid could have been rendered soluble by acid-base neutralization or ion exchange reactions.

The measured Al concentrations in the sulfur component, higher at the five western than at the four eastern sites, generally agree with higher predictions of dust fluxes by wind erosion of soils in major land resource areas (MLRA) of the U.S. Midwest (Gillette and Hanson, 1989). This result lends credibility to calculations of dust generation from a semiempirical model (Gillette and Passi, 1988) as a means of estimating fluxes of alkaline particles that may interact with acidic air pollutants. In areas of large acidic air pollution fluxes, these upward alkaline fluxes may determine the deposition fluxes of solubilized mineral elements to the surface.

B. Time Series Relationships in the Factors

Factors have now been defined that are combinations of the measured element concentration variables and when added together reproduce the original data quite well, that is, explain a high percentage of data variance. The amounts of each factor contributing to each of the 3,600 samples (around 400 per site) are indicated by factor scores. If the statistically resolved factors are regarded as representing distinct aerosol components, their variations in time and any relationships between factors should be linked to meteorological conditions that control their fluctuations. To help look for such links, the factor scores should be examined as time

Table 2–3. Multiple linear regression of elements versus factor scores at 9 SURE sites, October 1977. Variance explained by the regressions and percent (± SD) of concentration in constant term.

SURE site		S	Al	Si	Fe	K	Ca	Pb	Br	Cl	P	Average of S-Br
Duncan Falls, OH (4)	%R2	88.6	91.2	90.2	82.4	92.0	81.7	87.3	89.0	60.5	31.8	87.8
	%Const.	14.1	−3.6	5.0	−36.4	−7.0	25.2	−9.5	4.9	57.3	53.9	−0.9
	± SD	2.0	3.0	4.5	4.9	5.6	4.4	2.6	2.5	6.3	4.2	3.7
Fort Wayne, IN (7)	%R2	88.4	76.7	92.9	71.7	78.0	81.9	91.3	90.0	35.6	34.1	83.9
	%Const.	4.8	−3.3	−6.0	12.8	6.9	−14.9	−14.8	8.4	59.2	51.7	−0.8
	± SD	2.4	3.9	2.3	4.4	3.7	7.2	2.4	3.0	4.9	5.2	3.7
Rockport, IN (5)	%R2	97.7	86.2	96.3	87.8	83.9	88.8	93.5	93.3	0.6	53.4	90.9
	%Const.	2.6	10.3	−4.0	−5.1	−4.3	−2.0	−19.3	18.8	125.6	52.4	−0.4
	± SD	0.9	3.0	2.1	3.2	4.1	4.4	2.7	2.6	22.1	3.4	2.9
Lewisburg, WV (9)	%R2	95.5	77.7	94.1	83.8	83.5	68.6	82.9	86.0	8.8	13.2	84.0
	%Const.	2.5	12.4	6.8	−11.5	−3.4	−12.9	−13.0	17.5	60.5	69.9	−0.2
	± SD	1.3	3.3	1.7	3.3	2.9	5.8	3.4	3.2	8.1	4.0	3.1

Giles County, TN (6)	%R2	98.5	84.8	93.6	93.2	69.7	86.5	91.9	90.5	17.1	33.4	88.6
	%Const.	1.0	14.7	−5.9	−22.0	16.6	−13.1	−10.5	18.2	65.9	69.8	−0.1
	± SD	0.8	3.6	2.8	2.9	5.7	5.5	2.7	2.9	4.0	4.8	3.4
Montague, MA (1)	%R2	99.3	96.4	93.2	80.4	80.9	90.8	95.3	96.1	46.3	70.3	91.6
	%Const.	4.7	7.4	−5.0	−16.8	17.7	−8.1	−16.9	10.2	24.9	52.6	−0.9
	± SD	0.6	1.2	1.7	3.6	2.9	2.6	2.4	1.8	4.9	1.9	2.1
Scranton, PA (2)	%R2	96.9	95.3	98.0	95.3	87.5	77.5	85.0	89.8	34.2	38.5	90.7
	%Const.	3.4	8.7	−3.8	−8.0	−1.2	2.1	−11.1	9.1	27.6	46.2	−0.1
	± SD	1.3	2.2	1.8	2.8	6.0	4.9	3.0	2.5	6.5	4.1	3.1
Indian River, DE (3)	%R2	97.9	91.3	93.8	87.7	83.6	99.5	77.7	78.9	10.5	17.6	88.8
	%Const.	8.4	−10.6	−7.9	−24.0	27.9	−4.4	−32.8	15.7	169.6	84.8	−3.5
	± SD	1.0	3.7	2.8	4.5	2.9	0.5	5.8	4.4	22.9	4.0	3.2
Research Triangle Park, NC (8)	%R2	98.7	90.0	91.8	84.5	74.2	96.3	95.4	95.1	4.9	20.4	90.8
	%Const.	5.7	11.7	−2.3	−29.3	12.0	2.4	−4.8	12.1	131.7	76.1	0.9
	± SD	0.6	2.8	2.4	3.4	3.9	1.7	2.6	2.6	19.9	4.9	2.5
Averages	%R2	95.7	87.7	93.8	85.2	81.5	85.7	88.9	89.9	24.3	34.7	88.6
	%Const.	5.2	5.3	−2.6	−15.6	7.2	−2.9	−14.7	12.8	80.3	61.9	−0.7
	± SD	1.2	3.0	2.5	3.7	4.2	4.1	3.1	2.8	11.1	4.1	3.1

Table 2-4. Sulfur factor elemental concentrations, ng m^{-3}, and percent in each factor.

	Western group, 5 sites					Eastern group, 4 sites			
SURE site	Duncan Falls, OH (4)	Fort Wayne, IN (7)	Rockport, IN (5)	Lewisburg, WV (9)	Giles County, TN (6)	Montague, MA (1)	Scranton, PA (2)	Indian River, DE (3)	Research Triangle Park, NC (8)
Factor	(F4/4)	(F3/3)	(F3/3)	(F3/3)	(F3/3)	(F4/4)	(F4/4)	(F3/4)	(F4/4)
cS	2,136.16	2,681.90	3,488.72	2,252.19	3,120.58	1,331.61	2,157.87	1,681.66	2,439.23
%S	59.27	89.28	88.85	73.35	92.11	56.46	73.89	73.72	72.21
cAl	388.82	363.59	190.66	112.10	32.64	24.70	51.89	35.57	71.50
%Al	49.38	69.31	21.06	19.59	5.38	6.83	8.59	7.77	11.73
cSi	429.61	613.10	72.82	126.44	<60.00	136.40	80.13	92.93	332.50
%Si	16.18	41.25	2.68	9.79	<4.00	19.21	5.06	8.91	24.53
cFe	318.06	20.41	166.43	72.15	53.05	45.56	21.85	51.72	110.93
%Fe	47.31	4.52	25.73	27.54	15.62	25.03	5.75	20.43	39.65
cK	69.53	50.01	89.12	30.14	32.74	17.42	41.79	−5.13	27.16
%K	21.56	21.24	28.30	19.77	19.25	14.85	10.63	−2.51	15.01
cCa	−122.29	−240.46	−346.10	−61.07	−276.76	16.63	87.89	25.56	20.67
%Ca	−15.97	−22.07	−22.53	−10.69	−30.64	13.23	29.63	10.47	11.04
cPb	37.67	42.92	53.09	30.88	41.02	73.98	38.71	33.40	87.21
%Pb	24.19	22.71	22.65	26.64	26.43	31.54	26.11	24.24	23.61
cBr	−1.04	−4.65	−7.88	−1.03	−2.89	3.70	−0.73	−2.64	−3.89
%Br	−4.07	−14.43	−15.47	−4.66	−10.03	8.71	−3.03	−6.92	−6.04
cCl	52.23	119.87	−113.95	90.25	71.52	73.16	80.80	−1148.30	−101.90
%Cl	18.88	49.03	−35.26	32.25	25.77	33.95	42.09	−85.71	−25.35
cP	107.56	212.89	88.17	49.03	<20.00	53.35	97.29	<13.00	56.09
%P	22.99	45.92	16.42	11.45	<4.00	13.96	24.12	<4.00	12.06
Si/Al wt.	1.105	1.686	0.382	1.128	0.840	5.523	1.544	2.612	4.650
± SD	0.170	0.089	0.227	0.165	1.119	0.536	0.360	0.303	0.795

Table 2-5. Aluminum-rich factor elemental concentrations, ng m^{-3}, and percent in each factor.

	Western group, 5 sites					Eastern group, 4 sites			
SURE site	Duncan Falls, OH (4)	Fort Wayne, IN (7)	Rockport, IN (5)	Lewisburg, WV (9)	Giles County, TN (6)	Montague, MA (1)	Scranton, PA (2)	Indian River, DE (3)	Research Triangle Park, NC (8)
Factor	(F1/4)	(F1/3)	(F1/3)	(F1/3)	(F1/3)	(F3/4)	(F1/4)	(F1/4)	(F1/4)
cS	344.45	~40.10	291.95	567.66	94.34	312.10	126.97	125.65	501.08
%S	9.56	~1.33	7.44	18.49	2.78	13.23	4.35	5.51	14.83
cAl	476.87	183.01	520.42	359.05	398.87	235.03	437.07	401.76	433.53
%Al	60.56	34.89	57.47	62.76	65.70	64.98	72.37	87.82	71.11
cSi	698.12	846.73	2,290.10	910.95	1,328.74	333.23	1,486.67	817.26	877.52
%Si	26.29	56.96	84.14	70.57	81.08	46.93	93.88	78.38	64.73
cK	67.19	108.95	224.03	99.16	115.20	12.31	118.38	94.92	88.84
%K	20.83	46.27	71.14	65.03	67.73	10.50	30.11	46.49	49.10
cCa	467.83	1,192.17	1,406.53	417.07	912.08	32.57	73.59	46.90	47.72
%Ca	61.08	109.42	91.55	73.00	100.98	25.90	24.81	19.21	25.49
cFe	429.80	195.20	417.62	197.64	276.88	59.68	350.73	224.58	156.89
%Fe	63.93	43.25	64.55	75.44	81.52	32.79	92.25	88.71	56.07
cPb	<2.00	23.35	31.13	21.95	17.73	45.61	2.04	8.24	4.66
%Pb	<1.00	12.36	13.23	18.94	11.42	19.45	1.37	5.98	1.26
cBr	1.78	4.18	9.78	2.34	5.40	9.75	0.07	1.37	4.73
%Br	6.97	12.98	19.20	10.59	18.75	22.97	0.31	3.59	7.35
cCl	<6.00	<4.00	<50.00	<20.00	12.97	63.70	<10.00	<100.00	<60.00
%Cl	<2.00	<2.00	<15.00	<7.00	4.67	29.56	<5.00	<10.00	<15.00
cP	55.88	49.52	144.89	53.81	150.74	101.30	56.12	43.27	89.36
%P	11.94	10.68	26.99	12.56	26.76	26.50	13.91	12.89	19.22
Si/Al wt.	1.464	4.627	4.400	2.537	3.331	1.418	3.401	2.034	2.024
± SD	0.091	0.194	0.101	0.074	0.083	0.031	0.048	0.041	0.045

Table 2-6. Lead factor elemental concentrations, ng m^{-3}, and percent in each factor.

	Western group, 5 sites					Eastern group, 4 sites			
SURE site	Duncan Falls, OH (4)	Fort Wayne, IN (7)	Rockport, IN (5)	Lewisburg, WV (9)	Giles County, TN (6)	Montague, MA (1)	Scranton, PA (2)	Indian River, DE (3)	Research Triangle Park, NC (8)
Factor	(F3/4)	(F2/3)	(F2/3)	(F2/3)	(F2/3)	(F1/4)	(F2/4)	(F2/4)	(F2/4)
cS	537.24	137.31	45.24	174.82	139.33	229.08	430.35	129.75	104.68
%S	14.91	4.57	1.15	5.69	4.11	9.71	14.74	5.69	3.10
cAl	−53.79	<10.00	101.61	29.78	86.68	30.30	31.64	<8.00	15.65
%Al	−6.83	<2.00	11.22	5.21	14.28	8.38	5.24	<2.00	2.57
cSi	594.30	115.01	467.23	166.35	378.98	52.98	<15.00	<30.00	−20.74
%Si	22.38	7.74	17.17	12.89	23.13	7.46	<1.00	<3.00	−1.53
cK	69.19	60.17	15.28	28.31	<6.00	18.16	88.31	31.20	<2.00
%K	21.45	25.55	4.85	18.57	<3.00	15.48	22.47	15.28	<1.00
cCa	173.92	299.95	506.23	289.24	385.95	3.03	72.04	23.35	−4.54
%Ca	22.71	27.53	32.95	50.63	42.73	2.41	24.28	9.56	−2.43
cFe	109.10	178.13	95.89	22.27	84.43	41.92	<15.00	<15.00	37.89
%Fe	16.23	39.47	14.82	8.50	24.86	23.03	<5.00	<6.00	13.54
cPb	129.96	150.63	195.44	78.13	112.69	119.36	119.07	136.40	317.71
%Pb	83.45	79.71	83.38	67.39	72.60	50.89	80.32	99.00	86.02
cBr	23.07	29.96	39.47	16.95	21.06	18.91	22.11	30.01	53.63
%Br	90.26	93.00	77.50	76.60	73.12	44.54	91.75	78.55	83.25
cCl	24.90	−18.26	<30.00	<24.00	10.24	10.07	54.21	<300.00	−65.07
%Cl	9.00	−7.47	<10.00	<10.00	3.69	4.67	28.23	<25.00	−16.18
cP	42.61	−38.61	22.81	26.01	23.42	14.67	61.52	16.91	−14.16
%P	9.11	−8.33	4.25	6.07	4.16	3.84	15.25	5.04	−3.04
Si/Al wt.	−11.048	−23.229	4.598	5.586	4.372	1.748	−0.806	−2.009	−1.326
± SD	3.436	50.058	0.556	1.676	0.507	0.119	0.609	2.470	0.896
Br/Pb wt.	0.178	0.199	0.202	0.217	0.187	0.158	0.186	0.220	0.169
± SD	0.005	0.005	0.004	0.007	0.004	0.003	0.006	0.008	0.003

Table 2-7. Second dust factor elemental concentrations, ng m^{-3}, and percent in each factor.

	Western Group	Eastern group, 4 sites			
SURE site	Duncan Falls, OH (4)	Montague, MA (1)	Scranton, PA (2)	Indian River, DE (3)	Research Triangle Park, NC (8)
Factor	(F2/4)	(F2/4)	(F3/4)	(F4/4)	(F3/4)
cS	78.22	374.36	105.98	152.41	141.82
%S	2.17	15.87	3.63	6.68	4.20
cAl	~3.70	44.77	30.65	76.96	17.93
%Al	~0.47	12.38	5.08	16.82	2.94
cSi	801.38	222.65	102.78	197.92	197.38
%Si	30.17	31.36	6.49	18.98	14.56
cK	139.16	48.61	149.36	26.25	43.41
%K	43.15	41.45	38.00	12.86	23.99
cCa	53.37	83.72	57.05	159.18	118.79
%Ca	6.97	66.59	19.23	65.20	63.47
cFe	60.14	65.46	28.80	28.42	56.07
%Fe	8.95	35.96	7.58	11.23	20.04
cPb	2.09	35.10	4.94	<8.00	−22.73
%Pb	1.34	14.97	3.33	<6.00	−6.15
cBr	0.50	5.76	0.45	3.47	2.17
%Br	1.97	13.57	1.85	9.09	3.37
cCl	48.97	14.90	<2.00	77.97	106.33
%Cl	17.70	6.91	<1.00	5.82	26.45
cP	9.70	11.81	<4.00	−14.10	−20.14
%P	2.07	3.09	<1.00	−4.20	−4.33
Si/Al wt.	~217.	4.973	3.353	2.572	11.010
± SD	~159.	0.244	0.225	0.306	4.019

Table 2-8. Earth crust concentrations of elements and chemical equivalent weights.

Element	EC (ppm)	Atomic weight	Ionic charge	Equivalent weight
S	260	32.06	2−	16.03
Al	81,300	26.98154	3+	8.994
Si	277,200	28.086	4+	7.022
Ti	4,400	47.90	4+	11.975
Fe	50,000	55.847	3+	18.616
K	25,900	39.098	1+	39.098
Mg	20,900	24.305	2+	12.152
Ca	36,300	40.08	2+	20.04
Mn	950	54.9380	2+	27.469
Zn	70	65.38	2+	32.69
Cl	130	35.453	1−	35.453
P	1,050	30.97376	3−	10.325

series. As an example, consider one of the nine sites, Fort Wayne, Indiana (site 7), where soil dust was apparently rather prevalent.

In Figure 2–3 are plotted the absolute factor scores of three factors that represent Al-rich dust, automotive lead aerosol, and the mainly sulfur component, factors 1, 2, and 3, respectively. The vertical scale is proportional to concentration in air, and points are plotted every 2 hours. The horizontal scale is marked in days that begin at midnight, where dates <1 or >31 correspond to days in September or November 1977. Temporal variations of the three factors are complex, although upon close scrutiny some features can be seen. For instance, it seems that peak concentrations often occur in late afternoon for the Al-rich dust, but less often for lead aerosol and only rarely for sulfur. In addition to this apparently diurnal fluctuation, more gradual rises and falls in concentration are apparent for all three components.

The main temporal features are more efficiently portrayed by periodograms, shown in Figure 2–4, that help bring out the interplay of different frequencies of variation in each time series. (Alternatively, suspected periodicities, for example, diurnal, could be tested individually by composite data plots, such as by hour of the day, although such a procedure is inefficient for complex time series.) Periodograms are readily obtained using microcomputer software (STSC, 1987) by Fourier analysis in which different sine waves are calculated that sum to the time series data. The ordinate is a measure of the amplitude of a sine wave frequency plotted along the abscissa, so that a peak in this power spectrum indicates a relatively important frequency in the time series. Frequencies are given in cycles/sampling interval of 2 hours, up to the Nyqvist frequency of 0.5, equivalent to the shortest resolvable period of twice the sampling step length, that is, 4 hours. A diurnal period corresponds to 0.08333 on the abscissa, and longer periods are to its left. Diurnal variation is easily seen to be strong in Al dust, weaker in lead aerosol, and very weak in sulfur.

Table 2–9. Acid-base balance of sulfur factor at nine SURE sites, October 1977; chemical equivalent ratios to sulfur and their standard deviations.

SURE site		Al/S	Si/S	Fe/S	K/S	Ca/S	Pb/S	Br/S	Cl/S	P/S
Duncan Falls, OH (4)	Ratio ± SD	0.324	0.459	0.128	0.013	−0.046	0.003	0.000	0.011	0.078
		0.012	0.070	0.008	0.002	−0.007	0.000	0.000	0.002	0.008
Fort Wayne, IN (7)	Ratio ± SD	0.242	0.522	0.007	0.008	−0.072	0.002	0.000	0.020	0.123
		0.010	0.022	0.004	0.001	−0.016	0.000	0.000	0.001	0.010
Rockport, IN (5)	Ratio ± SD	0.097	0.048	0.041	0.010	−0.079	0.002	0.000	−0.015	0.039
		0.010	0.028	0.004	0.001	−0.011	0.000	0.000	0.007	0.006
Lewisburg, WV (9)	Ratio ± SD	0.089	0.128	0.028	0.005	−0.022	0.002	0.000	0.018	0.034
		0.009	0.014	0.002	0.000	−0.007	0.000	0.000	0.003	0.007
Giles County, TN (6)	Ratio ± SD	0.019	0.020	0.015	0.004	−0.017	0.002	0.000	0.010	−0.002
		0.009	0.025	0.002	0.001	−0.009	0.000	0.000	0.001	−0.010
Montague, MA (1)	Ratio ± SD	0.033	0.234	0.029	0.005	0.010	0.009	0.001	0.025	0.062
		0.003	0.010	0.002	0.000	0.001	0.000	0.000	0.002	0.004
Scranton, PA (2)	Ratio ± SD	0.043	0.085	0.009	0.008	0.033	0.003	0.000	0.017	0.070
		0.006	0.016	0.002	0.002	0.003	0.000	0.000	0.001	0.006
Indian River, DE (3)	Ratio ± SD	0.038	0.126	0.026	−0.001	0.012	0.003	0.000	−0.309	0.005
		0.010	0.021	0.003	−0.001	0.000	0.000	0.000	−0.044	0.007
Research Triangle Park, NC (8)	Ratio ± SD	0.052	0.311	0.039	0.005	0.007	0.006	0.000	−0.019	0.036
		0.008	0.020	0.002	0.001	0.001	0.000	0.000	−0.010	0.010

Table 2-10. Acid-base balance of the aluminum dust factor at nine SURE sites, October 1977; chemical equivalent ratios to sulfur and their standard deviations.

SURE site		Al/S	Si/S	Fe/S	K/S	Ca/S	Pb/S	Br/S	Cl/S	P/S
Duncan Falls, OH (4)	Ratio ± SD	2.468	4.627	1.074	0.080	1.086	0.000	0.001	−0.011	0.252
		0.191	0.445	0.086	0.010	0.086	0.001	0.000	0.008	0.036
Fort Wayne, IN (7)	Ratio ± SD	8.135	48.212	4.192	1.114	23.783	0.090	0.021	−0.022	1.918
		5.242	31.017	2.701	0.717	15.307	0.058	0.014	−0.051	1.278
Rockport, IN (5)	Ratio ± SD	3.177	17.908	1.232	0.315	3.854	0.016	0.007	0.034	0.771
		0.167	0.883	0.064	0.017	0.199	0.002	0.000	0.043	0.053
Lewisburg, WV (9)	Ratio ± SD	1.127	3.664	0.300	0.072	0.588	0.006	0.001	0.006	0.147
		0.051	0.142	0.013	0.003	0.032	0.001	0.000	0.009	0.025
Giles County, TN (6)	Ratio ± SD	7.536	32.516	2.527	0.501	7.734	0.029	0.011	0.062	2.481
		0.825	3.482	0.274	0.056	0.847	0.004	0.001	0.022	0.318
Montague, MA (1)	Ratio ± SD	1.342	2.438	0.165	0.016	0.083	0.023	0.006	0.092	0.504
		0.033	0.072	0.010	0.002	0.005	0.002	0.000	0.008	0.023
Scranton, PA (2)	Ratio ± SD	6.135	26.732	2.379	0.382	0.464	0.002	0.000	0.019	0.686
		0.744	3.232	0.288	0.055	0.066	0.002	0.000	0.017	0.114
Indian River, DE (3)	Ratio ± SD	5.699	14.849	1.539	0.310	0.299	0.010	0.002	−0.205	0.535
		0.396	1.024	0.108	0.022	0.020	0.004	0.001	−0.407	0.071
Research Triangle Park, NC (8)	Ratio ± SD	1.542	3.998	0.270	0.073	0.076	0.001	0.002	−0.069	0.277
		0.037	0.094	0.008	0.003	0.003	0.001	0.000	−0.030	0.030

Table 2-11. Sulfur and aluminum concentrations and chemical equivalent ratios of aluminum + silicon + iron to sulfur in sulfur-rich, F(S), and aluminum-rich, F(Al), aerosol components from factor analysis of element concentration data at nine SURE sites, October 1977.

SURE site	Concentration, ng m^{-3} Sulfur		Aluminum		Equivalent Ratio (Al + Si + Fe)/S ± SD	
	F(S)	F(Al)	F(S)	F(Al)	F(S)	F(Al)
Duncan Falls, OH (4)	2,136	344	389	477	0.911 ± 0.071	8.169 ± 0.492
Fort Wayne, IN (7)	2,682	40	364	183	0.771 ± 0.024	60.539 ± 31.573
Rockport, IN (5)	3,489	292	191	520	0.186 ± 0.030	22.317 ± 0.901
Lewisburg, WV (9)	2,252	568	112	359	0.245 ± 0.017	5.091 ± 0.151
Giles County, TN (6)	3,121	94	33	399	0.054 ± 0.027	42.579 ± 3.589
Montague, MA (1)	1,332	312	25	235	0.296 ± 0.011	3.945 ± 0.080
Scranton, PA (2)	2,158	127	52	437	0.137 ± 0.017	35.246 ± 3.329
Indian River, DE (3)	1,682	126	36	402	0.190 ± 0.023	22.087 ± 1.103
Research Triangle Park, NC (8)	2,439	501	72	434	0.402 ± 0.022	5.810 ± 0.101

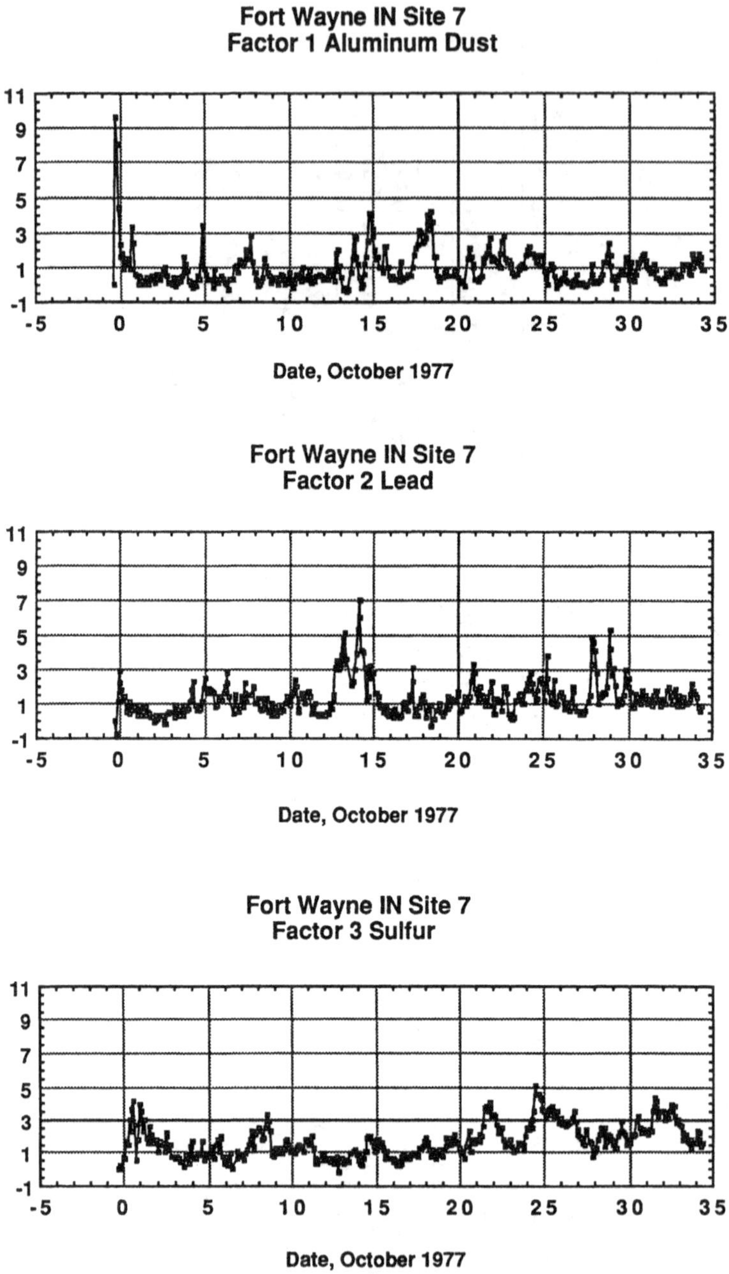

Figure 2–3. Absolute factor scores at Fort Wayne, IN, plotted as time series, late September to early November 1977, for the three factors—Al-rich dust, lead aerosol, and a mainly sulfur factor—that account for 84% of the variance of eight measured elemental concentration variables. A zero score means the factor was not present. The added zero sample (see text) is the first point plotted.

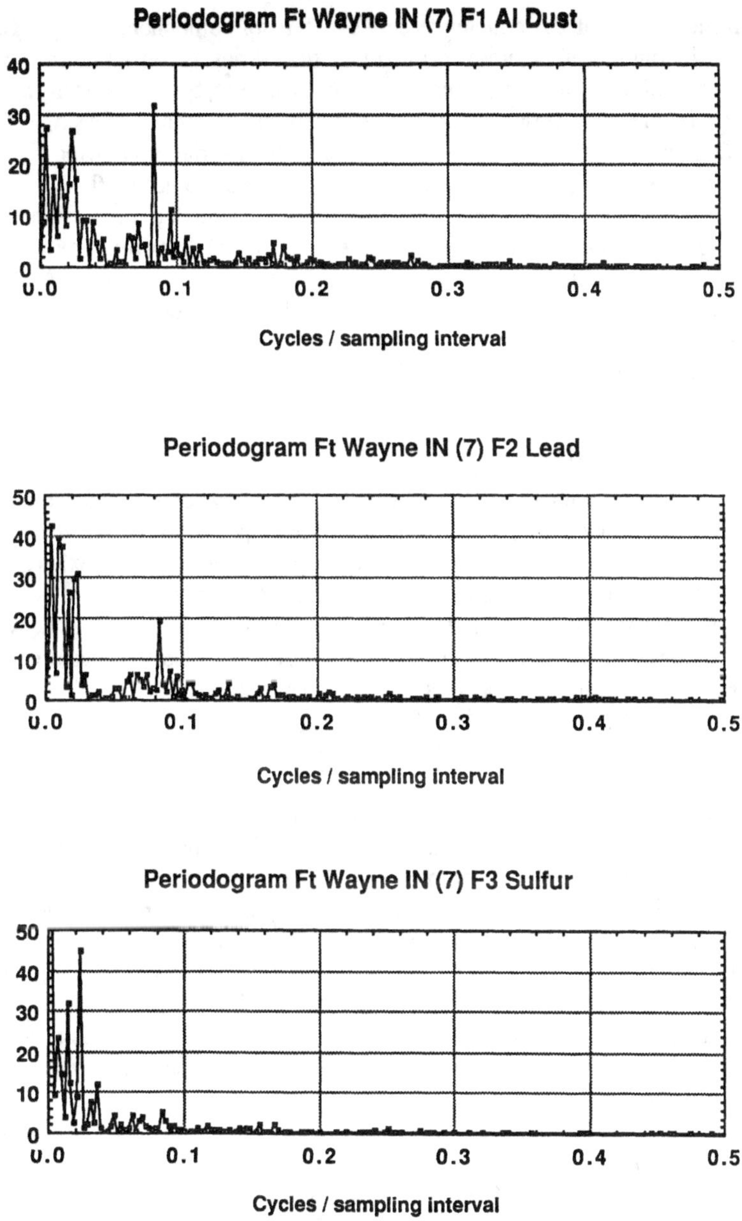

Figure 2–4. Periodograms of absolute factor score time series for the three factors of Figure 2–3 at Fort Wayne, giving the relative importance of a Fourier component on the ordinate and its frequency on the abscissa for a 2-hour sampling interval. A 24-hour period corresponds to 0.08333 cycle/sampling interval, and longer periods lie to its left.

An impression of the coherence between aerosol components can be gained from periodograms of the products of pairs of time series, given in Figure 2–5. For Al dust paired with lead, the diurnal frequency is present, but not as strong as other frequencies present. These appear to be relatively stronger than in the single-factor periodograms of Figure 2–4. They may be the result of variations in wind and other weather conditions that, although not documented for this investigation, may have controlled the transport of some of the dust and lead aerosol to the sampling site. The association suggests that some dust was generated on dusty roads by vehicular traffic. Some coherence between sulfur and Al dust appears for periods of several days, but that between sulfur and lead also for shorter periods. Thus, the mainly sulfur component may have some linkage to airborne soil dust but apparently more to automotive traffic.

C. Context of the Results

Although in the biochemical literature Al is rated low among trace metals in biological importance (Frieden, 1984), it has been reported to be toxic to a variety of bacteria as well as to fish through gill hyperplasia that leads to anoxia (Ochiai, 1987), and Al is prominent in the literature on the effects of acidic deposition on soils (Tabatabai, 1985) and freshwater ecosystems (Schindler, 1988). Water-soluble aluminum is reported to occur in Adirondack lakes of eastern North America at micromolar concentrations, higher than considered to be natural (Driscoll and Newton, 1985). Whereas SO_4^{2-} is attributed to wet and dry deposition of acidic sulfate air pollution, the high aquatic concentrations of Al have been considered to be due to ion exchange interactions of clays and humic substances with acidified water (Cronan et al., 1986; Walker et al., 1988). Nevertheless, although these interactions occur, "the terrestrial reactions which release aluminum are still not completely understood" (Schindler, 1988, note 63).

The results of this chapter suggest an additional possible source of soluble Al in surface waters of air-polluted regions, direct deposition from the atmosphere of soil dust aerosol that has already undergone dissolution by strong acidic attack while airborne. Such a process apparently has not been reported in the acidic deposition literature, although oceanographers have suggested dustfall as contributing to soluble Al concentrations measured in surface seawater (Orians and Bruland, 1986).

Soil minerals that contribute to ambient aerosol are known to be generally alkaline, and low pH can enhance the reaction of acids with bases. Acidic gases are calculated to diffuse quickly to particle surfaces, and accommodation coefficients, a measure of sticking probability, are high (Mozurkewich, 1986). Gas phase and liquid phase models (Calvert, 1983) predict that SO_2 should be oxidized and ultimately form H_2SO_4 liquid films on aerosol surfaces, either in cloud-free air or after cloud droplet evaporation. At ambient relative humidities, the pH of a surface film may be low (Ferek et al., 1983), and solubilization of even rather resistant mineral aerosol particles could take place within a few days of residence time in the atmosphere. Also HNO_3 may be present, although owing to its high vapor pressure acidic solutions could resist its uptake or retention. In atmospheres with excess

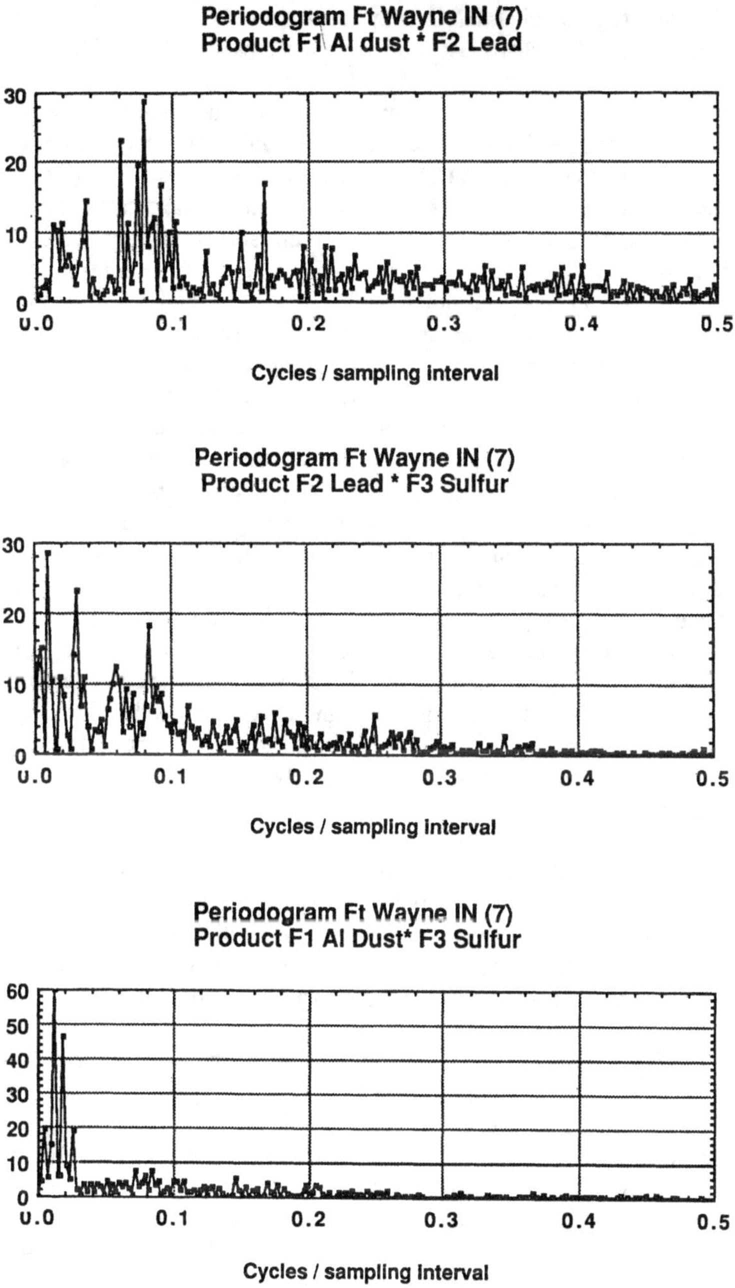

Figure 2–5. Periodograms of the product of pairs of absolute factor score time series of Figure 2–3 at Fort Wayne for three factor pairs.

aerosol alkalinity over acids, HNO_3 may add to the potential for airborne mineral dissolution, although perhaps less so if an acidic aerosol coating of H_2SO_4 should form.

The reaction of strongly acidic solutions with alkaline minerals may be more rapid with some than with others. On the ground dilute acids in rainwater may be expected to be neutralized by $CaCO_3$ more rapidly than by dissolution or even ion exchange reactions with alumino-silicate clays. If so, a simple calculation shows that acidic deposition of 50 mEq m^{-2} yr^{-1} onto soil containing 0.1% $CaCO_3$ would be completely neutralized by the $CaCO_3$ in less than 1 mm soil depth per year. Thus, Ca-containing and other alkaline constituents of soils may interfere with the mobilization of soluble Al in watersheds through mineral dissolution or cation exchange with acidic rain. However, presolubilization in the atmosphere could supply dissolved Al to the surface without this interference.

For present purposes, the important airborne Al concentration is not total aerosol Al but that portion that may have been rendered soluble by reaction with H_2SO_4. For example, this may be from dust that has reached high enough altitude to find SO_2 dispersed from tall stack emissions, as may be expected by soil erosion in strong winds. This component of Al must be distinguished from other Al-containing dust that is present at a sampling site, for instance, by the multivariate statistical approach carried out here. Using this approach some Al has been found within a mainly sulfur component and interpreted to represent an internal mixture of soil mineral with H_2SO_4 and their reaction products. If measured S represents SO_4^{2-} from H_2SO_4 and exceeds Al and other potentially alkaline mineral constituents in this component, and atmospheric conditions were favorable, the reaction may have gone appreciably toward complete solubilization of the mineral.

From the concentrations in air of elements contained in a resolved component, one can try to estimate their fluxes to the surface and judge whether they are sufficiently large to be of potential importance to watersheds. For an approximate mean flux, mean aerosol concentration may be multiplied by an approximate deposition velocity. Dry deposition velocities for sulfate aerosol are well under 1 cm sec^{-1}, but wet and dry deposition both share in the total of acidic deposition (cf. Calvert, 1983). Thus, without intending to arrive at a precise estimate, for nine sites in Table 2–4, one can compromise by multiplying the mean sulfur component S concentration, 2.4 ± 0.7 μg m^{-3}, by 1 cm sec^{-1} and obtain a mean S flux of about 500 equivalents per hectare per year (50 mEq m^{-2} yr^{-1}). This is not far from U.S. Environmental Protection Agency sulfur emissions inventory fluxes and acidic deposition fluxes typical for the region.

In Table 2–9 the sulfur component chemical equivalent ratios Al^{3+}/SO_4^{2-} range from 0.32 to 0.02 and average 0.10. If the aerosol S concentration in the sulfur component can be used as a guide to the deposition flux of S, and aerosol composition an indicator of wet as well as dry deposition, the Al deposition flux from the sulfur component should average 10% of this, 50 equivalents ha^{-1} yr^{-1} (5 mEq m^{-2} yr^{-1}), with midwestern fluxes being greater than those in the eastern region. Much of this may be Al that has been rendered soluble in the atmosphere.

The magnitude of the Al flux from the sulfur component to the surface is quite large, especially as midwestern and eastern parts of the United States are not considered to be particularly dusty. In order for a comparable soluble Al flux to result from chemical reactions in acidified soils, this simple line of reasoning suggests that an appreciable fraction of all acidic deposition must result in clay mineral Al mobilization. This percentage could be less if HNO_3 and other pollution acids also contributed to ion exchange processes that release soil Al, but the presence of other reactive inorganic and organic bases in soils may interfere with these processes. Thus, the reaction of pollution SO_2 or H_2SO_4 with soil minerals while airborne may in fact account for an important fraction of the soluble Al found in surface waters.

IV. Discussion

As a further test of this possibility, although at best semiquantitative, one may compare aerosol concentrations with 26-month average compositions of 20 Adirondack lakes that have been impacted by acidic precipitation (Driscoll and Newton, 1985), shown in Table 2–12. (Drainage basin areas of the lakes, not included in the publication, were kindly provided by Charles Driscoll.) Two, Barnes and Little Echo, are seepage lakes with inputs mainly by deposition onto the lake surfaces, whereas the rest are drainage lakes with inputs mainly from surrounding watersheds. Lake and basic areas, lake volumes and depths, as well as the concentrations of sulfate, calcium, and soluble aluminum are given. Assuming steady-state conditions and that lake basin sulfate is that contained mainly in the lake itself, and assuming a sulfate deposition flux of 50 mEq m^{-2} yr^{-1}, the mean residence time of sulfate can be estimated. It averages 1.2 years with a broad range of 0.042 to 6.36 years (1.96 and 4.37 years for the two seepage lakes), suggesting that lake water sulfate concentrations may reflect atmospheric deposition averaged over weeks to years of time. It may therefore be reasonable to attempt to compare average lake water soluble Al^{3+}/SO_4^{2-} ratios with the October 1977 averages measured in the atmosphere.

Lake SO_4^{2-} averages 129.6 μEq L^{-1} (SD 10%), whereas Al^{3+} is more variable and averages 5.5 μmol L^{-1} or 16.5 μEq L^{-1} (SD 100%). (The two seepage lakes average less: SO_4^{2-} = 65 and 78 and Al^{3+} = 3.3 and 3.6 μEq L^{-1}.) The range of equivalent ratio Al^{3+}/SO_4^{2-} = 0.007 to 0.44, average 0.122 and geometric mean 0.072 (SD × 3.1); for the two seepage lakes equivalent Al^{3+}/SO_4^{2-} = 0.051 and 0.046. At the Massachusetts and eastern Pennsylvania aerosol sampling sites, located on different sides of the Adirondacks, the October 1977 equivalent Al/S = 0.033 and 0.043, average 0.04, in the sulfur component, within a factor of two agreement with the lake average and perhaps fortuitously closer to the two seepage lake ratios. It would, of course, be desirable to have aerosol data over a full year or more. However, for now it seems that direct dry or wet deposition of acid-solubilized mineral Al with sulfate from the atmosphere to a watershed could account quite well for much of the soluble Al present in these lakes. It is conceivable that the same may be true for other areas of eastern North America that are impacted by acidic precipitation.

Table 2-12. Adirondack lake composition.[a]

Lake	Area, 10^4 m^2		Lake vol	Av lake	SO_4	Ca	Al		Basin SO_4	Res Time	Al/SO_4	Al/SO_4
	Lake	Basin	(10^4 m^3)	depth (m)	(mEq/m^3)	(mEq/m^3)	($mmol/m^3$)	pH	(mEq/m^2)	(yr [50])	(eq/eq)	(log_{10})
Arbutus	50.0	317.0	146.0	2.9	141.0	152.0	1.0	6.2	64.9	1.299	0.021	−1.672
Barnes	5.1	5.1	7.7	1.5	65.0	30.0	1.1	4.7	98.1	1.963	0.051	−1.294
Big Moose	520.0	9,555.0	3,600.0	6.9	140.0	93.0	8.9	5.1	52.7	1.055	0.191	−0.720
Black	32.0	373.0	199.0	6.2	130.0	191.0	0.3	6.8	69.4	1.387	0.007	−2.160
Bubb	20.0	264.0	42.0	2.1	131.0	108.0	1.8	6.1	20.8	0.417	0.041	−1.385
Cascade	40.0	689.0	172.0	4.3	139.0	159.0	2.8	6.5	34.7	0.694	0.060	−1.219
Clear	73.0	567.0	660.0	9.0	139.0	157.0	0.8	7.0	161.8	3.236	0.017	−1.763
Constable	23.0	1383.0	45.0	2.0	149.0	98.0	10.5	5.2	4.8	0.097	0.211	−0.675
Darts	144.0	10,749.0	415.0	2.9	139.0	97.0	7.6	5.2	5.4	0.107	0.164	−0.785
Heart	11.0	18.0	54.0	4.9	106.0	119.0	0.6	6.4	318.0	6.360	0.017	−1.770
Little Echo	0.8	1.0	2.8	3.5	78.0	36.0	1.2	4.3	218.4	4.368	0.046	−1.336
Merriam	7.8	277.0	11.5	1.5	137.0	58.0	19.0	4.5	5.7	0.114	0.416	−0.381
Moss	45.0	1,248.0	272.0	6.0	141.0	146.0	2.2	6.4	30.7	0.615	0.047	−1.330
Otter	10.0	409.0	22.0	2.2	138.0	88.0	5.0	5.5	7.4	0.148	0.109	−0.964
Rondaxe	92.0	14,185.0	220.0	2.4	134.0	112.0	4.4	5.9	2.1	0.042	0.099	−1.007
Squash	3.9	170.0	7.2	1.8	131.0	65.0	19.2	4.6	5.5	0.111	0.440	−0.357
Townsend		150.0			154.0	95.0	9.9	5.2			0.193	−0.715
West	12.0	184.0			111.0	94.0	6.6	5.2			0.178	−0.749
Windfall	1.5	399.0	4.9	3.3	141.0	143.0	5.6	5.9	1.7	0.035	0.119	−0.924
Woodruff	18.1	811.0	34.0	1.9	147.0	430.0	1.0	6.9	6.2	0.123	0.020	−1.690
Average				3.6	129.6	123.6	5.5	5.7	61.6	1.232	0.122	−1.145
± SD				2.1	22.3	81.3	5.6	0.8	85.3	1.706	0.121	0.485
Geometric Mean												0.072
× SD												(× 3.1)

[a] From Driscoll and Newton (1985).

Deposition of acid-solubilized Al in the respiratory tract may also occur while breathing in such areas. The October 1977 average concentrations of 0.39 to 0.025 $\mu g\ m^{-3}$ of aerosol Al in the sulfur factor at the nine sites, assuming daily inhalation by a human of 10 m^3 and 50% aerosol deposition efficiency, could deposit from 2 to 0.1 μg per day of soluble Al. Animal experiments with short exposures to high concentrations of aluminum sulfate (Drummond et al., 1986) suggest that prolonged exposures of humans to the measured concentrations in air could lead to detectable respiratory effects. The physiological fate as well as immediate effects of soluble Al after respiratory deposition are important to determine.

In addition to the results presented here, other data sets of aerosol composition measurements have been studied in order to ascertain whether air pollution sulfur is present in internal mixture with soil mineral elements. The data include remote parts of the Asia-Pacific region (Winchester and Wang, 1989). These remote data generally indicate some atmospheric sulfur uptake, approaching rather than exceeding a limit of sulfuric acid neutralization by alkaline mineral constituents. In contrast, the eastern North American data generally indicate an excess of sulfur over mineral constituents, implying excess acidity in polluted air over that required for complete mineral dissolution.

To summarize, in spite of the complexity of atmospheric processes in eastern North America, only three or four factors are needed to account for some 90% of the variance of a month of measured data at each of the nine sites from Indiana to Massachusetts. At all of these, a portion of aluminum is associated with aerosol sulfur in a statistically resolved component of the aerosol mixture in ambient air. Compared to the average concentrations of S in the sulfur component, those of Al are not small. Chemical equivalent ratios in the sulfur component of Al^{3+}/SO_4^{2-} average about 10%, with Al being relatively higher in the U.S. midwest within agricultural areas with highest predicted dust fluxes from soil erosion.

Therefore, it can be argued that soil dust particles tend to collect air pollution sulfur as aqueous sulfuric acid particle coatings. At ambient relative humidities, these coatings may assume molal concentrations and pH low enough to solubilize clay minerals. Because at all sites S exceeds Al + Si + Fe on an equivalent basis, excess sulfuric acid may coat dust particles more than that required for acid-base neutralization and dissolution. At sites in Massachusetts and eastern Pennsylvania, where average equivalent Al/S = 0.04 in the sulfur component, approximate agreement is found with 26-month average compositions of 20 Adirondack lakes, geometric mean equivalent Al/S = 0.072 (SD × 3.1). Direct dry or wet deposition of acid solubilized mineral Al with sulfuric acid from the atmosphere to a watershed could account for much of the soluble Al as well as sulfate in these lakes of the eastern United States that are impacted by acidic precipitation.

Acknowledgments

Aerosol composition measurements at the nine northeastern U.S. sites were made at Florida State University for a study supported in part by the U.S. Environmental

Protection Agency. I first employed the procedures for statistical analysis of the data using microcomputer software while on sabbatical leave at the National Bureau of Standards, Center for Analytical Chemistry, Gaithersburg, Maryland, and at the National Oceanic and Atmospheric Administration, Geophysical Monitoring for Climate Change laboratories in Boulder, Colorado, and subsequently for research funded by U.S. National Science Foundation grant ATM 8601967. Their support is greatly appreciated.

References

Barnard, W. R., G. J. Stensland, and D. F. Gatz. 1986. Water Air Soil Pollut 30:285–293.
Calvert, J., panel chairman. 1983. *Acid deposition processes in eastern North America.* National Academy Press, Washington, D.C.
Cronan, C. S., W. J. Walker, and P. R. Bloom. 1986. Nature 324:140–143.
Driscoll, C. T., and R. M. Newton. 1985. Environ Sci Technol 19:1018–1024.
Drummond, J. G., C. Aranyi, L. J. Schiff, J. D. Fenters, and J. A. Graham. 1986. Environ Res 41:514–528.
Fan, Song-Miao. 1985. Acidity of atmospheric aerosols in North Florida. M.S. Thesis, Dept. of Oceanography, Florida State University, Tallahassee.
Ferek, R. J., A. L. Lazrus, P. L. Haagenson, and J. W. Winchester. 1983. Environ Sci Technol 17:315–324.
Frieden, E. 1984. *In* E. Frieden, ed. *Biochemistry of the Essential Ultratrace Elements.* Plenum, New York.
Gillette, D. A., and K. J. Hanson. 1989. J Geophys Res 94:2197–2206.
Gillette, D. A., and R. Passi. 1988. J Geophys Res 93:14,233–14,242.
Johansson, T. B., R. E. Van Grieken, J. W. Nelson, and J. W. Winchester. 1975. Anal Chem 47:855–859.
Lindberg, S. E., G. M. Lovett, D. D. Richter, and D. W. Johnson. 1986. Science 231:141–145.
Lindberg, S. E., and R. R. Turner. 1988. Water Air Soil Pollut 39:123–156.
Ma Ci-Guang, Ge Ji-Rong, Li Min, and Shen Ji. 1987. Arid Environmental Monitoring 1(2):12–17 (in Chinese).
Mason, B., and C. B. Moore. 1982. *Principles of geochemistry*, 4th ed. Wiley, New York. Pp. 46–47.
Mozurkewich, M. 1986. Aerosol Science and Technology 5:223–236.
Nelson, J. W. 1977. *In* T. G. Dzubay, ed. *X-ray fluorescence analysis of environmental samples,* 19–34. Ann Arbor Science Publishers, Ann Arbor, Mich.
Ochiai, E. I. 1987. *General principles of biochemistry of the elements.* Plenum, New York.
Orians, K. J., and K. W. Bruland. 1986. Earth Planet Sci Lett 78:397–410.
Schindler, D. W. 1988. Science 239:149–157.
STSC. 1987. *Statgraphics, statistical graphics system by Statistical Graphics Corporation, user's guide.* STSC, Inc., Rockville, Md.
Tabatabai, M. A. 1985. CRC Critical Reviews in Environmental Control 15(1):65–110.
Tang, I. N., H. R. Munkelwitz, and J. G. Davis. 1976. J Aerosol Sci 9:505–511.
Thurston, G. D., and J. D. Spengler. 1985. Atmos Environ 19:9–25.
Walker, W. J., C. S. Cronan, and H. H. Patterson. 1988. Geochim Cosmochim Acta 52:55–62.
Winchester, J. W., and Wang Ming-Xing. 1989. Tellus 41B:323–337.

Source-Receptor Relationships for Atmospheric Trace Elements in Europe

Jozef M. Pacyna*

Abstract

Source receptor relationships are presented for trace elements emitted from various sources in Europe and measured at a number of remote locations. Quantitative and qualitative data for the major sources of trace elements in Europe are presented, and the emission data have been used in long-range transport models to calculate trace element concentrations and deposition patterns. These estimates are verified by measurements, and a few examples are given. It was concluded that the estimates of trace element emissions can be related to the air concentrations measured at remote locations. Statistical methods were found useful to assess the source category of trace elements measured at remote locations. However, it has been proven much harder to localize the sources. To do so, it is necessary to use these techniques together with the information on the meteorological situation, particularly on air mass trajectories and/or synoptic configurations, and accurate emission inventories.

I. Introduction

During the last few decades there has been a growing interest in studying the behavior of trace elements in the environment. Major sources of trace element emissions to the environment have been recognized, and quantitative assessments have been attempted. As the toxic hazards of various elements to human health and to the environment became recognized, the main emphasis in many European countries was placed on establishing the methods to control these hazards in the neighborhood of industrial sources. Legislative measures dealing specifically with the control of chemicals including trace elements have been enacted in many countries.

In the 1970s, however, high concentrations of anthropogenic trace elements were observed also in various remote regions. Unexpectedly high atmospheric

*Norwegian Institute for Air Research, P.O. Box 64, 2001 Lillestrøm, Norway.

turbidities in the Alaskan Arctic in the early 1970s (Shaw and Wendler, 1972) motivated chemical analysis in the whole Arctic region, and, particularly in winter, high concentrations of Pb, V, Mn, Cd, Ni, and other man-made elements were measured (e.g., Rahn and McCaffrey, 1979; Larssen and Hanssen, 1979; Barrie et al., 1981; Heintzenberg et al., 1981). Enhanced concentrations of Zn, Cu, In, W, Sb, Pb, Cd, and other atmospheric trace elements were also observed in other remote areas as the Antarctic atmosphere (e.g., Maenhaut et al., 1979) and over the eastern equatorial Pacific (Maenhaut et al., 1983). Adams and others (1980) have reported on high enrichment factors of trace elements in aerosols measured at the Chacaltaya Mountain in Peru and at Jungfraujoch in the Swiss Alps. Air particles, collected at the summit of Whiteface Mountain, New York, far away from industrial sources, contained high concentrations of many anthropogenic elements (Husain Samson, 1979). There are also many remote places in Europe, where similar observations have been made (e.g., Peirson et al., 1975; Ronneau et al., 1978; Lannefors and Hansson, 1980). Characteristically, the high concentrations of anthropogenic trace elements at remote regions are episodic and show strong seasonal variations for a majority of elements. This leads to the conclusion that the occurrence of these pollutants at remote locations is due to long-range atmospheric transport from the source regions.

In many cases high concentrations of anthropogenic trace elements are accompanied by increased concentrations of sulfates, particularly in the fine particle fraction (<1.0 μm in diameter). Such similarities suggest that trace elements can be used to apportion other pollutants, as, for instance, the precipitation sulfate (Rahn and Lowenthal, 1984). However, one should not underestimate differences between the scavenging rates of particles of different size and solubility and the scavenging of water-soluble gases like SO_2 and HNO_3 (Hidy, 1984).

In the following, source-receptor relationships are presented for trace elements emitted from various sources in Europe and measured at a number of remote locations. Quantitative and qualitative data for the major sources of trace elements in Europe are presented, and the emission data have been used in long-range transport models to calculate trace element concentrations and deposition patterns. These estimates are verified by measurements, and a few examples are given. Various techniques can then be used to assess the contribution of different sources to the trace element concentrations in the air and their deposition at a given receptor point.

II. Emission Patterns of Atmospheric Trace Elements in Europe

The first step in all source-receptor modeling is to prepare an accurate and complete set of emission data. In Europe a few attempts have been made to establish an emission data base for trace elements on a regional scale. The atmospheric emissions of Cd from various sources in the European community have been estimated by Rauhut (1978), van Wambeke (1979), van Enk (1979), and Hutton (1982) with a major goal to use them in migration models, which stimulate the pathways of Cd through the environmentt and give average Cd concentration levels in the atmosphere, soil, and water. A Cd emission survey was also a subject of study at the OECD Air Management Policy Group as a part of their work on the control of specific toxic substances in the atmosphere (1984). Cadmium, often referred to as the typical element of industrial origin in the atmosphere, has also been studied in individual European countries. Emission surveys for FRG have had the longest history (e.g., various UBA studies: 1977, 1981, 1982; VDI, 1975, 1984; Sartorius et al., 1977). A major aim of these studies was to present an emission survey that could be used by policy makers to control the state of the environment. Emission surveys for Cd have also been prepared for the United Kingdom (Hutton and Symon, 1986), the Netherlands (e.g., Kendall et al., 1985), and Sweden (e.g., NSV, 1982).

The Pb emissions have also been estimated in Europe. The main source of Pb is gasoline combustion, and the emissions are often calculated by multiplying the gasoline consumption by the Pb content of gasoline in a given country. For other trace elements, far less information is available about national atmospheric emissions in the European countries.

Among source categories, coal combustion in electric power plants and industrial boilers is one of the major sources of atmospheric trace elements and as such a subject of emission inventories (e.g., CEC, 1981; Brumsack and Heinrichs, 1984; Braun et al., 1984).

The emission surveys mentioned above were very valuable for the development of control strategies in Europe, but less applicable for modeling the long-range transport of air pollutants. A first preliminary survey of the atmospheric emissions of 16 trace elements from all major sources in Europe was prepared for the base year 1979/1980 (Pacyna, 1984). The estimates were based on emission factors calculated separately for each of the European countries and statistics on the consumption of raw materials and the production of various industrial goods. An *emission factor* is defined as the statistical average or a quantitative estimate of the rate at which a pollutant is released to the atmosphere as a result of an activity, such as combustion or industrial production, divided by the level of that activity. Thus, the emission factor relates the quantity of pollutants emitted to some indicator such as production capacity, rate of raw material usage, or quantity of fuel burned. The emission factors are often based on either stack measurements or material balances, including information about raw material characteristics, collector removal efficiencies, and industrial technology profile.

The 1979/1980 survey has caused some discussion with respect to the accuracy of the emission figures for specific source categories and was modified (Pacyna, 1985). The major emphasis was placed on (1) the concentrations of trace metals in raw materials, (2) the production technology employed in various industries, and (3) the type and efficiency of control installations. The emission factors in Figure 3–1 (Pacyna and Münch, 1987) illustrate differences between the European countries in this respect. For combustion sources, the emission factors differ, often by one order of magnitude. The total emissions of several trace elements from

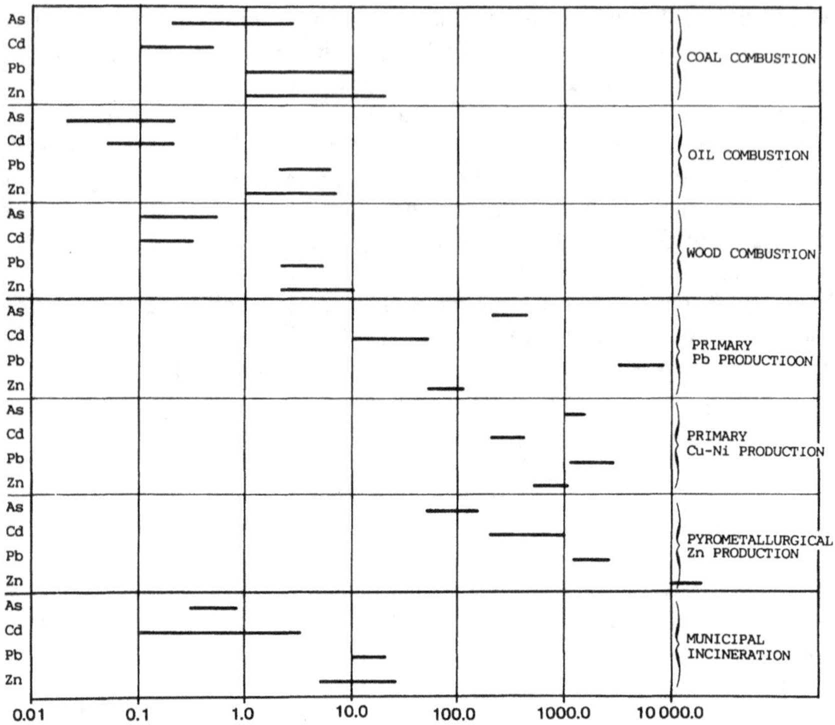

Figure 3–1. Emission factors for trace elements from industrial sources in Europe.

combustion processes, smelter operations, and other industrial processes in Europe are given in Table 3–1. The emission estimates were based on the emission factors from Figure 3–1. The elements Be, Co, Hg, Mo, Sb, and Se are emitted chiefly from the combustion of coal, whereas Ni and V are released mainly from oil combustion. Smelters and secondary nonferrous foundries release the largest amounts of As, Cd, Cu, and Zn. The two elements Cr and Mn are emitted mainly from iron, steel, and ferro-alloy production. Finally, Pb enters the environment primarily as a result of gasoline combustion.

Source-receptor models require emission data to be spatially distributed. This means that the total emissions for a given country need to be allocated to administrative or geographical units or certain grid cells. The spatial distributions of the As, Cd, Pb, Sb, and V emissions in Europe are shown in Figure 3–2. The grid system of 150 km by 150 km is the same as used previously in the UN ECE EMEP program for SO_2 emissions in Europe (e.g., Saltbones and Dovland, 1987), presented in Figure 3–3. Source emission patterns of trace elements in Figure 3–2 show significant differences that may be useful in developing "fingerprints" for source apportionment models. The similarity of the emission patterns of some trace elements and SO_2 suggests that trace elements may serve as indicators of the

Table 3-1. Trace element emissions from major anthropogenic sources in Europe in 1980 (in $t \cdot y^{-1}$).

Element	Combustion processes	Smelter operations	Other industrial processes	Total
As	720	3,420	150	4,290
Be	50	—	—	50
Cd	310	660	190	1,160
Co	2,000	—	—	2,000
Cr	2,780	—	16,100	18,880
Cu	4,920	8,660	1,920	15,500
Hg	220	80	40[b]	340
Mn	2,500	300	14,880	17,680
Mo	850	—	—	850
Ni	13,750	1,780	470	16,000
Pb	77,650[a]	6,170	16,220	100,040
Sb	270	10	100	380
Se	370	10	30	410
V	28	—	—	28,730
Zn	7,730	54,270	16,360	78,360
Zr	1,700	—	—	1,700

[a] Including 74300 $t \cdot y^{-1}$ from gasoline combustion.

[b] Very likely that this amount is far underestimated.

origin of acidic aerosols measured at remote receptors. However, this suggestion should be taken with some caution as SO_2 behaves differently from the rather unactive trace elements during long-range transport.

III. Long-Range Transport Models for Trace Elements in Europe

The development of emission surveys in Europe has made it possible to relate enhanced concentrations of trace elements in remote aerosols and deposition to their sources. As these increased concentrations were often observed only during some short periods, called *episodes*, their origin has been discussed using air mass trajectories.

At the beginning of the 1980s, extensive programs started to study the long-range transport of trace elements to remote regions in Scandinavia. It is not surprising that these studies started in Scandinavia, given the synoptic situations in Europe. Generally, Europe is situated in the westerlies, which means that the annual mean air flow passes from western Europe into eastern Europe and Scandinavia in a northeasterly direction along the Baltic Sea. In addition, it means that when easterly winds bring the polluted air masses out over the Atlantic, these air masses sooner or later will be brought back to Europe and generally in a

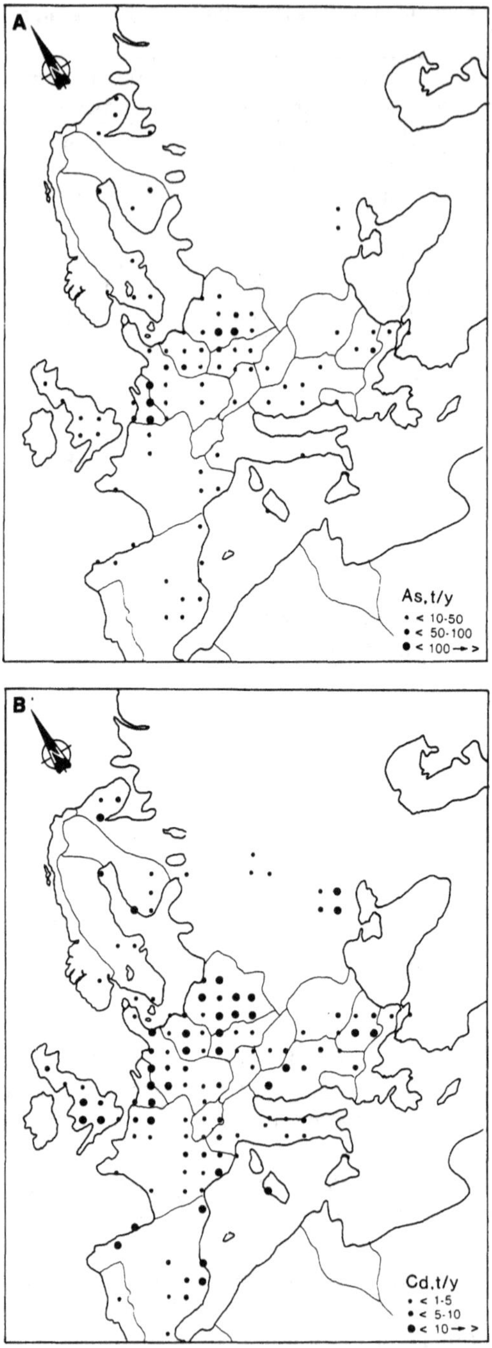

Figure 3–2. Spatial distributions of the As (A), Cd (B), Pb (C), Sb (D), and V (E) emissions in Europe within the EMEP grid of 150 km by 150 km.

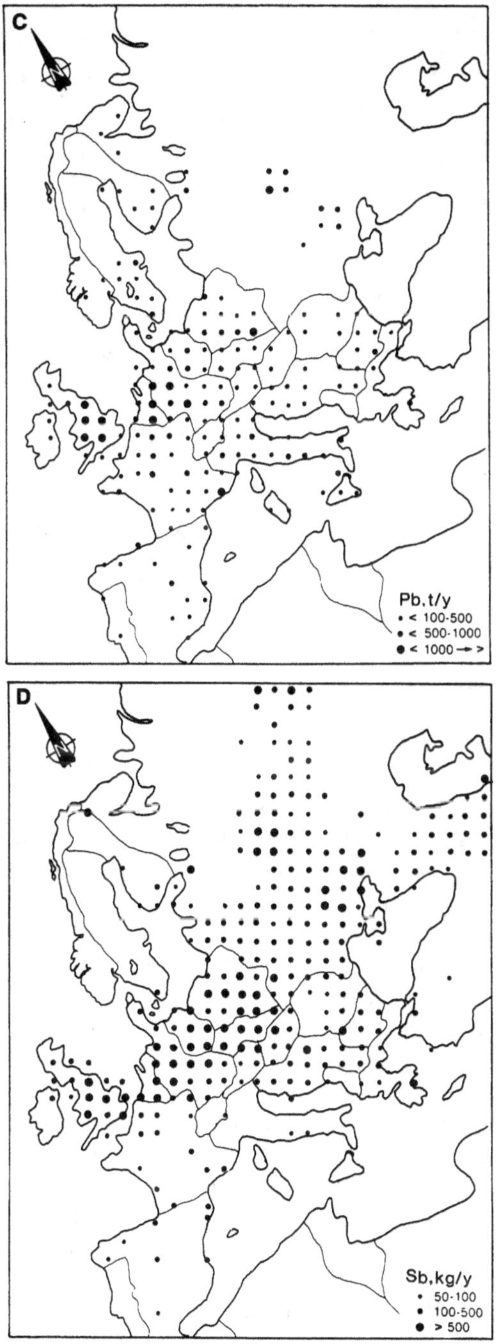

Figure 3-2. (*Continued*)

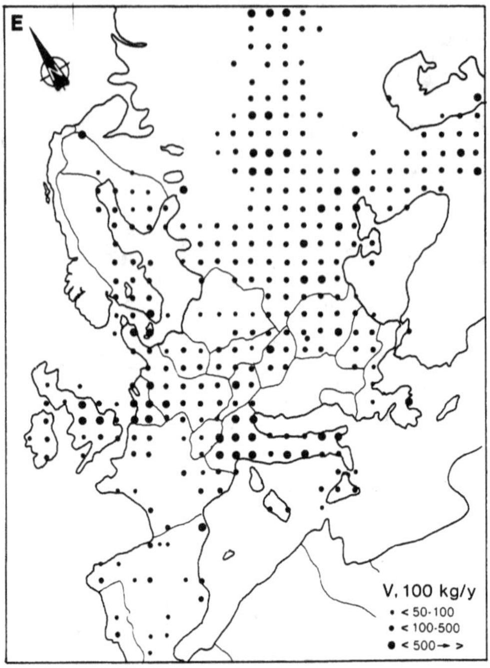

Figure 3-2. (*Continued*)

northeasterly direction toward the west coast of Scandinavia. A recent discussion of the EMEP data (Pacyna, 1987a) illustrates this situation. A correlation analysis of the sulfate concentrations measured at different EMEP stations showed correlation coefficients in the range of 0.8 to 1.0 between observations at stations in central Europe and observations a few days later in Scandinavia.

Lannefors and Hansson (1980) concluded that the concentrations of many trace elements in southern Scandinavia were one order of magnitude higher in air masses that had passed over the European continent compared to concentrations in air masses from the North Atlantic region. This also applies to particulate sulfates. The enrichment factors of S, V, Ni, Cu, Zn, Pb, and to some extent Mn relative to Ti were high during the episodes of long-range transport of polluted air and showed a strong seasonal variation, with the highest concentrations during winter, except for S (Lannefors et al., 1983). Most of the mass of trace elements was found in the accumulation mode of particles (0.1 to 2.5 μm EAD)*, but the authors indicated that there were some differences in the size distribution. According to Martinsson and others (1984), size distribution evaluations indicated that the components Cu, Ni, V, Pb, S, Zn, K, and Mn, listed in order of increasing mass

*EAD = equivalent aerodynamic diameter.

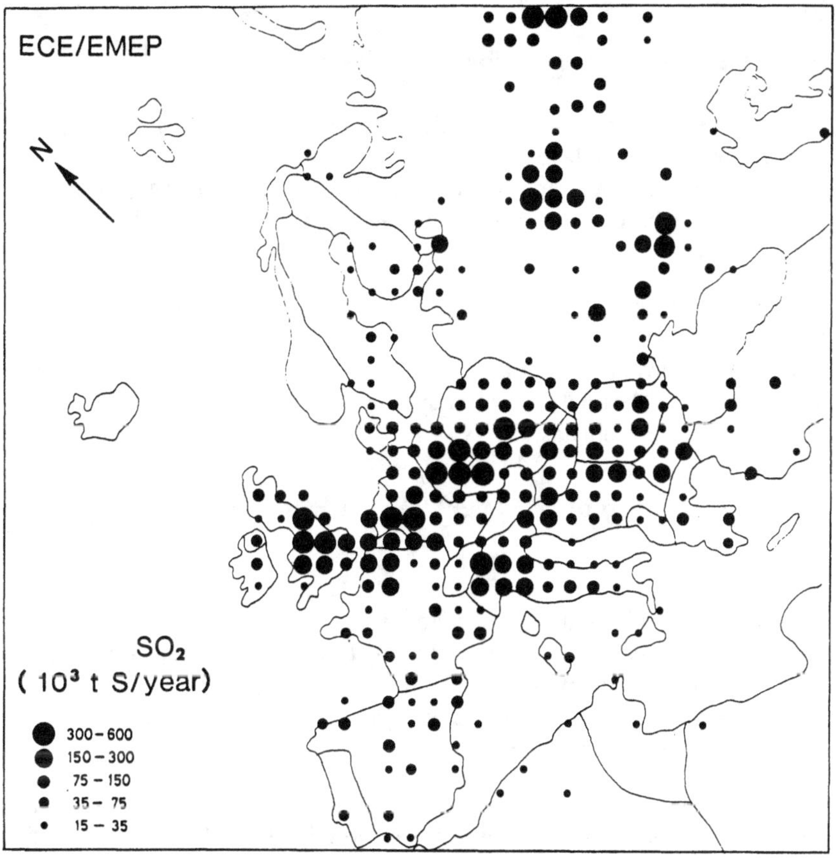

Figure 3–3. Spatial distribution of the SO_2 emissions in Europe within the EMEP grid of 150 km by 150 km (on the basis of data by Saltbones and Dovland, 1987).

median aerodynamic diameter, mainly originated from high-temperature sources. Lannefors and others (1983) concluded that foreign sources contribute on the order of half to three-quarters to the yearly average concentrations of S, Ni, and Pb as found in southern Sweden, and about a third in the case of V.

A first model of the trace element transport with air masses to Scandinavia (Pacyna et al., 1984a) was developed soon after the European emission survey was completed. The model is similar to that used in the OECD program on long-range transport of sulfur compounds in Europe (OECD, 1979; Eliassen and Saltbones, 1983). The following mass balance equation was used in the model:

$$\frac{dq}{dt} = (1 - \alpha) \cdot \frac{Q}{h} - k \cdot q \qquad (1)$$

where

q = trace element concentration in air, ng · m^{-3}
t = time, s
Q = trace element emission per unit area and time, ng m^{-2} s^{-1}
h = height of mixing layer, m
k = decay rate for the trace element considered (wet and dry deposition), s^{-1}
α = fraction of trace element emission deposited in the same grid element as it is emitted

The mass balance equation was integrated along the calculated trajectories. Because Equation 1 is linear in the concentration q, the contributions from emissions at individual positions along a trajectory can be considered separately and added to give the total concentration. For a trajectory consisting of N sections or timesteps Δt, one obtains:

$$q(N\Delta\, t) = q(\text{o}) \quad e^{-k\,N\,\Delta t} + \sum_{i=0}^{N-1} \; (1-\alpha)\,\frac{Q_i\,\Delta t}{h}\,e^{-k(n-i)\,\Delta t} \qquad (2)$$

where

$q(N\Delta\, t)$ = trace element concentration at the end of the trajectory, ng m^{-3}
$q(\text{o})$ = trace element concentration at the start of the trajectory, ng m^{-3}
Q_i = trace element emission in the i-th grid, ng m^{-2}s^{-1}
N = number of trajectory timesteps
Δt = length of timestep, s

The above model has been used to explain the origin of aerosols at three stations in southern Scandinavia (Birkenes in Norway, Rörvik in Sweden, and Virolahti in Finland), measured during episodes of pollution transport in the period February to June, 1980 (Pacyna et al., 1984b). The measured and calculated concentrations of various elements during one of the episodes (April 14 through 16) are shown in Figure 3–4 for Birkenes and Rörvik. The peaks of the Cu, V, Ni, Mn, and Mo concentrations in Virolahti appeared 2 to 3 days later. It should be mentioned that the SO_4^{-2}, NO_3^-, and Cl^- concentrations followed the concentrations of the the trace elements and particularly the concentrations of As, Pb, and V (Pacyna et al., 1984b). The 850 mb trajectories for the period April 14 through 16, 1980, indicated that the air masses had passed over industrial regions of the Netherlands, the Federal Republic of Germany, and the German Democratic Republic before reaching Birkenes and Rörvik. Agreement within a factor of two between measured and estimated concentrations of trace elements (Figure 3–4) was observed for most of the episodes during the measurements. It is interesting to discuss possible reasons that could have caused the difference. Three major groups of uncertainties should be taken into account: (1) the emission estimates, (2) the deposition processes, and (3) the meteorological data. The uncertainties of the emission estimates have partly been discussed. The completeness of emission inventories for sources such as waste incineration, industrial application of metals,

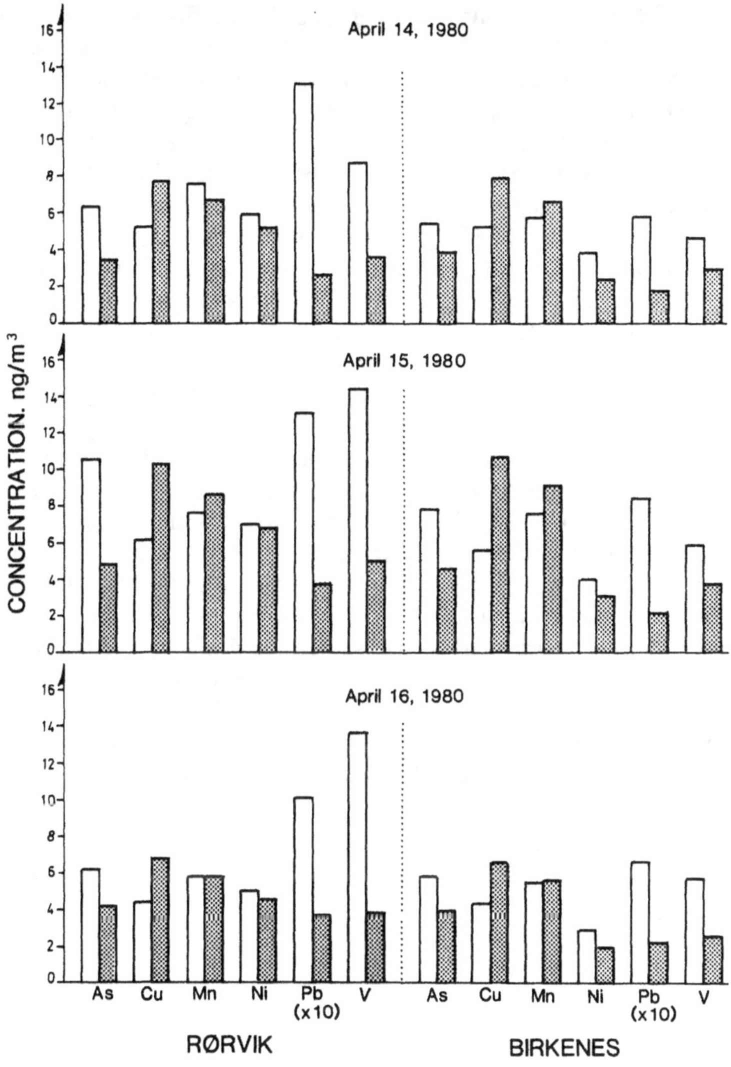

Figure 3–4. Measured (open bars) and calculated concentrations of trace elements (solid bars) at Birkenes and Rörvik in April 1980 (Pacyna, 1986a). The Pb concentrations should read as multiplied by 10. (Copyright © 1986 by D. Reidel Publishing Company. Reprinted by permission.)

and combustion of fuels in residential burners is often doubtful because of a lack of information on the chemical composition of the input material (e.g., municipal and industrial wastes and residual oil), control equipment, and sometimes location (important for the spatial distribution of the emission estimates). Besides, the existing emission inventories for trace elements give the sum of the fine and coarse fractions of the emitted particles, whereas only the first is important when long-range transport is studied. This may result in some overestimations of the expected concentrations at remote locations for trace elements such as Cu, Mn, and Ni. Then again, a part of the elements in the coarse fraction of particles seems to be easily deposited after emission, particularly in the emission area itself. This local deposition has been a subject of source-receptor studies in industrial regions. Measurement around copper and lead smelters and coal-fired power plants in Poland (e.g., Glowiak et al., 1977; Pacyna, 1980) concluded that up to 15% of the mass of the emitted trace elements may be deposited in the same area as it is released (the area of 150 by 150 km). This local deposition factor varies substantially, and only 5% was assumed in the OECD project on sulfur transport in Europe (OECD, 1979). Key factors in this connection are (1) physical-chemical properties of trace elements (e.g., volatility, tendency to condense on fine particles), (2) temperature of exhaust gases, (3) height of a stack, and (4) meteorological conditions. Also the grid size of the model used may be important. However, a major portion of emitted trace elements is transported over long distances and a part of these pollutants is removed by dry and wet deposition.

Dry deposition velocities vary substantially for trace elements, depending on the particle size and the roughness of the surface where the deposition takes place, for example, crops, grass, soil, snow, or forests. The meteorological conditions are also important, particularly wind velocity and stability. Ranges of dry deposition velocities for some trace elements are shown in Figure 3–5. Wet deposition seems to be more important than dry deposition for most trace elements and should not be neglected in modeling their transport. Cawse (1980) concludes that wet deposition can contribute 80% to 100% to the total deposition of Pb, Zn,

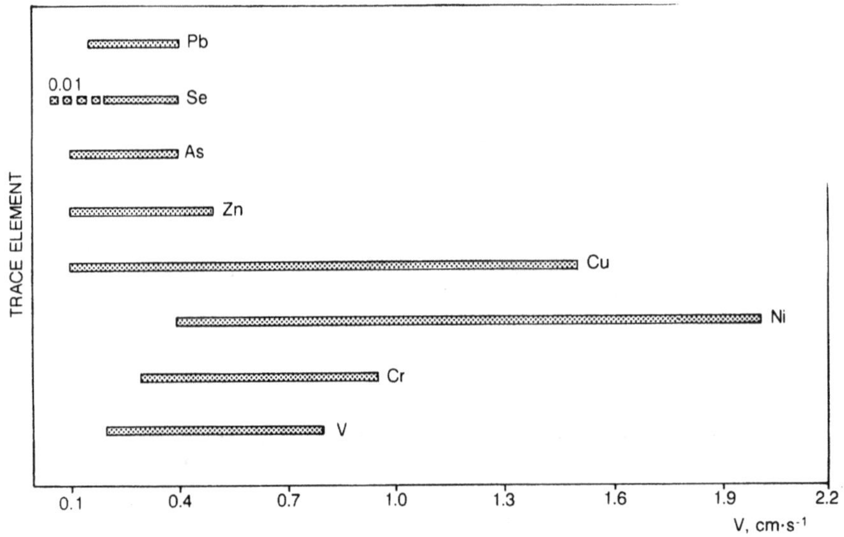

Figure 3–5. Average deposition velocities of trace elements (in cm · s^{-1}).

Cu, and Co, and 60% to 80% for Ni, As, Sb, Cr, and Se. Conversely, up to 65% of Mn and V seems to be dry deposited. A recent study by Michaelis (1987) does not, however, confirm the above suggested wet and dry contributions for Cr and Pb. This latter study suggests only 40% to 50% for Pb and 10% to 30% for Cr as wet contributions, and up to 60% for V. The likely explanation of these differences is that the assessment by Cawse (1980) is on the basis of measurements in rural land areas, whereas Michaelis conducted his measurements on the shore of the North Sea. This illustrates how important but difficult it is to assume deposition velocities when modeling the atmospheric transport of trace elements.

Even more difficult is assessing to what extent inaccuracies in trajectory computations may contribute to the total uncertainties of the source-receptor models presented above. Recently Miller (1987) has presented a review on the use of back air trajectories in interpreting atmospheric chemistry data. Calculation from on-site meteorological measurements is still the only way to determine the time path of a particle in the atmosphere. Miller (1987) presented advantages and disadvantages of different types of trajectories, namely, dynamic (isobaric and isentropic), kinematic, and mixed. He concluded that, despite its limitations, the single, back-trajectory analysis method is very useful in the interpretation of atmospheric chemistry data, but no total assessment of the uncertainties of this method is available.

The trajectory model based on the mass balance Equation 1 was also used to study the long-range transport of trace elements to the Norwegian Arctic (Pacyna et al., 1985). An emission inventory for 16 elements from 12 major source regions in the USSR was prepared (NILU, 1984) and used together with the European survey presented earlier in this work. The model was used to calculate concentrations during the March 1983 episode of air pollution transport to the Arctic. An agreement between estimates and measurements within a factor of two to three was obtained for Sb, Pb, Se, V, As, and Cr.

A more advanced model, but with the same emission data, has been developed by Krell and others (1986). Their three-dimensional trajectory model based on the Monte Carlo method was used to study the long-range transport of Pb over the North Sea. The same model was previously used to study the long-range transport of S compounds in Europe. The results showed that the calculated and measured monthly mean deposition and surface concentrations of Pb over land (two stations in Denmark, two in the Federal Republic of Germany, and two in the United Kingdom) differ by a factor of two.

Summarizing, the measured air concentrations and deposition of trace elements at remote locations in Europe and the Arctic can be related to the emission estimates with the help of trajectory models. The source-receptor relationships presented in this section apply mainly to episodes of pollution transport and show an agreement between measurements and estimates by a fairly constant factor of ± 2 for most of the trace elements considered. Different values of this magnitude are often encountered in simpler situations on local scales where the pollution sources have been better characterized (e.g., Kowalczyk et al., 1982).

It should be mentioned that in this comparison a predicted mixed-layer average concentration is related to a ground-level observation. However, during the episodes of pollution transport, the lower troposphere at remote locations often seems to be homogeneously mixed, as shown by the aircraft measurements in the Arctic (Ottar et al., 1986).

Characteristically, S behaves similarly to the trace elements, and the models of trace element transport within air masses presented in this section were also used to study the transport of sulfur.

IV. Application of Statistical Methods in Source-Receptor Studies

Statistical methods such as multiple regression analysis, cluster analysis, time series analysis, principal component or factor analysis, and discriminant analysis have found several applications in studies of the origin of aerosols (e.g., Gaarenstrom et al., 1977; Raabe, 1978; Finzi et al., 1979; Horowitz and Barakat, 1979; Henry and Hidy, 1979; Hopke et al., 1980; Thurston and Spengler, 1981; Lowenthal and Rahn, 1985). They were also used to assess the impact of metal emissions in Europe on the concentrations measured in Scandinavia and the Arctic.

A sector analysis of daily mean concentrations from the measurements at Birkenes, Rörvik, and Virolathi in 1980 (already mentioned in the previous section) has been performed, and the sectorial contributions of Pb and S are shown in Figure 3–6 as an example (Pacyna et al., 1984b). Generally, the two southern sectors, which can be called the *European sectors,* dominated at Birkenes and Rörvik. The ratio of the average concentrations in the northern sectors at Birkenes to the average concentrations in the southern sectors can be used to assess roughly the Norwegian and the foreign contributions. The same ratio calculated for Rörvik may separate the Swedish and foreign contributions; however, an effect of emission sources in Norway cannot be excluded. The Norwegian relative contributions on fine mode particles at Birkenes and the relative contributions of Swedish fine mode particles at Rörvik are presented in Table 3–2 for some anthropogenic pollutants. For most of them, the contribution from the European emissions dominate. The northeastern section was, however, of equal importance as the southern sectors at Virolahti, and the emission sources from the USSR seem to affect the concentrations in this sector.

Although the above sector analysis is a useful method to assess the major direction of pollution transport to a given receptor, the method is of little help to localize the emission sources. In order to relate the measured concentrations of trace elements at the three stations discussed above to the emissions in some potential regions, sector analysis has been performed by considering the elemental composition of each sector at each station (Pacyna et al., 1984b). The element concentrations relative to the Se concentrations were calculated for trace elements with more than 60% of mass in the fine fraction of the particules. The percentage contributions of trace element concentrations in the fine fraction of particles to the

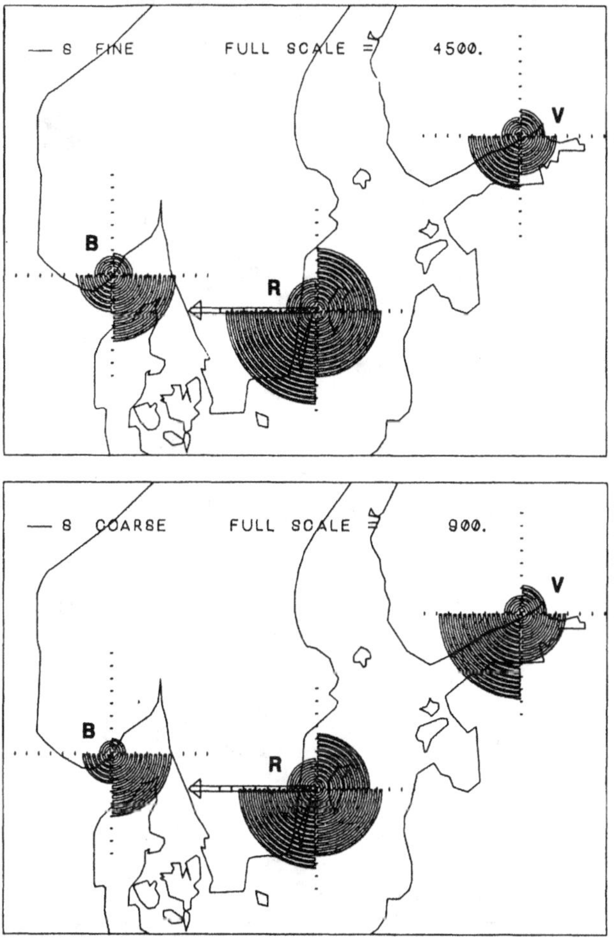

Figure 3–6. Sectoral concentrations of S and Pb in fine and coarse fractions of particles at Rörvik (R), Virolathi (V), and Birkenes (B).

concentrations in both the fine and coarse fractions are shown in Figure 3–7. The calculations of the trace element concentrations relative to the Se concentrations are given in Table 3–3 for two sectors: southeast and southwest. The ratios for individual trace elements within a sector are similar for all three stations, supporting the hypothesis of a similar history of the air masses. However, it should be noted that these ratios are built on average concentrations measured in the fine fraction of particles. To distinguish between eastern and western European aerosols affecting receptors in Scandinavia, only concentrations during episodes of long-range transport from a given area can be used. Based on the meteorological information (850 mb trajectories) and daily measurements at Birkenes and Rörvik, the elemental tracers were calculated for the eastern and western European

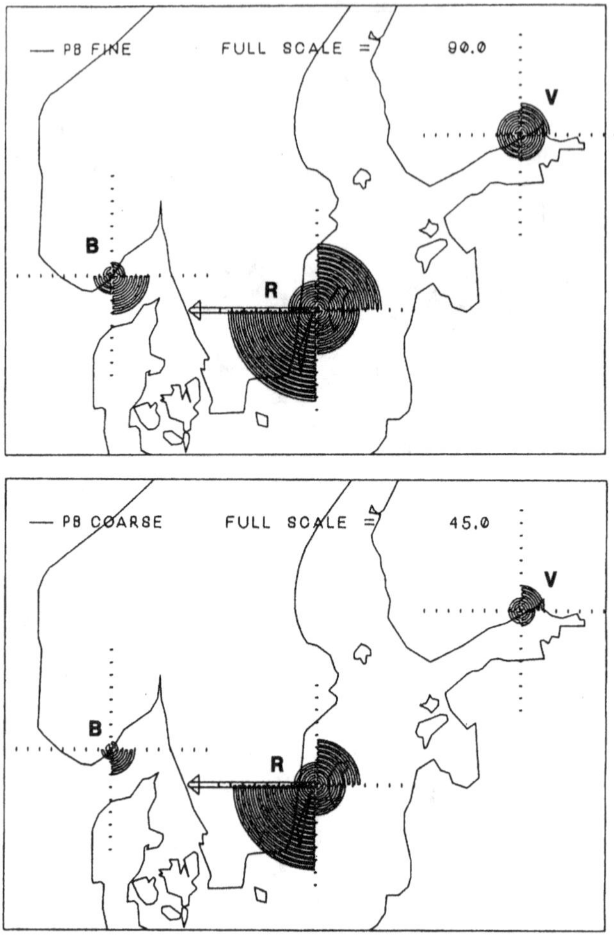

Figure 3-6. *(Continued)*

Table 3-2. Norwegian fine mode (<2 μm) relative contributions of totals (%) at Birkenes and Swedish fine mode relative contributions of total at Rörvik.

Element	Birkenes (this work)	Rörvik (this work)	Rörvik (Martinsson et al., 1984)
V	35	43	40
Mn	52	29	10
Ni	50	40	45
Zn	30	38	10
Pb	30	38	—
As	30	—	—
Cr	27	75	—
Cd	—	39	—

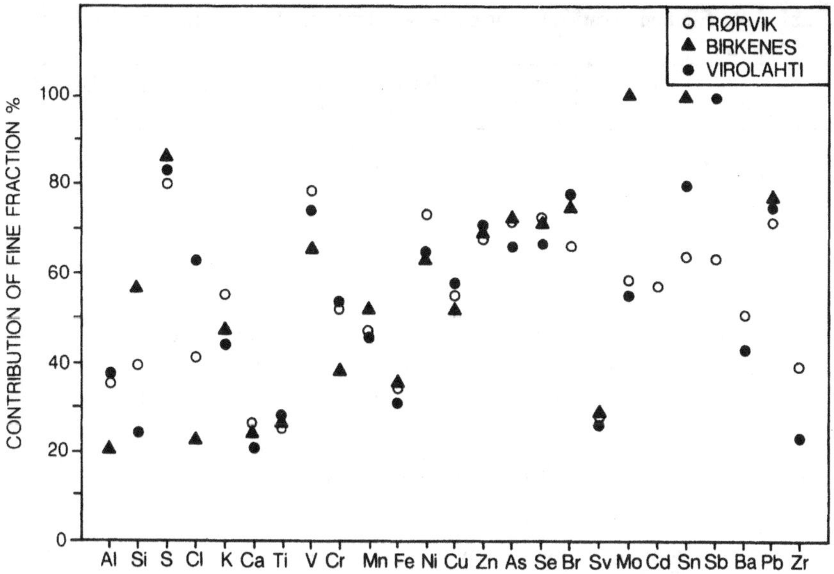

Figure 3–7. Percentage contribution of the trace element concentrations in fine fraction of particles to the trace element concentrations in both fine and coarse fractions measured at Rörvik, Birkenes, and Virolahti.

aerosols separately, and the results are shown in Table 3–4. The data show that the difference between tracers for the western and eastern European aerosol is up to fourfold, which is not very large, considering the uncertainties of tracer estimates.

Similar estimates of elemental tracers for the total concentrations (fine plus large particles) resulted in even smaller differences between the eastern and western European aerosols, at most twofold. Thus, total aerosols seem to be less suitable tracers than fine fraction aerosols.

Table 3-3. The ratios of the trace element concentrations to the Se concentrations measured in the fine fraction of particles at Birkenes (B), Rörvik (R), and Virolahti (V) for the southeastern and southwestern sectors.

Elemental ratios	SE sector			SW sector		
	B	R	V	B	R	V
S/Se[a]	9.0	10.2	11.0	6.4	5.7	6.8
V/Se	20.0	25.4	28.5	8.4	9.0	12.0
Ni/Se	12.8	9.8	14.1	4.9	6.0	6.2
Cu/Se	8.4	7.0	9.0	4.0	5.3	5.0
Zn/Se	130	140	160	90	110	120
As/Se	9.2	9.1	7.5	6.0	7.5	6.1
Pb/Se	120	140	160	100	110	150
Cr/Se	1.6	1.2	2.0	1.6	1.5	2.2

[a]The S/Se ratio should be multiplied by 1,000.

Table 3-4. Elemental ratios for the eastern and western European aerosol at Birkenes and Rörvik.

Ratio[a]	Western European (10 samples)[b]	Eastern European (11 samples)[c]	Moscow + Urals (11 samples)
Cr/Se	2.0	3.5	5.0
Cu/Se	6.1	3.3	7.9
As/Se	6.0	7.0	27.1
V/Se	6.7	25.0	7.1
Mo/Se	0.8	1.3	1.0
Sn/Se	4.1	6.3	7.1

Pacyna (1986a).

[a] Based on concentrations in particles <2 μm EAD.

[b] Including the United Kingdom, France, the Netherlands, Belgium, Luxemburg, and Federal Republic of Germany.

[c] Including German Democratic Republic, Poland, Czechoslovakia, Hungary, and the following republics of the USSR: Lithuania, Latvia, Ukraine, and White Russia.

A tracer system was suggested by Rahn and others (1982) already in 1981 and 1982, when they used the noncrustal Mn/V ratio to search for midwestern aerosols in the northeastern United States. Later, the Mn/V tracer became a matter of discussion in the literature (e.g., Wolff, 1982; Kneip et al., 1983; Husain et al., 1983; Hidy, 1984). A major limitation of this technique appears in its application in areas where large amounts of V are emitted. Recently Rahn and Lowenthal (1984) have developed a tracer technique that uses seven elements (As, Sb, Se, Zn, In, noncrustal Mn, noncrustal V) to characterize the aerosol. Particularly interesting is the application of this new tracer technique to the Arctic aerosol (Lowenthal and Rahn, 1985). In the first step, a principal component analysis was used to isolate the major types of aerosol present in a given sample. Then a group of trace elements was assigned to the factor representing the anthropogenic component and used to build a tracer system. In Table 3–5 elemental composition of this factor at Barrow, Alaska, during the winter 1979–1980 is compared to the elemental composition at Ny Ålesund, Spitsbergen, during various seasons (with Sb as a reference element). There is rather good agreement between these sets of data, suggesting the same origin of the air pollution measured at both locations. In order to localize the potential emission sources, Lowenthal and Rahn (1985) have used a discriminant analysis to construct source groups for each of the signatures in Table 3–5. As discriminant analysis cannot resolve mixtures of sources, they used chemical element balance apportionment. Using the above techniques, Lowenthal and Rahn (1985) suggested that 70% of the anthropogenic pollutants at Barrow stems from the USSR sources. This fits very well the results obtained for Ny Ålesund (Ottar et al., 1986).

Source-receptor relationships for trace elements in the Arctic were also a subject of studies applying other statistical techniques. Some of them are based on principal component analysis. Heidam (1986) studied the elemental composition of the Greenland aerosol by means of factor analysis of the logarithmic concen-

Table 3-5. Elemental ratios estimated by Rahn and Lowenthal (1984) at Barrow, Alaska, compared to the ratios calculated in this work for Ny Ålesund, Spitsbergen.

Elemental ratio to Sb	Barrow, Alaska, winter 1979–1980	Ny Ålesund Winter 1983	Ny Ålesund Winter 1983	Ny Ålesund Winter 1986
V_x[a]	4.2	8.1	6.8	3.5
Mn_x[b]	5.0	7.2	4.1	2.2
Zn	46.0	56.0	41.0	38.0
As	8.9	8.2	7.0	4.0
Se	0.54	1.9	3.1	2.2
In ($\times 10^3$)	—	27.0	32.0	15.0

[a] V_x = noncrustal V.

[b] Mn_x = noncrustal Mn.

trations. The analysis showed that the atmospheric aerosol can be described by four or five different source-related and statistically independent components. However, Heidam (1986) could not identify source areas. Knowing the chemical composition of the aerosol at two locations—Ny Ålesund, Spitsbergen, and Vardø in northern Norway—Pacyna (1986b) was able to show that the same air mass had passed both locations. Some differences of the chemical composition of the aerosols at the two locations could be explained by deposition en route, as there are no additional source regions between these stations. An analysis of the linear dependencies between the concentrations of the various elements at Vardø and at Ny Ålesund was carried out with the use of correlation coefficients, arranged in a matrix. The analysis of the correlation matrix was performed by principal component analysis. In addition canonical correlation analysis was performed. This method measures the interdependence of two groups of variables and searches for similar linear structures. The analysis proved that the trace elements measured at Ny Ålesund and Vardø had the same origin.

Finally a cluster analysis of individual particle data from Ny Ålesund proved to be useful for the chemical characterization of samples (Saucy et al., 1987). The results showed that the aerosol at Ny Ålesund in the autumn of 1984 consisted primarily of Si-, Cl-, and S-rich particles, with considerable differences in the composition of the coarse and fine fractions.

An interesting combination of pattern recognition technique and principal component analysis has been presented by Martinsson and others (1984). The method was used to study the influence of aerosol long-range transport to Scandinavia on the basis of cascade impactor measurements. The results indicate that contributions of foreign sources to the trace element concentrations in Sweden vary from 50% to 90%, depending on the element.

Summarizing, the above discussed statistical methods are very useful to assess the source category (e.g., anthropogenic activity, soil dust) of trace elements measured at various remote locations. However, it has proven much harder to localize the sources. To do so, it is necessary to use these techniques together with information on the meteorological situation, particularly on air mass trajectories and/or synoptic configurations, and accurate emission inventories. One should be aware, however, of the uncertainties of the trajectory computations, as discussed earlier.

V. Recommendations for Further Research

The major conclusion from the research on source-receptor relationships for trace elements so far is that the estimates of trace element emissions can be related to the air concentrations measured at remote locations. Further explanation of this relationship by investigating in more detail the emission processes and the long-range transport of trace elements is the general recommendation for future work.

The qualitative assessment of emission inventories on a regional scale is now reasonably satisfactory, but this is not the case for the quantitative assessment. More measurements are needed to provide the information on the emission rates of trace elements from various technical processes and the efficiency of control equipment employed for a given source category. Of special importance is the trace element distribution on various-size particles, focusing on the fine particles (<2 μm diameter) emissions. Generally, the relative solubility of trace elements increases with decreasing particle size. This tendency is obviously not related to the particle size distribution alone. Lindberg and Harriss (1983) have pointed out that the connection between solubility and particle size may be related to the increased surface area to volume ratio for fine particles, characteristic surface properties of these metals, and the increased free H^+, organic carbon, and strong acid anion concentrations in the fine particle fraction. This may reflect the relationship between trace elements and acidic compounds. More attention in future work should also be given to the chemical form of the emitted trace elements, as this is a key factor for their solubility (e.g., Pacyna, 1987b).

Development of source-receptor modeling is needed and should be considered together with a question about the accuracy required in source-receptor relationships. The major goal of the future source-receptor models for trace elements should be to help the decision-makers develop control strategies of not only trace elements but first of all acidic deposition precursors. In this connection, the research on atmospheric transport of trace elements should be carried out in close relation to the research on atmospheric transport of sulfur and nitrogen compounds, oxidants, and the like. The role of trace elements as signatures of the acidic deposition precursors within the air masses is also a part of this research. Any statistical signatures calculated so far need to be related to the trace element emissions (or their ratios) with the help of information on the meteorological conditions.

Acknowledgment

The author thanks Dr. Brynjulf Ottar for helpful discussions.

References

Adams, F. C., M. J. van Craen, and P. J. van Espen. 1980. Environ Sci Technol 14:1002–1005.

Barrie, L. A., R. H. Hoff, and S. M. Daggupaty. 1981. Atmos Environ 15:1407–1419.

Braun, H., H. Vogg, G. Halbritter, K.-R. Bräutigam, and H. Katzer. 1984. Recycling International 1:1–2.

Brumsack, H., and H. Heinrichs. 1984. Environ Technol Lett 5:7–22.

Cawse, P. A. 1980. Inorganic Pollution and Agriculture (Ministry of Agriculture, Fisheries and Food, London) 326.

CEC. 1981. *Protection in the environment.* Ispra Rept. 3921. Commission of the European Communities (CEC). Joint Research Centre, Ispra, Italy. 68 p.

Eliassen, A., and J. Saltbones. 1983. Atmos Environ 17:1457.

Finzi, G., P. Zanetti, G. Fronza, and S. Rinaldi. 1979. Atmos Environ 13:1249–1255.

Gaarenstrom, P. D., S. P. Perone, and J. L. Mayers. 1977. Environ Sci Technol 15:1421–1427.

Glowiak, B. J., R. Palczynski, and J. M. Pacyna. 1977. *US environmental quarterly report HASL-321*. Health and Safety Laboratory, US ERDA, New York. 18 p.

Heidam, N. Z. 1986. *In* J. O. Nriagu, and C. I. Davidson, eds. *Toxic metals in the atmosphere,* 267–294, John Wiley & Sons, New York.

Heitzenberg, J. H.-C. Hansson, and H. O. Lannefors. 1981. Tellus 33:162–171.

Henry, R. C., and G. M. Hidy. 1979. Atmos Environ 13:1581–1596.

Hidy, G. M. 1984. JAPCA 34:518–531.

Hopke, P. K., R. E. Lamb, and D. F. S. Natusch. 1980. Environ Sci Technol 14:164–172.

Horowitz, J., and S. Barakat. 1979. Atmos Environ 13:811–818.

Husain, L., and P. J. Samson. 1979. J Geophys Res 84:1237–1240.

Husain, L., and J. Webler, and E. Canelli. 1983. JAPCA 33:1185.

Hutton, M. 1982. *Cadmium in the European community*. MARC Tech Rept 26. MARC, London. 99 p.

Hutton, M., and C. Symon. 1986. Sci Total Environ 57:129–150.

Kendall, P. M. H., C. F. P. Bevington, and D. J. Pearse. 1985. *Atmospheric cadmium emission and deposition in the Netherlands.* Metra Consulting BV, Kockengen, Netherlands. 51 p.

Kneip, T. J., P. J. Lioy, and G. T. Wolff. 1983. Environ Forum 2:9.

Kowalczyk, C. S., G. E. Gordon, and S. W. Rheingrover. 1982. Environ Sci Technol 16:79–90.

Krell, U., J. Lehmhaus, and E. Roeckner. 1986. *Proc WMO conf on air pollution modelling and its application*. Leningrad, USSR.

Lannefors, H. O., and H.-C. Hansson. 1980. Studies in Environmental Science 8:159–164.

Lannefors, H. O., H.-C. Hansson, and L. Granat. 1983. Atmos Environ 17:87–101.

Larssen, S., and J. E. Hanssen. 1979. *Proc WMO symp. on the long-range transport of pollutants*. Sofia, Bulgaria.

Lindberg, S. E., and R. C. Harriss. 1983. J Geophys Res 88:5091–5100.

Lowenthal, D. H., and K. A. Rahn. 1985. Atmos Environ 19:2011–2024.

Maenhaut, W., H. Raendonck, A. Selen, R. van Grieken, and J. W. Winchester. 1983. J Geophys Res 88:5353–5364.

Maenhaut, W., W. H. Zoller, R. A. Duce, and G. L. Hoffman. 1979. J Geophys Res 84:2421–2431.

Martinsson, B. G., H.-C. Hansson, and H. O. Lannefors. 1984. Atmos Environ 18:2167–2182.

Michaelis, W. 1987. *Proc. 16th NATO inter techn meeting on air pollution modelling and its applications*. Lindau, FRG.

Miller, J. M. 1987. *In The use of back air trajectories in interpreting atmospheric chemistry data: A review and bibliography*. NOAA Tech Mem ERL ARL-155. National Oceanic and Atmospheric Administration, Silver Spring, Md. 28 p.

NILU 1984. *Emission sources in the Soviet Union*. NILU TR 4/84. Norwegian Institute for Air Research, Lillestrøm, Norway. 30 p.

NSV. 1982. *Monitor 1982*. NSV Rept. Naturvårdsverket, Solna, Sweden. 176 p.

OECD. 1979. *The OECD programme on long-range transport of air pollutants. Measurements and findings*. Organization for Economic Co-operation and Development, Paris.

OECD Air Management Policy Group. 1984. *Control of specific toxic substances in the atmosphere: Cadmium*. Organization for Economic Co-operation and Development, Paris.

Ottar, B., J. M. Pacyna, and T. C. Berg. 1986. Atmos Environ 20:87–100.

Pacyna, J. M. 1980. *Coal-fired power plants as a source of environmental contamination by trace metals and radionuclides*. Post-doctorate Habilitation Thesis. University of Wroclaw, Wroclaw, Poland. 169 p.

Pacyna, J. M. 1984. Atmos Environ 18:41–50.

Pacyna, J. M. 1985. *Spatial distributions of the As, Cd, Cu, Pb, V, and Zn emissions in Europe within a 1.5^0 grid wet*. NILU OR 60/85. Norwegian Institute for Air Research, Lillestrøm, Norway. 22 p.

Pacyna, J. M. 1986a. Water Air Soil Pollut 30:825–835.

Pacyna, J. M. 1986b. *Source-receptor relationships for air pollutants in the Arctic*. NILU OR 3/86. Norwegian Institute for Air Research, Lillestrøm, Norway. 46 p.

Pacyna, J. M. 1987a. *Proc 3d EMEP workshop in data analysis and presentation*. Cologne, FRG.

Pacyna, J. M. 1987b. *In* T. C. Hutchinson, and K. M. Meema, eds. *Lead, mercury, cadmium and arsenic in the environment* (SCOPE 31), 69–88. John Wiley & Sons, Chichester, England.

Pacyna, J. M., and J. Münch. 1987. *Proc 6th inter conf heavy metals in the environment*. New Orleans, La.

Pacyna, J. M., B. Ottar, J. E. Hanssen, and K. Kemp. 1984b. *The chemical composition of aerosols measured in southern Scandinavia*. NILU OR 66/84. Norwegian Institute for Air Research, Lillestrøm, Norway.

Pacyna, J. M., B. Ottar, U. Tomza, and W. Maenhaut. 1985. Atmos Environ 19: 857–865.

Pacyna, J. M., A. Semb, and J. E. Hanssen. 1984a. Tellus 36B:163–178.

Peirson, D. H., P. A. Cawse, L. Salmon, and R. S. Cambray. 1975. Nature 241:252–256.

Raabe, O. G. 1978. Environ Sci Technol 12:1162–1167.

Rahn, K. A., and D. L. Lowenthal. 1984. Science 223:132–139.

Rahn, K. A., D. H. Lowenthal, and F. N. Lewis. 1982. *Elemental tracers and source areas of pollution aerosol in Narragansett, RI*. Tech. Rept. University of Rhode Island, Narragansett, R.I.

Rahn, K. A. and R. J. McCaffrey. 1979. *Proc. WMO symp. on the long-range transport of pollutants*. Sofia, Bulgaria.

Rauhut, A. 1978. *Survey of industrial emission of cadmium in the European Economic Community*. ENV/223/74/E. CEC, Brussels.

Ronneau, C., J. L. Navarre, P. Priest, and J. Cara. 1978. Atmos Environ 12:877–881.
Saltbones, J., and H. Dovland. 1987. *Emissions of sulphur dioxide in Europe in 1980 and 1983*. EMEP/CCC-Rept 1/86. Norwegian Institute for Air Research, Lillestrøm, Norway.
Sartorius, S., B. Seifert, and F. Vahrenholt. 1977. Luft 37:11.
Saucy, D. A., J. R. Anderson, and P. R. Buseck. 1987. Atmos Environ 21:1649–1657.
Shaw, G. E., and G. Wendler. 1972. *Proc. conf. on atmospheric radiation*. Fort Collins, Colo.
Thurston, G. D., and J. D. Spengler. 1981. *Proc. EPA-SIAM conf. environmentrics 81*. Alexandria, Va.
UBA. 1977. *Luftqualitätskriterion für cadmium*. Berichte 4/77. Umweltbundesamt, Berlin.
UBA, 1981. *Cadmium-Bericht. Ein Beitrag zum Problem der Umweltbelastung dürch nicht: oder schwer abbaubare Stoffe-dargestellt am Beispiel Cadmium*. Umweltbundesamt, Berlin.
UBA, 1982. *Anhörung zu Cadmium. Protokoll der Sachverständigenanhörung, Berlin.* Umweltbundesamt, Berlin.
van Enk, R. H. 1979. *The pathway of cadmium in the European Community*. EUR/6626/EN. CEC, Brussels.
van Wambeke, L. 1979. *Measures taken by the European communities in respect of cadmium between 1977 and 1979 and inventory of emissions into the environment.* ENV/322/79/FR. CEC, Brussels.
VDI. 1975. *Heavy metals as air pollutants: Lead, zinc and cadmium*. VDI Berichte 203. Verein Deutscher Ingenieure. Düsseldorf, FRG.
VDI. 1984. *Schwermetalle in der Unwelt*. UFOPLAN No. 10403186. Verein Deutscher Ingenieure. Düsseldorf, FRG. 483 p.
Wolff, G. T. 1982. Science 216:1172.

The History of Atmospheric Deposition of Cd, Hg, and Pb in North America: Evidence from Lake and Peat Bog Sediments

Stephen A. Norton,* Peter J. Dillon,† R. Douglas Evans,‡ Gregory Mierle,† and Jeffrey S. Kahl*

Abstract

Lake and ombrotrophic peat bog sediments record increases in the concentrations and accumulation rates of Cd, Hg, and Pb for most of temperate North America for the last 100 years. These increases are largely related to the burning of coal, smelting of nonferrous metals, the transportation industry, and the industrial production of chlorine.

Modern atmospheric fluxes of Cd in central North America are about 1,000 × background fluxes; accumulation rates for Cd in sediments have increased two to three times above background, beginning about 100 years ago. The anthropogenic component of the total Cd flux in recent sediment is about 0.1 to 0.2 $mg/m^2/yr$.

Global-scale Hg pollution of the atmosphere is suggested by concentrations of Hg in northern hemisphere air that are double the Hg content of southern hemisphere air. The accumulation rates of Hg in sediment have approximately doubled over the last 100 years. However, these accumulation rates are approximately an order of magnitude less than those for Cd.

Modern increases in Pb concentrations are ubiquitous for all lakes examined thus far in North America. Input is from multiple sources and thus the timing of increased accumulation rates in sediment varies across the continent. Typical modern accumulation rates reach maxima of 20 to 30 $mg/m^2/yr$, or 100 × that of Cd and 1,000 × that of Hg. Recent decreases in atmospheric Pb are reflected in decreases in the accumulation rate of Pb in both lake and peat bog sediment in eastern North America.

---

*Department of Geological Sciences, University of Maine, Orono, ME 04469, USA.

†Ontario Ministry of the Environment, Dorset Research Centre, Dorset, Ontario POA 1E0, Canada.

‡Trent Aquatic Research Centre, Trent University, Peterborough, Ontario K9J 7B8, Canada.

I. Introduction

Air pollution attributable to the burning of fossil fuels has existed in North America for more than 100 years. Prior to 1885, wood was the principal source of thermal energy. The mix of fossil fuels consumed has changed since then from dominantly coal (until about 1950) to oil (to the present), with significant amounts of natural gas since about 1950 (Husar, 1986). Associated with the burning of these fuels are the emissions of SO_x and NO_x, the principal constituents of acidic precipitation. Also associated with fossil fuel consumption are both atmospheric and direct aquatic emissions of trace metals contained in the fuels themselves. For metals that are readily volatilized during fossil fuel consumption (e.g., Hg), the main emissions are to the atmosphere. Such elements range widely in character during emission and atmospheric transport. Elements emitted largely as volatile gases (e.g., Hg and As) may have relatively long residence times in the atmosphere (Lindberg, 1987). Others, such as Fe and Al, are more refractory and have much shorter residence times, because of their association with coarser particles (Torrey, 1978; Babu, 1975; Ainsworth and Rai, 1987).

Through time, different types and sources of coal and oil have been utilized for energy production, each with its own characteristic metal emission chemistry (Ainsworth and Rai, 1987). Additionally, other industrial activities add to the atmospheric burden of metals. Most important of these is the smelting of metals (especially Pb, Zn, Cu, and Ni), a process that involves emissions of not only the metals of interest in this chapter (Cd, Hg, Pb) but also other trace metals mineralogically associated with the ore minerals. Thus As, Cd, Hg, Ni, and others are released in significant quantities from the roasting of sulfide ores of almost any metal. Other industries release metals associated with production streams or waste streams (e.g., airborne Hg from chlor-alkali plants). The dominant modern source of Pb is the burning of leaded gasoline, a practice that started in earnest in the 1930s, peaked in the 1970s, and now is reduced to about 25% of peak values (Royal Society of Canada, 1986).

Emissions of metals from activities not associated with the generation of electricity as well as emissions from fossil fuel electric power-generation stations are not distributed evenly across the continent. Thus, inventorying of atmospheric emissions of metals and resultant deposition must be done on a regional, rather than continental, scale. Source-receptor relationships for trace metals have now been regionally modeled and validated for Europe (Pacyna, this volume) but not for North America.

Although measurements of metals in air and precipitation have been conducted for a variety of purposes, few synoptic-scale *and* long-term studies have been conducted on the metal content of air, precipitation, or deposition to natural surfaces. Several summaries of the metal chemistry of air and precipitation have appeared recently (e.g., Galloway et al., 1982; Jeffries, volume 4 of this series; Pacyna, this volume; Lindberg and Turner, 1988). One broad conclusion from these syntheses is clear: Anthropogenic activity has increased the flux of certain trace metals to the earth's surface by up to several orders of magnitude. Surveys of collection surfaces receiving this deposition give further evidence for anthropogenic influence on the atmospheric flux of trace metals. For example, collection and analysis of moss from Norway, Sweden, and Finland show a clear spatial relationship between sources of pollution, changes in emission, and the chemistry of living mosses that receive their nutrients entirely from the atmosphere (Ruhling et al., 1987; SNV, 1987; see also a review by Puckett, 1986). Groet (1976) conducted a similar survey in the northeastern United Stated in 1973, as did Percy (1983) in the Maritime Provinces of Canada, but unfortunately neither has been repeated. Surveys of soil chemistry of forest land indicate a correspondence between emissions sources and the accumulation of trace metals in soil (Hanson and Norton, 1982; Johnson et al., 1982; Steinnes, 1988). Such surveys are restricted to the last two decades and are thus unable to yield information about the longer-term atmospheric pollution and deposition of trace metals. The surveys are also not sufficiently well calibrated to enable the reconstruction of actual atmospheric deposition rates, only spatial and short-term trends.

In the last 15 years, numerous studies of the chemistry of sediment cores from lakes not receiving direct discharge of metals have demonstrated a long-term history of increased accumulation of metals (see reviews of Coleman, 1985; Alderton, 1985; and Norton, 1986a, 1986b). Initially this pollution was identified by increases in the concentration of trace metals suspected of being of atmospheric origin. With the recent availability of absolute chronology for the sediment cores through the use of ^{210}Pb dating (Appleby and Oldfield, 1978) and varves (Renberg, 1979), it has become possible to reconstruct the accumulation rate of pollutants in the sediment. However, for a variety of reasons, the accumulation rates of trace metals in sediment are generally not equivalent to atmospheric deposition rates. Processes that alter the correspondence of atmospheric deposition rates and sediment accumulation rates of trace metals include sediment focusing (Hakanson, 1977; Dearing, 1983); transport of metals of interest from the watershed to the lake with subsequent sedimentation (Dillon et al., 1988; Norton, 1989); decreased deposition in the lake due to acidification of the water column (Campbell et al., 1985); mobilization from the sediment as a result of acidification of the water column (Kahl and Norton, 1983); bioturbation or other types of sediment mixing (Davis, 1974); diagenesis of the sediment related to sulfate reduction (e.g., Carignan and Tessier, 1988); and, of course, any direct discharges of metals to the lake or its watershed. The faithfulness of the sediment record in recording the changing atmospheric record is thus dependent on many factors, some of which are specific to the watershed and some of which are specific to the chemistry of the metal under consideration.

Cores from aggrading ombrogenic peat bogs also have been utilized as recorders of atmospheric deposition (e.g., Schell, 1987; Norton, 1987). These records also suffer from imperfect recording of accumulation rates from the atmosphere for reasons including the following: inaccuracies of dating the peat because of possible chemical mobility of ^{210}Pb or variable contributions of dry deposition of material over small areas (Norton, 1987); highly variable redox conditions causing mobilization and localization of certain elements (Clymo, 1965, 1987); zones of immobilization of certain elements because of sulfate reduction below the water table (Bayley and Schindler, 1987); and control over the chemical mobility of certain elements because of the high concentrations of dissolved organic matter (Gorham et al., 1985).

Comparisons between sediments from closely spaced lakes and ombrogenic bogs (Norton and Kahl, 1987a) suggest that only relatively immobile elements, most notably Pb, may yield sediment accumulation trends that are similar in pattern to atmospheric deposition. All elements are subject to mechanical focusing or postsedimentation redistribution by chemical, biological, or other physical means.

This chapter reviews the precipitation chemistry and lake and peat sediment chemistry of three metals emitted to the atmosphere in significant amounts as a result of anthropogenic activity: Cd, Hg, and Pb. These three metals have contrasting source terms, atmospheric residence times, and chemical mobility. Emissions of Hg (especially) and Pb are linked to the consumption of coal; Cd and Pb are emitted from nonferrous smelting; the most important source of atmospheric Pb is automobile emissions. All are thus related spatially with emissions associated with acidic rain, SO_x, and NO_2. Each of the three elements has known toxic effects and each behaves differently in the environment. Mercury, the least abundant in the environment, appears to be quite mobile in certain environments because of transformations involving gaseous phases as well as complexing with dissolved organic matter. It is subject to biomagnification, with a significant repository in natural systems being fish. Lead, on the contrary, is the most abundant of the three elements but is subject to biominification. Because of this property and because of the relative insolubility of Pb, ecological damage from Pb in natural systems has not been demonstrated even though it accumulates to considerable concentrations in soils and sediment. Cadmium has greater mobility in water than Pb, is mobilized under acidic conditions, and may cause direct effects to biota.

II. Methods

The concentrations of Cd, Hg, and Pb are determined in tissue, soils, and sediment with little difficulty. Consequently, the concentrations of these elements in various solid environmental compartments have been well characterized. From these data it is possible to make inferences about the aqueous fluxes of elements between environmental compartments.

However, considerable difficulties have plagued the accurate assessment of the flow of Cd, Hg, and Pb through the atmosphere and through terrestrial and aquatic ecosystems. Only in the last 10 to 15 years has it been possible to measure accurately Pb and Cd in precipitation and surface waters without contamination problems. Accurate measurements of Hg in solutions are largely restricted to the last 10 years.

Initial characterization and interpretation of sediment cores from lakes consisted of demonstrating increasing concentrations of metals of interest in increasingly "young" sediment. Approximate chronology was established with pollen or stratigraphic markers of known age (e.g., for Pb, Davis and Norton, 1978; for Cd, Kemp et al., 1978; for Hg, Thomas, 1973). The use of ^{137}Cs for dating, which was popular in the 1970s, enabled, with certain assumptions, the estimation of average accumulation rates for sediment components. The use of ^{137}Cs for chronology in lakes with low sedimentation rates has proven problematic (Davis et al., 1984) and has been replaced by dating with ^{210}Pb (popularized by Appleby and Oldfield, 1978) and varves (Renberg, 1979). These latter two methods typically yield internally consistent ages for the last 100 to 150 years of sediment accumulation. This chronology, in combination with complete chemistry, has enabled the calculation of gross accumulation rates where bulk chemistry has been determined (e.g., the methods of Buckley and Cranston, 1971, used by Kahl and Norton for data in this chapter) or the accumulation rate of elements based on concentrations extracted by acid leaching (commonly aqua regia) from solid sediment (used by Dillon, Evans, and Mierle for data in this chapter). Tessier and others (1979) developed a procedure for the sequential extraction of metal species from sediment. Similar techniques have been reviewed by Rapin and others (1986), who discussed the differences between the operationally defined metal concentrations. White and Gubala (1989) have determined selected extractable components of the total sediment chemistry and have been able to demonstrate important changes in the state in which elements are delivered to and stored in the sediment. These studies have shed light on mechanisms of transport, processes related to lake acidification, and postsedimentation alteration of the sediment (diagenesis). Concentrations of metals determined by these different analytical methods, combined with age determinations of the intervals of sediment, yield significantly different accumulation rates for immobile elements (e.g., Pb) but substantially equivalent rates for those elements that exist in the sediment largely as acid-soluble material (e.g., Zn) (Table 4–1).

This chapter focuses on lakes that receive no direct discharge of metals either to the watershed or the lake. Ideally one would like to reconstruct atmospheric deposition rates for metals from sediment accumulation rates. However, for reasons discussed above, it is highly likely that most chemical profiles in sediment are not simply related to atmospheric deposition rates. The one possible exception is Pb. There are two sources of trace metals deposited in the sediment: material derived from the soils and bedrock of the watershed, and anthropogenic contributions that may be deposited directly on the lake, or on the watershed and then transported to the lake, to be deposited in the sedimentary record. Two approaches have been used to differentiate the anthropogenic component from the autochthonous component. Dillon and Evans (1982) subtracted the average trace metal content (concentration) of preindustrial sediment from the trace element content of the sediment column deposited under the influence of atmospheric pollution. This approach can be done on a gross scale to assess the integrated anthropogenic contribution at the coring site or on any sampling interval to develop a continuous anthropogenic flux to the sediment, if dating of sediments is available. If multiple

Table 4–1. Interlaboratory comparison of the analysis of standard sediment by different methods.

Laboratories	Organic	Pb (ppm)	Zn (ppm)	Cu (ppm)	Cd (ppm)	Ni (ppm)	Al (%)	Mg (%)	Ca (%)	K (%)	Na (%)	Ti (%)	Fe (%)	Mn (%)
K.J.[a]	12.1	35	140	15	0.56	99							4.9	0.26
M.V.[a]		42	144	27	0.63	77	3.10	0.72	0.266	0.33		0.016	5.06	0.24
K.T.[a]		60	144	17	2.2	56	3.08	0.70	0.248	0.44		0.017	4.98	
S.N.[b]	11.7	72	173	7			7.41	1.08	0.343	1.82	0.88	0.48	5.98	0.30
R.B.[b]			166				7.67	1.16	0.33	1.69	0.86	0.50	5.83	0.30

[a] Acid extraction, aqua regia.

[b] Total digestion (Buckley and Cranston, 1971).

cores are available from many different locations, a whole lake burden can be calculated (Dillon and Evans, 1982; Evans et al., 1983). Norton and others (1989) have utilized the concentration ratio of the trace metal of interest to TiO_2 in preindustrial sediment to determine the autochthonous portion of the total content of trace metal for any sediment interval. Both of these methods for determination of the anthropogenic component have shortcomings but yield similar trends for changes in the anthropogenic flux of metals to lakes.

III. Results

A. Cadmium

Cadmium is enriched in the atmosphere of the northern hemisphere relative to major elements. This conclusion is based on many data but particularly on the concentration of Cd concentrations in ice and snow cores. The Cd/Al ratios found exceed those expected for normal crustal dust (Herron et al., 1977). However, these data have been challenged by Nriagu (1980) and Boutron (1980). Of particular concern is the issue of contamination of samples. For example, Bewers and others (1987) reject all Cd precipitation data for remote areas before 1979. They suggest that precipitation in remote areas should be in the concentration range of 0.1 to 1 ng/L, as found by Mart (1983). Batifol and Boutron (1984), based on data from snow in European glacial environments, suggest that there has been a doubling of Cd in the atmospheric flux since about 1945. LaZerte and others (in press) finds concentrations in precipitation ranging from 10 to 260 ng/L, with a mean of 100, for samples collected recently near Dorset, Ontario. Chan and others (1986) report mean concentrations of 70 to 100 ng/L for 36 stations across Ontario and average annual wet deposition of about 0.07 mg/m^2 of Cd. Thus it appears that the concentration may range over three orders of magnitude between "clean areas" and "polluted areas." Presumably, this translates into increased atmospheric deposition fluxes ranging up to 1,000 $\times$ background in polluted areas that were originally in "clean" areas.

Sediment data support this conclusion. Evans and others (1983) found sediment concentrations of Cd for lakes in the Haliburton, Ontario, area of 1 to 6 ppm dry weight. Data from one lake are shown in Figure 4–1. Based on 52 cores from 10 lakes, the average ratio of maximum concentration to background concentration was 3.43 $\pm$ 1.2. This ratio depends on (1) the magnitude of the increased flux from the atmosphere, (2) the background concentration, which varies considerably from lake to lake, and (3) the gross sedimentation rate. The whole lake burdens (calculated as described above) ranged from 6 to 10 mg/m^2, most of which has accumulated in the last 50 years. This loading is comparable to the wet deposition on an annual basis. Wong and others (1984) give concentration profiles for nine remote lakes in Algonquin Park, Ontario. Concentrations in recent sediment are typically triple those of background sediment (Figure 4–2). The accumulation rates for total Cd (approximately 75% of which is anthropogenic) for the nine lakes range from 0.07 to 0.75 mg/m^2/yr, with an average of 0.27. These values are

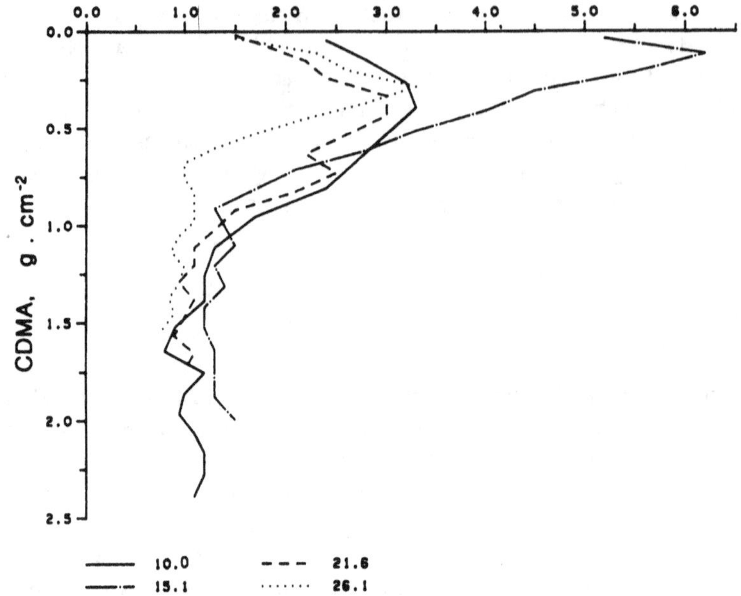

Figure 4–1. The concentration of Cd (μg g · dry mass^{-1}) in multiple cores from Clear Lake, Ontario (from Evans et al., 1983). CDMA is cumulative dry mass. The four curves are data for cores taken at four depths.

comparable to those obtained from a 50-year averaging of the data from Evans and others (1983). Similar enrichments have been observed for lakes in the Adirondack Mountains, New York (Heit et al., 1981; Galloway and Likens, 1979). Various studies indicate similar enrichments for sediments in the Great Lakes; however, these lakes also receive direct discharges of Cd from anthropogenic activity within the watersheds, and comparison is difficult. Nriagu and others (1979) estimated that the fluxes to sediment of the eastern and western basin of Lake Erie have been 4 and 7 mg/m^2/yr, with sewage contributing significantly to this total. These values are comparable to the entire lake burden value for those lakes studied by Evans and others (1983). Christensen and Chien (1981), also using ^{210}Pb-dated cores, indicated strong enrichment in sediment younger than 1860 in two cores from Lake Michigan. They calculated an anthropogenic component of the total flux for Cd of 0.12 and 0.34 mg/m^2/yr for cores from the central basin and Green Bay, respectively. The length of time for these fluxes was unspecified, but the lower value corresponds to estimates for the present atmospheric flux in the Dorset, Ontario, area.

The degree to which sedimentary fluxes of Cd reflect atmospheric fluxes is not well established but can be estimated for the Haliburton, Ontario, area. There, annual atmospheric fluxes average 0.045 mg/m^2. Assuming that this flux has been maintained for 50 years, it is clear that the sediment burden on a whole lake basis (6 to 10 mg/m^2) can be accounted for only if there are significant contributions

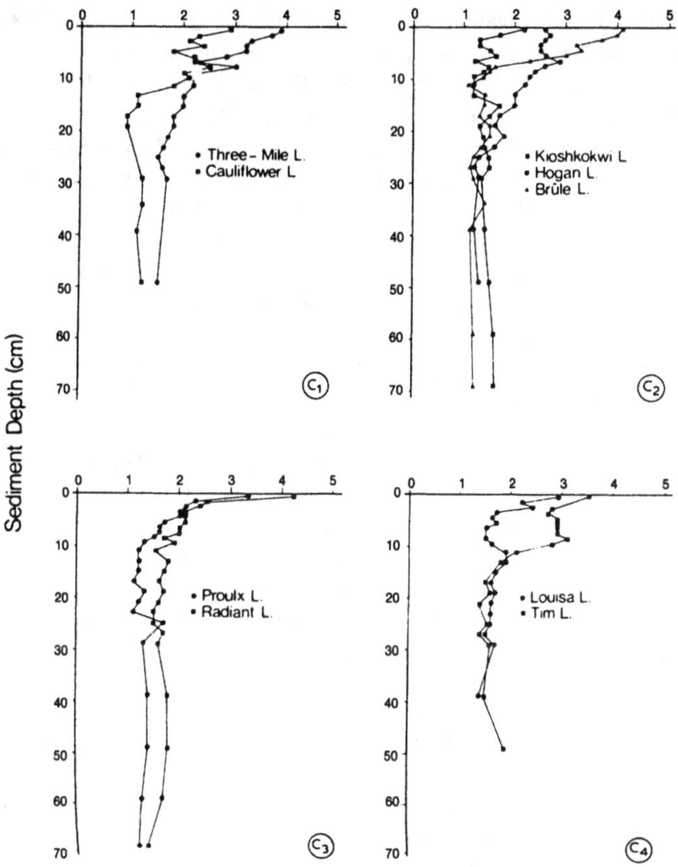

Figure 4–2. The concentration of Cd ($\mu g\ g^{-1}$) in single cores from nine lakes in Ontario (from Wong et al., 1984). Concentration based on dry mass.

from the watershed to the lake. Consistent with this, Evans (unpublished) found mean concentrations of Cd for eight headwater streams ranged from 50 to 180 ng/L, comparable to values in precipitation. We interpret these elevated values in sediment to be related to surface water acidification in the area. Stephenson and Mackie (1988) surveyed 64 central Ontario lakes and found concentrations that ranged from <2 to 587 ng/L; the higher concentrations were close to Sudbury, Ontario. Further, they found a strong negative correlation between the pH of the lake waters and the Cd concentration. LaZerte and others (1989) determined that in Plastic Lake, Ontario, an acidified catchment-lake system, even an acidified lake may act as a trap for Cd. Lindberg and Turner (1988) estimated that 8% to 29% of atmospherically deposited Cd was exported from selected watersheds in the southeastern United States. Thus, the flux of anthropogenic Cd to the sediment will be a function of atmospheric inputs coupled with altered mobility associated with acidification in the watershed. It is highly probable that Cd initially stored in the terrestrial part of the watershed may be released later during soil acidification (Mayer and Lindberg, 1985). The resultant flux to the lake may be significantly elevated over values expected from concurrent atmospheric deposition alone, even assuming 100% yield of the modern atmospheric flux from the watershed. Clearly, the net accumulation rate of Cd in lake sediments may not necessarily parallel trends in atmospheric deposition, although the onset of increased accumulation rates probably coincides with the beginning of atmospheric deposition. If one assumed that atmospheric deposition is of the order of 0.1 ng/cm^2, a figure associated with a background Cd concentration of 1 ng/L in precipitation, the whole lake burdens discussed above would require hundreds to thousands of years to accumulate, depending on the watershed yield of Cd and the efficiency of retention of Cd in the lake.

Few studies are available from ombrogenic bogs to clarify the historical atmospheric inputs of Cd. Glooschenko and Capoblanco (1982) analyzed peat at two different depths from ombrotrophic bogs in northern Ontario and found that concentrations of Cd were elevated about 100% in surface peats, relative to background. However, sectioning of the peat was coarse, dating was not performed, and the data were highly variable. Therefore, changes in accumulation rates cannot be estimated. Schell (1987) reported a fivefold increase in the accumulation rate of Cd for a ^{210}Pb-dated core from a bog in Pennsylvania. Modern accumulation rates were 1.5 $mg/m^2/yr$, about an order of magnitude higher than exists in the Haliburton area, known for high Cd concentrations in precipitation (see above). The Pennsylvania values are probably high because of a significant autochthonous component in the total burden present in the core. This interpretation is supported by the high background Pb accumulation rates that are at least an order of magnitude higher than any other values reported for remote lakes. This bog must be slightly minerotrophic.

B. Mercury

Several natural processes contribute to the global atmospheric burden of Hg. Oceanic and soil degassing are probably the most important (McCarthy et al., 1970; Johnson and Braman, 1974; Kim and Fitzgerald, 1986). Lesser sources include volcanoes (Varekamp and Buseck, 1981; Siegel and Siegel, 1978) and emissions from ore deposits (Lindberg, 1987). The Hg is generally released to the atmosphere as Hg vapor and persists predominantly in that form (Jonasson and Boyle, 1972; Johnson and Braman, 1974; Kim and Fitzgerald, 1986), with estimated residence times ranging from months to years (Lindberg, 1987).

Anthropogenic inputs of Hg to the atmosphere constitute a significant fraction, perhaps a third, of the total burden (Andren and Nriagu, 1979). The concentration of Hg north of 10° north latitude averages 1.5 ng/m^3 over the open oceans and averages 2.9 ng/m^3 at a coastal site in Rhode Island (Fitzgerald et al., 1983, 1984). Systematic monitoring of Hg in air at inland sites has not been reported, but concentrations of 1 to 10 ng/m^3 appear to be representative for nonurban sites

(Schroeder, 1982). In contrast, mid-Pacific concentrations of Hg between 10 and 20° south latitude average 1 ng/m^3 (Fitzgerald et al., 1984). Atmospheric concentrations of Hg in the northern hemisphere are approximately twice those of the southern hemisphere.

The elevated concentrations of Hg in the atmosphere of the northern hemisphere are reflected in the chemistry of ice cores from Greenland. Initial studies by Weiss and others (1971, 1975) were probably flawed by sample contamination and analytical detection limits. However, the results of Applequist and others (1978) showed that the Hg in precipitation has doubled in the last two decades in comparison to the eighteenth century.

Paleolimnological data for Hg are sparse for North America. Thomas (1973) and Kemp and others (1978) found Hg enrichment in the sediments of Lakes Huron and Superior, but these lakes also receive considerable effluent from industrial activities within their watersheds. Thus, quantitatively linking accumulation rates in the sediment to atmospheric fluxes was not possible. Schmitt and Sasseville (1978) report a doubling of the concentration of Hg in recent undated sediment for four remote lakes in northern Maine. Rada and others (1989) indicate an increase in the concentration of Hg in sediment in 11 lakes in Wisconsin (10 of which were seepage lakes) of up to a factor of 2.8. The timing of the increase is coincident with that of Pb (see below) for the region. This temoral coincidence in conjunction with the fact that 10 of 11 lakes are seepage lakes suggests that the increases are related to atmospheric deposition directly to the lake surface and that the Hg is related to the same source, probably coal burning. Heit and others (1981) report detailed chemistry for sediment from two remote lakes in New York. They reported a doubling of the concentration of Hg in recent sediment. However, accurate accumulation rates cannot be determined because of the use of ^{137}Cs for dating. Ouellet and Jones (1983) report data for 26 lakes, 11 of which were described as having no watershed disturbances. They used ^{137}Cs dating for chronology and placed the onset of metal pollution at about 1950. However, the time scale for Hg pollution is comparable to that of Pb (as it was in Heit et al., 1981), the onset for which was in the late nineteenth century (see, e.g., Appleby and Oldfield, 1979) based on ^{210}Pb dating. Meger (1986), also using ^{137}Cs dating, reported a post-1880 average flux for Hg to a remote lake on the Minnesota-Ontario border of 0.17 mg/m^2/yr, twice the pre-1880 value (Figure 4–3). Evans (1986) studied sediment from 14 remote lakes in south-central Ontario, using the multiple core technique to assess Hg whole-lake burdens. He found elevated concentrations in younger (but undated) sediment, typically an increase of 100% over background. Whole-lake burdens averaged 0.79 mg/m^2, about an order of magnitude less than for Cd. If the anthropogenic burdens determined by Evans (1986) for Ontario lakes are distributed over 50 years, the calculated annual accumulation rates range from about 0.01 to 0.03 mg/m^2/yr. For the entire lake set ($n = 14$), there was no relationship between the anthropogenic lake burden and the watershed:lake area ratio, although there was a significant relationship ($p < 1\%$; $n = 5$; $r = 0.96$) for a regional subset of these lakes. These data suggest that the relative importance of direct input from the atmosphere and input from the

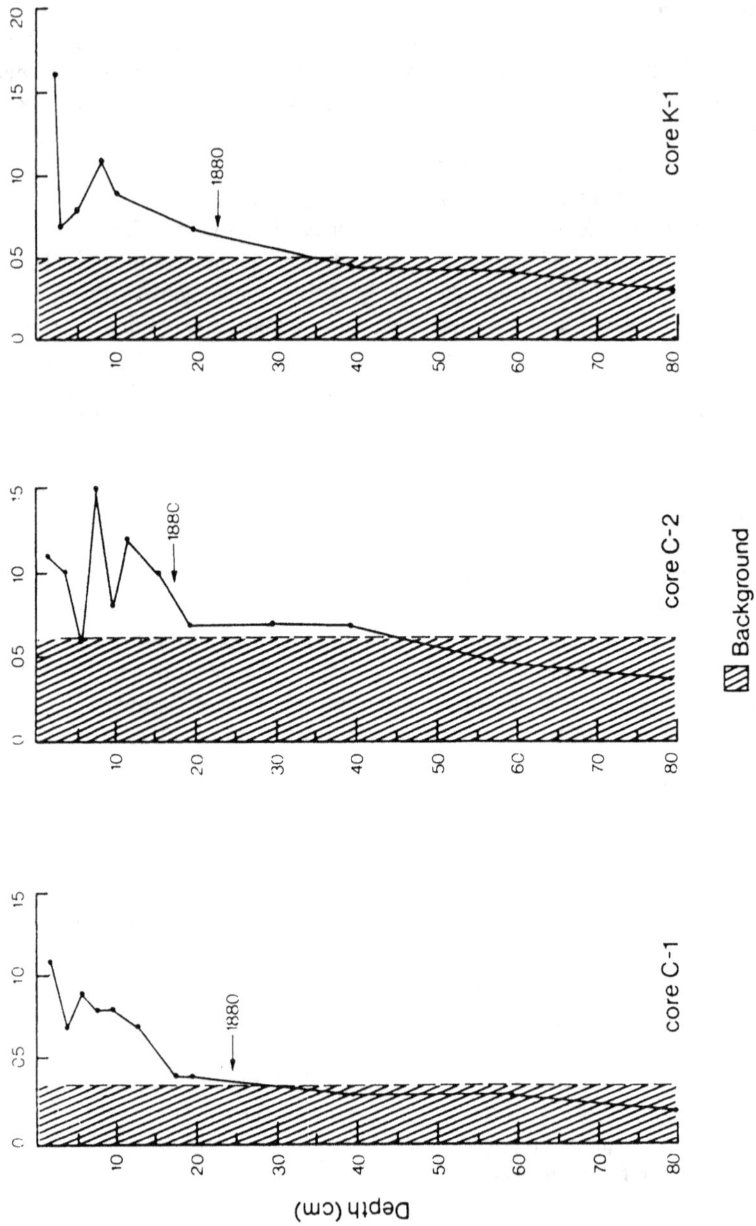

Figure 4–3. The concentration of Hg ($\mu g\ g^{-1}$) in cores from Crane Lake, Mn (from Meger, 1986). Concentration based on dry mass.

watershed varies regionally, either because retention by the watershed varies with geologic and edaphic factors or because deposition varies. There is so little data with regard to regional deposition patterns of Hg that the latter possibility cannot be evaluated. Data on the former possibility are also sparse, but there is some evidence that relates to the transport of Hg from the drainage basins to the lakes.

First, extensive lake sediment surveys have been conducted in Sweden (Johansson, 1985) and Finland (Rekolainen et al., 1986; Tolonen and Jaakkola, 1983). The results indicate sharply increased accumulation of Hg in sediments in this century, particularly in areas known to have historically high emissions from local chlor-alkali plants. Although these emissions have been substantially reduced in the last 10 to 15 years, there is not a parallel decrease in the accumulation rate of Hg in sediment (Finnish studies) or concentration (Swedish studies). Part of this could be related to a lag associated with focusing of sediment. Most of this is probably related to slow release of the accumulated Hg from the watershed, a mechanism proposed for a system in the United States by Lindberg and Turner (1977). The release rate is apparently increased with acidification and also promoted by high concentrations of dissolved organic matter. The latter is particularly important in Finland.

Second, soil organic matter has a high affinity for Hg (Krenkel, 1973). Consequently, Hg deposited on land is probably initially strongly adsorbed and retained until it is either degassed or mobilized in the aqueous phase by dissolved organic matter. Mercury applied to soil columns over a 19-week period remained entirely in the upper 10-cm layer or was lost, presumably by degassing to the atmosphere (Hogg et al., 1975). Mierle (1989), studying three watersheds in Ontario, estimated that 81% to 91% of the total wet Hg deposited annually was retained in the watershed. Despite this high retention efficiency, the watersheds supplied 43% of the Hg input to the lakes. Retention was lowest in the watershed with a large fraction of wet, lowland soil. During the summer, this watershed yielded runoff water with high concentrations of dissolved organic carbon and total Hg up to 25 ng/L. This study suggests that humic and fulvic acids can mobilize Hg from watersheds and increase the flux to lakes. Clearly, factors that affect the production and release of humic matter (e.g., topography, rainfall, and soil acidity) will affect the supply of Hg to lakes. Mercury may also be volatilized from the sediment, further complicating the reconstruction trends in atmospheric deposition of Hg.

Because of the high affinity of Hg for organic matter, peat bogs present an opportunity for evaluation of changing fluxes through time. Pheiffer Madsen (1981) reported accumulation rates for Hg from dated moss cores from two ombrotrophic bogs in Denmark. Background values (prior to 1850) averaged about 0.01 mg/m^2/yr. For complete retention of all atmospheric Hg in precipitation, this would correspond to about 10 ng/L in precipitation, assuming 1 m of rainfall/yr, certainly elevated above background values but reasonable for western Europe at that time. Values rose sharply in the last 50 to 60 years to 10 $\times$ background, comparable to values found in lake sediments in Scandinavia (see, e.g., Johansson, 1985).

Thus, the paleolimnological evidence for an increased atmospheric flux of Hg is strong in eastern North America and the Scandinavian countries. However, the timing of the increases are not well established. Also, the degree of parallelism between atmospheric fluxes and net sediment accumulation rates is probably poor because of differing retention by watersheds and lakes. Clear water systems should yield the best relationships because of a reduced influence of dissolved organic matter on the mobility of Hg.

C. Lead

In 1978, Nriagu stated that anthropogenic activity was responsible for up to 96% of the global Pb emissions. However, as analytical techniques have improved, estimates of background levels of Pb in precipitation and air have declined to less than the approximately 4% implied by Nriagu. Boutron and Patterson (1986) demonstrated that the anthropogenic contribution to the troposphere is about 99% of the total; the remainder is contributed by volcanoes (ca. 2.4×10^6 g/yr) and soil dust (2.8×10^6 g/yr). The residence time of Pb in the atmosphere is estimated by Nriagu (1986) to be about 10 days. Thus, the proportions of natural and anthropogenic Pb in precipitation should be in the same proportion as the emissions.

The principal contributors to the modern atmospheric burden of Pb are from the burning of gasoline with alkyl Pb additive (estimated 70% of the total) and from smelters (25%) (Settle and Patterson, 1980). The use of alkyl Pb in the United States declined substantially from 13×10^{10} g/yr in 1979 to 5.7×10^{10} g/yr for 1983 (Eisenreich et al., 1986). The trend toward reduced use of Pb in gasoline has continued in North America. The reduction in emissions is paralleled by a reduction in concentrations of Pb in ambient air (EPA, 1986) to less than 50% of the maximum values achieved in the 1960s. Deposition of Pb has decreased as well (Eisenreich et al., 1986).

Information on deposition of Pb from the atmosphere has been reviewed many times. The Pb concentrations in air for rural and urban environments have been reviewed by Nriagu (1978). Precipitation data have been reviewed by Galloway and others (1982), Jeffries and Snyder (1981), Campbell and others (1985), Murozumi and others (1969), Ng and Patterson (1981), Boutron and Patterson (1986), Barrie and others (1987), and other investigators.

Just as for Cd ang Hg, long-term monitoring of Pb in precipitation has not been performed, and spatial coverage of North America in any one survey has not been extensive enough to delineate the pattern of deposition of Pb and the relationship of any pattern to emissions, except in the most general sense.

In the absence of long-term monitoring of the deposition of Pb, lake sediment chemistry has been used to reconstruct the history of atmospheric deposition of Pb. It is the trace metal most frequently reported on because of the ease of measurement, the strong signal-to-noise ratio, and the obvious linkage with the three major sources of air pollution by metals: energy production, nonferrous smelting, and transportation. Two lines of evidence support the conclusion that the input of Pb to aquatic systems results primarily from atmospheric deposition. First, retention of Pb in terrestrial systems is typically very high (Andren et al., 1975; Swanson and Johnson, 1980; Turner et al., 1985; Lindberg and Turner, 1988). Lead deposited on terrestrial systems accumulates in the organic soil layers and to a lesser extent in mineral soil layers (Friedland et al., 1984; Friedland and Johnson, 1985; Turner et al., 1985). Typical leakage values range from 0% to 5% (Turner et al., 1985; Schut et al., 1986; Andren et al., 1975; Lindberg and Turner, 1988). For lake systems with a large watershed:lake area ratio, this flux of Pb could be a significant fraction of the total input reaching the lake (Borg and Johansson, 1988). Second, Dillon and Evans (1982), using multiple cores (see Figure 4–4 for a data set for one representative lake), compared the total Pb burden for a group of eight lakes in Ontario. The lakes varied in area, depth, watershed area, water chemistry, and other factors. They found that all the lakes had approximately the same total anthropogenic Pb burden (around 680 mg/m^2; range = 608 to 768 mg/m^2). They concluded that most of the Pb in the sediment was derived from direct deposition from the atmosphere. If it is assumed that this amount has accumulated over approximately 100 years (Norton, 1986a), it is more than an order of magnitude more than would be expected from natural processes. Evans and Rigler (1985) used the same methodology in four other regions of eastern Canada and found that regional burdens were internally consistent.

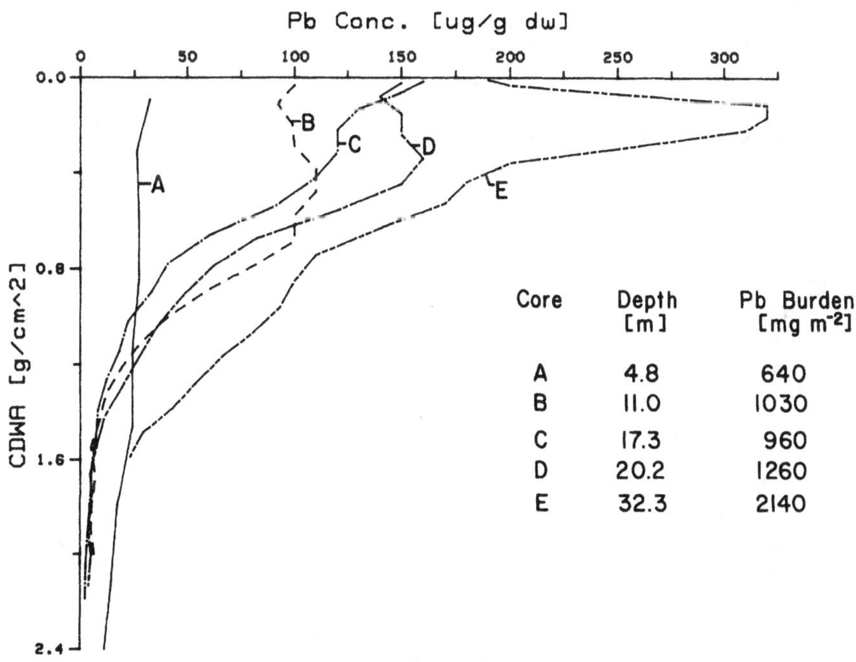

Figure 4–4. The concentration of Pb in multiple cores from Jerry Lake, Ontario (from Dillon and Evans, 1982). Concentration based on dry mass.

However, the average regional burden decreased to the north and east of southern Ontario, reaching a low value of 44 mg/m^2 in northern Quebec.

Early studies of the distribution of Pb in sediments centered on the Great Lakes. These studies are reviewed by Alderton (1985). Although it is likely that much of the Pb residing in the sediments is transported via the atmosphere, considerable direct surface inputs are probable in some localities. Thus, we focus our comments on more remote lakes that receive the bulk of their Pb via the atmosphere (Table 4–2).

A consistent picture has emerged for the eastern part of North America in terms of both the timing of atmospheric pollution and the approximate atmospheric flux. One of the first analyses of a ^{210}Pb-dated core from a relatively undisturbed lake in North America was reported by Wahlen and Thompson (1980). They calculated

Table 4-2. Selected studies of Pb in lake sediment for remote lakes in North America.

Author	Area	Lakes (n)	Dating
Baron et al. (1986)	Colorado	4	^{210}Pb
Brewer (1986)	Maine	4	^{210}Pb
Davis et al. (1983)	Northern New England	6	^{137}Cs, ^{210}Pb
Dillon et al. (1986)	Ontario	4	^{210}Pb
Dixit (1986)	Ontario	6	^{210}Pb
Dixit et al. (1989)	Ontario	3	^{210}Pb
Evans et al. (1986)	Ontario		^{210}Pb
Evans and Dillon (1982)	Ontario	1	^{210}Pb
Evans and Rigler (1985)	Quebec	3	^{210}Pb
Galloway and Likens (1979)	New York	1	^{137}Cs
Hanson and Norton (1982)	Maine	2	^{137}Cs, ^{210}Pb
Hanson et al. (1982)	Maine	5	^{137}Cs, ^{210}Pb
Heit et al. (1981)	New York	2	^{137}Cs
Johnson (1987)	Ontario	14	^{210}Pb
Johnson et al. (1988)	Ontario	4	^{210}Pb
Johnston et al. (1982)	Maine	3	^{137}Cs, ^{210}Pb
Kahl et al. (1984)	Maine, New Hampshire	3	^{137}Cs, ^{210}Pb
Kahl and Norton (1983)	Maine	2	^{210}Pb
Norton (1983)	Colorado	4	^{210}Pb
Norton (1984)	New York	10	^{210}Pb
Norton (1986a, 1986b)	United States	14	^{210}Pb
Norton (1987)	Maine	1	^{210}Pb
Norton et al. (1982)	N. New England	5	^{137}Cs, ^{210}Pb
Norton and Kahl (1987b)	Wyoming	4	^{210}Pb
Norton et al. (1989)	United States	30	^{210}Pb
Nriagu et al. (1982)	Ontario	7	^{210}Pb
Ouellet and Jones (1983)	Quebec	11	^{137}Cs
Wahlen and Thompson (1980)	New York	2	^{210}Pb
Williams (1980)	Maine, New Hampshire	6	^{137}Cs
Wong et al. (1984)	Ontario	10	^{210}Pb

that maximum flux of excess Pb (the anthropogenic component) for Sylvan Lake in New York as 160 mg/m^2/yr, quite large compared to values reported below (Table 4–3). However, the calculated unsupported ^{210}Pb was at least a factor of three higher than would result from just atmospheric loading, suggesting sediment focusing of the ^{210}Pb and stable Pb (see below). In northern New York, initial increases in the concentrations and accumulation rates for Pb in 18 lakes (Norton et al., in review; Norton, 1984) occur from about 1850 to 1875, based on ^{210}Pb-dated cores. (Earlier studies—Galloway and Likens, 1979; and Heit et al., 1984, among them—suggested similar enrichments but the chronologies, based on ^{137}Cs, are incorrect.) Background concentrations and accumulation rates range considerably in these single core studies (Figure 4–5). Anthropogenic Pb accumulation rates range from about 10 to 60 mg/m^2/yr. These values are comparable to those reported for soil accumulations in the northeastern United States by Friedland and others, 1984. The maxima accumulation rates in sediments typically occur between 1960 and 1980 (Table 4–3), suggesting some correspondence between atmospheric concentrations and sediment accumulation of Pb. Total integrated anthropogenic Pb ranges from 221 to 2,100 mg/m^2. The high end of the range is comparable to profundal cores from Dillon and Evans (1982). The lower end would yield much lower whole-lake burdens, even in adjacent lakes. This suggests that the results of Dillon and Evans may not be applicable for lakes with different hydrologic characteristics, especially where bypassing of sediment is important. Most of the variation in the New York data set appears to be controlled by sediment focusing rather than variation in the atmospheric deposition rate. This is supported by the strong relationship between the integrated unsupported ^{210}Pb and integrated anthropogenic Pb in the cores from 18 lakes in the Adirondack Mountains, New York (Figure 4–6).

In sediments from lakes in Ontario, concentrations increased between about 1880 and 1900 (Dixit et al., 1989; Nriagu et al., 1982; Wong et al., 1984) to maxima in recent sediment. Dixit (1986) reports that the onset of anthropogenic Pb deposition for six Sudbury, Ontario, lakes ranges from 1895 to 1924, somewhat earlier than regions to the east and southeast. Site-specific total Pb accumulation rates reach maxima of as much as 40 mg/m^2/yr with a range spanning an order of magnitude. Total integrated whole-lake anthropogenic Pb burdens for 20 lakes in southeastern Ontario range from 312 to 768 mg/m^2 (Evans and Rigler, 1985; Dillon and Evans, 1982). Further northeast in the Laurentide Park of Quebec, burdens are slightly lower ($n = 3$, range = 272 to 399 mg/m^2). In the Schefferville area, 400 km north of the park, burdens range from 31 to 59, an order of magnitude less. The onset of anthropogenic Pb pollution in the remote Schefferville region is in the 1850s, based on means of 22 cores in each of two lakes (Evans and Rigler, 1985). This region has had virtually no Pb emissions. The results from these studies are strong evidence for the widespread anthropogenic influence on the atmospheric deposition rates and sediment accumulation rates for Pb.

In New England, the history of elevated deposition of Pb is comparable to New York and eastern Canada (Figure 4–7). As for the Adirondack Mountain region of New York, single core accumulation rates for Pb are highly variable, depending on

Table 4-3. Characteristics of the anthropogenic flux of Pb to 18 lakes in the Adirondack Mountains, New York, based on ^{210}Pb-dated cores.

Lake	Integrated unsupported ^{210}Pb in lake sediment (pCi/cm^2)	Maximum anthropogenic accumulation rate for Pb ($\mu g/cm^2/yr$)	Integrated anthropogenic Pb in lake sediment ($\mu g/cm^2$)	Sediment accumulation rate, 100-year average ($mg/cm^2/yr$)
Brooktrout[a]	10.36	2.1	107	9.3
Deep	8.1	1.9	83	6.5
Fourth	10.41	2.3	112	17.0
Jerseyfield	12.32	3.3	166	9.4
Merrian	9.94	3.7	82	6.1
Panther	21.3	4.8	208	12.9
Sagamore	14.01	4.0	114	20.1
Silver	11.1	2.2	76	7.6
T	9.61	2.4	117	13.5
U. Wallface	7.2	3.2	99	7.2
Arnold[b]	7.1	1.5	58	20.9
Bear	4.4	1.2	48	6.7
Big Moose	19.1	5.9	210	18.0
Clear	17.2	3.0	138	8.8
Deep	10.7	2.7	100	7.5
Little Echo	2.6	0.8	22	3.1
Merriam	9.7	3.2	86	5.3
Queer	18.6	5.6	178	13.0
U. Wallface	7.3	1.2	68	5.0
West	7.3	1.8	97	9.6
Windfall	17.7	3.8	137	12.8

[a] From Norton (1984).

[b] From Norton et al. (1989b).

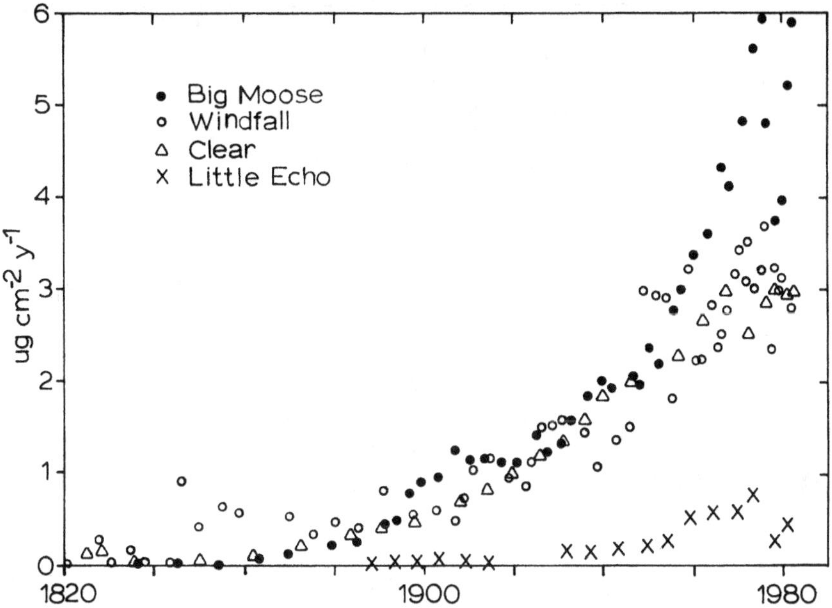

Figure 4–5. Accumulation rates of anthropogenic Pb for four lakes in the Adirondack Mountains, NY.

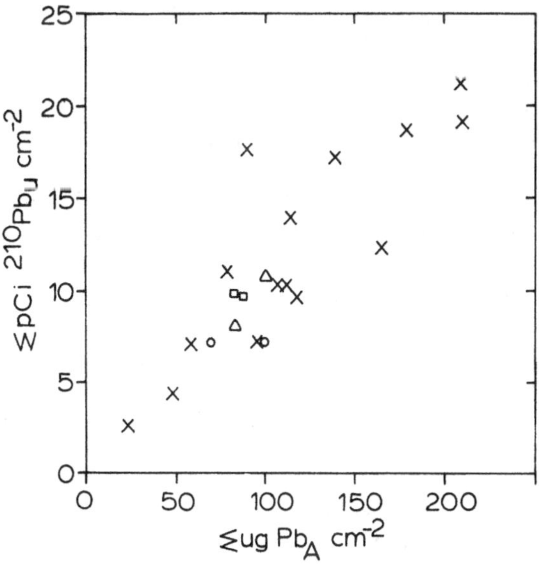

Figure 4–6. Integrated unsupported ^{210}Pb versus integrated anthropogenic Pb for single cores from 18 lakes in the Adirondack Mountains, NY. Open symbols indicate different studies at the same lake; circles, Upper Wallface; triangles, Deep; squares, Merriam. See Table 4–3 for additional data.

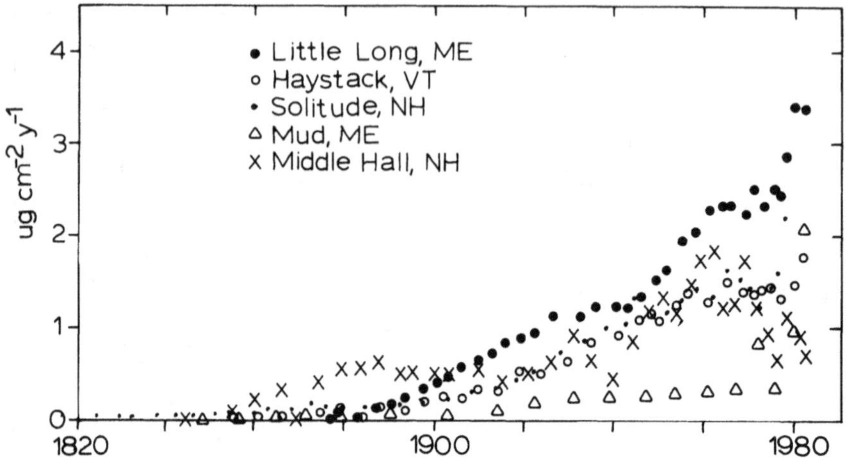

Figure 4–7. Accumulation rates of anthropogenic Pb for five lakes in New England.

the amount of focusing of sediment rather than variation in the atmospheric input. For example, four lakes located within 2 km of each other in coastal eastern Maine have very different accumulation rates for total Pb. Mud and Little Long Ponds (the two extremes) are only 200 m apart but have Pb accumulation rates that differ by a factor of ten (see Figure 4–7). The vast majority of the total Pb is anthropogenic. Both integrated total Pb and anthropogenic Pb are in proportion to the unsupported integrated ^{210}Pb values. The chronology of the initial increases is the same (Norton et al., 1989; Brewer, 1986).

Recent sediment for many lake cores typically has lower Pb accumulation rates (see, e.g., Figures 4–5 and 4–7) and concentrations. Two explanations have been offered to explain this: changes in lake chemistry and decreases in atmospheric inputs. First, Dillon and others (1986) suggest that acidification of lake waters below pH 5.5 may result in the decreased sedimentation of Pb associated with particles. Leaching of Pb from already deposited sediment is unlikely because of the higher pH associated with the sediment-water interface microenvironment. Also microcosm experiments indicate that Pb is not mobilized from sediment until a much lower pH is reached in the water column (Kahl and Norton, 1983; Davis et al., 1982). Empirical evidence supports the suggestion of Dillon and others (1986). For example, Borg and Andersson (1984) found that in lakes with low dissolved organic carbon, the portion of the total Pb that was dialyzable increased as pH declined. Also, Effler and others (1985) and Dillon and others (1987) demonstrated that acidification of lake waters reduces the concentration of dissolved organic carbon (DOC). If this reduction is due to decreased export of DOC from the watershed, there would be reduced export of Pb to the lake as well because of the strong positive linkage of DOC and the concentration of Pb in surface waters (Swanson and Johnson, 1980; Turner et al., 1985; Schut et al., 1986; LaZerte et al., 1989).

The second explanation for the decrease in recent accumulation rates is related to the documented recent decrease in atmospheric deposition of Pb. This is a reasonable explanation for the nonacidic lake situations and may also explain some of the recent decline in acidic lakes. Binford and others (in review) found no relationship between pH and cumulative Pb or cumulative ^{210}Pb (for one core per lake only, however), unlike Dillon and others (1986).

In Minnesota, Michigan, and Wisconsin, accumulation rates for Pb increase slightly later than further east, but typically before 1900 (Norton et al., 1989) (Figure 4–8). Maximum accumulation rates and integrated anthropogenic single-core loads are comparable to values in the eastern United States and Canada.

Florida lakes for which data are available (Norton et al., 1989) are all seepage lakes. These cores are apparently mixed, as evidenced by normal integrated unsupported ^{210}Pb but very high mass accumulation rates (Norton et al., 1989), and thus their Pb stratigraphies are suspect. However, total anthropogenic Pb values (from single cores) are lower, as a group, than those of the upper Great Lakes states, New York, New England, and southeastern Ontario—ranging up to 940 but with an average of 500 mg/m^2. For both the Florida and Great Lakes states, the maxima for accumulation rates generally fall after 1980, significantly later than for New York and New England. This may be a reflection of increased energy production in these parts of the United States (perhaps compensating for decreased alkyl Pb in gasoline) while energy production is down 20% to 30% in the northeastern United States.

For all the studies reported above, ancillary data indicate that Pb is accompanied by other elements typically associated with coal consumption, such as Zn. Lakes

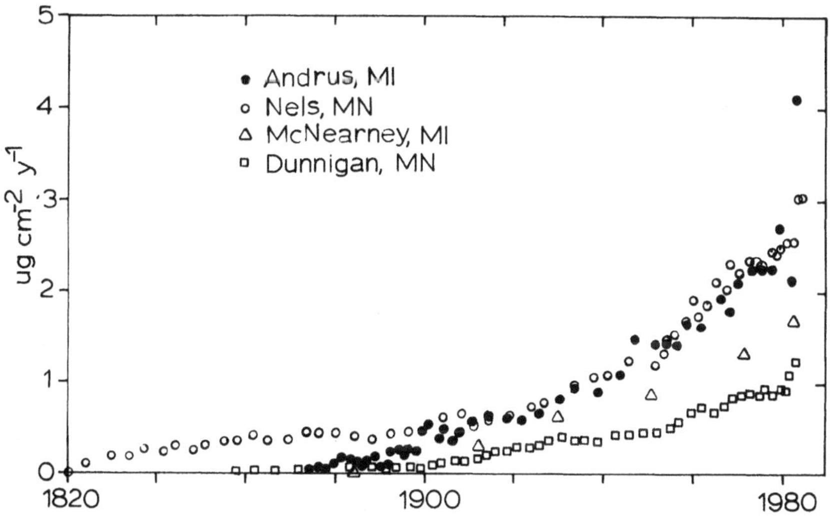

Figure 4–8. Accumulation rates of anthropogenic Pb for four lakes in the upper midwestern United States.

in the proximal windshed of smelters, such as those near Sudbury in southeastern Ontario, have other metals and metaloids as well, such as Se, As, Ni, and Co.

Comparatively few studies exist for lakes west of the Mississippi in the United States. Baron and others (1986) reported elevated concentrations and accumulation rates starting late in the nineteenth century for four lakes in Rocky Mountain National Park, Colorado. The lakes are remote, with absolutely no watershed development. These increases are attributed to dust related to mining activity in the region, inasmuch as the Pb is unaccompanied by appreciable amounts of other metals associated with fossil fuels. Virtually no coal was produced or consumed in that region prior to the twentieth century. Although accumulation rates approach values for eastern North America, accumulation rates are erratic (Figure 4–9). Several hundred kilometers to the northwest in the Wind River Mountains of Wyoming, Norton and Kahl (1987b) document similar increases for Pb starting in the later 1800s for three alpine cirque lakes. Initial increases for Pb are also unaccompanied by Cu or Zn and are thus attributed to mining activity in the region. Accumulation rates range up to 20 mg/m^2/yr, similar to values in eastern North America (Figures 4–5 and 4–7). Shirahata and others (1980) demonstrated a change in the isotopic composition of Pb in sediments dated to the late 1800s from a shallow (0.6 m), small (0.02 ha) alpine pond in the Sierra Nevada Mountains of California, and a slight increase in the accumulation rate of Pb (around 1 mg/m^2/yr). Last, Norton and Kahl (1988) report elevated Pb accumulation rates for alpine lakes in the Sierra Nevada Mountains in California. The major increases (up to 10 mg/m^2/yr) occur within the last 50 years and are apparently related

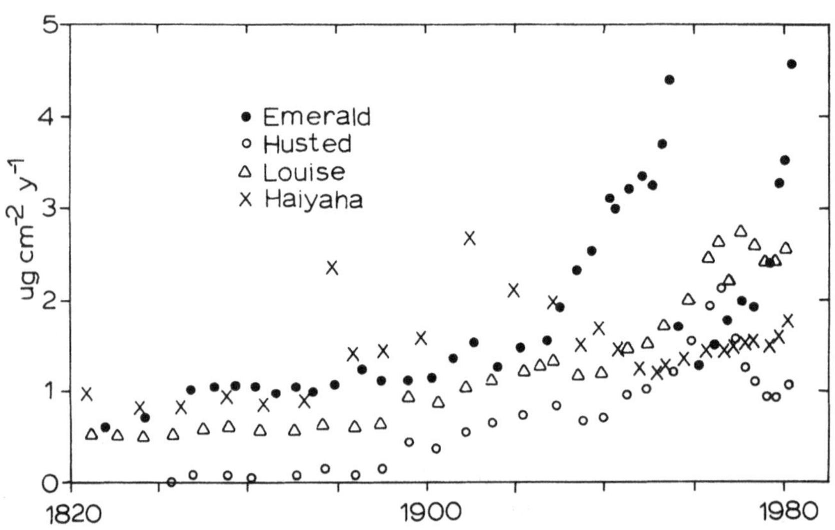

Figure 4–9. Accumulation rates of total Pb in single cores from four lakes in Rocky Mountain National Park, CO. The accumulation rate for Hiayaha is for the anthropogenic component.

primarily to the emissions from automobiles along the west coast of California (Figure 4–10). The other common metal pollutants associated with coal and oil consumption (Zn, Cu, and V) have not been detected in excess amounts.

Ombrotrophic peat bogs make ideal recorders of trends in atmospheric deposition of Pb. They receive all of their nutrients from the atmosphere, and their organic content is typically in excess of 95%, a great deal of which is *Sphagnum*, a strong cation exchanger (Clymo, 1987). All the studies below utilized ^{210}Pb dating for chronology and estimation of accumulation rates. Hemond (1980) was one of the first to estimate the atmospheric deposition of Pb from data in freshwater peat cores. He estimated values from Thoreau's bog in Massachusetts of about 43 $mg/m^2/yr$ for the period 1948 to 1977, with values declining to 10 or less for the period prior to 1881. The integrated anthropogenic Pb for the last 100 years is about 2,600 mg/m^2. This is two to three times the values reported by Norton (1987) and Norton and Kahl (1987b) for bogs in Maine. These values are consistent with surveys of the concentration of Pb in mosses in the northeastern United States (Groet, 1976). Background accumulation rates for Pb in a bog located in a maritime environment along coastal Maine have been estimated at 1 to 2 $mg/m^2/yr$ from both hummock and hollow cores by Norton and Kahl (1987a). This may be an overestimation caused by downward migration of anthropogenic Pb. Maximum deposition values for anthropogenic Pb are estimated at 40 and 20 $mg/m^2/yr$ for the hummock and hollow, respectively, in closely spaced cores. The discrepancy between the two accumulation rates is consistent with the inventory of unsupported ^{210}Pb in the cores. Hummocks had twice the inventory of hollows. The reverse of this was observed by Oldfield and others (1979a) with four times as much unsupported ^{210}Pb in a hollow core versus a hummock core for a bog in Cumbria, England. Alternatively, Oldfield and others (1979b) found up to an order of magnitude more airborne magnetic particles in hummocks than in pools for bogs also located in Cumbria.

McCaffrey and Thomson (1980) determined from a salt marsh core from coastal

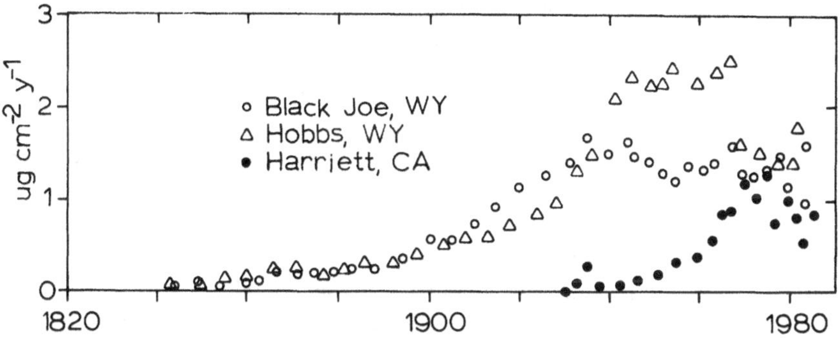

Figure 4–10. Accumulation rates of anthropogenic Pb in single cores from Harriett Lake in the Sierra Nevada Mountains, CA, and from Hobbs and Black Joe Lakes in the Wind River Mountains, WY.

Connecticut that anthropogenic Pb accumulation rates initially increased between 1860 and 1865, reaching values as high as 80 mg/m^2/yr and no lower than 40, since 1910.

IV. Summary

North America has been subjected to a history of atmospheric pollution associated with the burning of fossil fuels, smelting, transportation emissions, and other industrial activities. Although much concern has been expressed about acidic precipitation associated with emissions of NO_x and SO_x, the record of metal pollution is much clearer in terms of spatial and temporal trends because of their relatively conservative behavior in sediment systems.

Increased deposition of metals is recorded by ombrotrophic bog sediments and lake sediments. Various investigators have evaluated the deposition of metals using cores of accumulating sediment, dated by either ^{210}Pb or ^{137}Cs to evaluate the timing and the absolute accumulation rates. Differences in analytical methods for the metals, as well as mechanical and chemical processes in nature, render these surrogate metal deposition recorders imprecise in terms of the reconstruction of accurate atmospheric fluxes from measured net sedimentary fluxes. However, generalizations are possible, based on widespread data.

Atmospheric deposition of the three metals Cd, Hg, and Pb is elevated in those regions receiving higher than background amounts of sulfate, without exception. The atmospheric deposition rate for Cd in eastern Ontario and the upper midwestern United States is approximately three orders of magnitude higher than in remote parts of the northern hemisphere. Accumulation rates in modern sediments of remote lakes are in the range of 0.1 to 0.2 mg/m^2/yr, corresponding to Cd concentrations in precipitation of about 0.1 μg/L, assuming 1 m precipitation. The accumulation rates in lakes range over about an order of magnitude. This probably reflects sediment focusing in some lakes as well as differing amounts of transport of Cd from the watershed to the lake, depending largely on the pH of surface waters. Increased atmospheric deposition, as reflected by lake and bog sediments, started as early as 1860 and continues to the present.

Paleolimnological evidence for increased deposition of Hg is sparse but surprisingly consistent. Concentrations and accumulation rates of Hg in lake sediment have approximately doubled in eastern North America, with a time scale generally the same as that for Pb (see below). The doubling of sediment Hg accumulation rates parallels the increase of Hg in northern hemisphere air over values in the southern hemisphere. Studies in Sweden indicate that Hg continues to be exported from watersheds to lakes in amounts in excess of recent atmospheric deposition, indicating continued mobilization from organic soils.

Data for Pb in sediments are plentiful and give a very consistent picture for eastern North America. Increased atmospheric deposition of Pb was clearly evident in sediments as early as 1850 and is associated with the burning of coal. This source term was supplemented in the last 50 years by alkyl Pb in vehicle emissions. Sediment accumulation rates for Pb typically show an exponential increase until 1970 to 1985. Many sediment records show a recent decrease in accumulation rates, consistent with the sharp decrease in Pb emissions and the measured decrease in Pb in air. Maximum deposition rates measured in lakes cluster in the 20 to 30 mg/m^2/yr range, as do values measured in ombrotrophic peat cores. This suggests a comparable atmospheric deposition rate because of the immobile character of Pb. This rate corresponds to concentrations of about 20 μg/L in precipitation, if all the Pb is deposited as wet precipitation, which it is almost certainly not, even in lakes. Elevated accumulation rates of Pb in lake sediment, but associated with mining activity, are present in "pristine" western montane regions of the United States, suggesting that at least for the United States there may be no lake ecosystems that are untainted by atmospheric pollution of metals.

The increased concentrations and accumulation rates of Cd, Hg, and Pb in lake and peat bog sediments indicate that anthropogenic activity dominates the biogeochemical budgets for these elements in virtually all of temperate eastern North America and much of the west as well. The timing and magnitude of the changes in anthropogenic fluxes of Cd, Hg, and Pb to lake and bog sediments enable the imperfect reconstruction of trends in atmospheric deposition as well as permitting inferences about the changing status of the acidity of the watershed soils and lake water column.

Acknowledgments

Research leading to this paper was supported by the following—for Norton and Kahl: U.S. National Science Foundation (Grants #DEB-7922142, DEB7810641), U.S. Department of the Interior (Grants USOWRTAO48ME, RM-81-206GR [Man and the Biosphere]), U.S. Fish and Wildlife Service (Grant 1416000979040), U.S. Environmental Protection Agency (Grant APP-0119-1982), U.S. Forest Service (Grant 53-84M8-5-0040), the Electric Power Research Institute, and the State of Maine (Grant 916597 [U. S. DOE]); for Dillon and Mierle: Ontario Ministry of the Environment; for Evans: Natural Sciences and Engineering Research Council of Canada, Ontario Ministry of the Environment, and Environment Canada. Norton is grateful to Carolyn Moody for assistance in compiling data for lakes in the United States.

References

Ainsworth, C. C., and D. Rai. 1987. *Chemical characterization of fossil fuel combustion wastes*. Electric Power Research Institute, Palo Alto, Calif.

Alderton, D. H. M. 1985. *In Historical Monitoring,* 1–95, Monitoring and Assessment Centre, London.

Andren, A. W., S. E. Lindberg, and L. C. Bate. 1975. *Atmospheric input and geochemical cycling of selected trace elements in Walker Branch Watershed.* ORNL Environ Sc Div Publ 728, Oak Ridge National Laboratory, Oak Ridge, Tennessee. 68 p.

Andren, A. W., and J. O. Nriagu. 1979. *In* J. O. Nriagu, ed. *The biogeochemistry of mercury in the environment,* 1–21, Elsevier, Amsterdam.

Appelquist, H., J. K. Ottar, T. Sevel, and C. Hammer. 1978. Nature 273:657–659.

Appleby, P. G., and F. Oldfield. 1978. Catena 5:1–8.

Appleby, P. G., and F. Oldfield. 1979. Environ Sc Tech 13:478–480.

Babu, S. P. 1975. *Trace elements in fuel.* American Chemical Society, Washington, D.C. 216 p.

Baron, J., S. A. Norton, D. R. Beeson, and R. Herrmann. 1986. Can J Fish Aquat Sci 43:1350–1362.

Barrie, L. A., S. E. Lindberg, W. H. Chen, H. B. Ross, R. Arimote, and T. M. Church. 1987. Atmos Envir 21:1133–1135.

Batifol, P., and C. Boutron. 1984. Atmos Environ 18:2507–2515.

Bayley, S. E., and D. W. Schindler. 1987. *In* T. C. Hutchinson and K. M. Meema, eds. *Effects of atmospheric pollutants on forests, wetlands and agricultural ecosystems,* 531–548. Springer-Verlag, Berlin.

Bewers, J. M., P. J. Barry, and D. J. MacGregor. 1987. *In* J. O. Nriagu and J. Sprague, eds. *Cadmium in the aquatic environment,* 19–34. Wiley Interscience, New York.

Binford, M. W., S. A. Norton, and J. S. Kahl. 1989. *In* D. F. Charles and D. R. Whitehead, eds. *Paleolimnological investigation of recent lake acidification.* Junk Pub., The Hague.

Borg, H., and A. Andersson. 1984. Verh Internat Verein Limnol 22:725–729.

Borg, H., and K. Johansson. 1988. Unpublished manuscript.

Boutron, C. F. 1980. Nature 284:575.

Boutron, C. F., and C. C. Patterson. 1986. Nature 323:222–225.

Brewer, G. F. 1986. *Sulfur, heavy metal, and major element chemistry of sediments from four Eastern Maine ponds.* Unpublished M.S. Thesis, University of Maine. 139 p.

Buckley, D. E., and R. E. Cranston. 1971. Chem Geol 7:273–284.

Campbell, P. G. C., P. M. Stokes, and J. N. Galloway. 1985. *Acid deposition: Effects on geochemical cycling and biological availability of trace metals.* National Academy Press, Washington, D.C. 83 p.

Carignan, R., and A. Tessier. 1988. Geochim Cosmochim Acta 52:1179–1188.

Chan, W. H., A. J. S. Tang, D. H. S. Chung, and M. A. Lusis. 1986. Water Air Soil Pollut 29:373–389.

Christensen, E. R., and N.-K. Chien. 1981. Environ Sci Tech 15:553–558.

Clymo, R. S. 1965. J Ecol 53:747–758.

Clymo, R. S. 1987. *In* T. C. Hutchinson and K. M. Meema, eds. *Effects of atmospheric pollutants on forests, wetlands and agricultural ecosystems,* 513–530. Springer-Verlag, Berlin.

Coleman, D. O. 1985. *In Historical Monitoring,* 155–173, Monitoring and Assessment Centre, London.

Davis, A. O., J. N. Galloway, and D. K. Nordstrom. 1982. Limnol Oceanog 27:163–167.

Davis, R. B. 1974. Limnol Oceanog 14:466–488.

Davis, R. B., C. T. Hess, S. A. Norton, D. W. Hanson, K. D. Hoagland, and D. S. Anderson. 1984. Chem Geol 44:151–185.

Davis, R. B., and S. A. Norton. 1978. Pol Arch Hydrobiol 25:99–115.

Davis, R. B., S. A. Norton, C. T. Hess, and D. F. Brakke. 1983. Hydrobiologia 103:113–123.

Dearing, J. A. 1983. Hydrobiologia 103:59–64.

Dillon, P. J., and R. D. Evans. 1982. Hydrobiologia 91:121–130.

Dillon, P. J., H. E. Evans, and P. J. Scholer. 1988. Biogeochemistry 5:201–220.
Dillon, P. J., R. A. Reid, and E. deGrosbois. 1987. Nature 329:45–48.
Dillon, P. J., P. J. Scholer, and H. E. Evans. 1986. *In* P. G. Sly, ed. *Sediments and water interactions,* 491–499, Springer-Verlag, New York.
Dixit, S. S. 1986. *Algal microfossils and geochemical reconstructions of Sudbury Lakes.* Ph.D. Thesis, Queens University, Ontario. 190 p.
Dixit, S. S., A. S. Dixit, and R. D. Evans. 1989. Environ Sci Technol 23:110–115.
Effler, S. W., G. C. Schafran, and C. T. Driscoll. 1985. Can J Fish Aquat Sc 42:1707–1711.
Eisenreich, S. J., N. A. Metzer, and N. R. Urban. 1986. Environ Sci Technol 20:171–174.
EPA. 1986. *Air quality criteria for lead.* EPA 600/8-83/028 U.S. Environmental Protection Agency, Research Triangle Park, N.C.
Evans, H. E., P. J. Smith, and P. J. Dillon. 1983. Can J Fish Aquat Sci 40:570–579.
Evans, R. D. 1986. Arch Envir Contam Toxic 15:505–512.
Evans, R. D., and P. J. Dillon. 1982. Hydrobiologia 91:131–137.
Evans, R. D., and F. H. Rigler. 1980. Environ Sci Technol 14:216–218.
Evans, R. D., and F. H. Rigler. 1985. Water Air Soil Pollut 24:141–151.
Fitzgerald, W. F., G. A. Gill, and A. D. Hewitt. 1983. *In* C. S. Wong, E. Boyle, K. W. Bruland, J. D. Burton, and E. D. Goldberg, eds. *Trace metals in sea water,* 297–315. Plenum Press, New York.
Fitzgerald, W. F., G. A. Gill, and J. P. Kim. 1984. Science 224:597–599.
Friedland, A. J., and A. H. Johnson. 1985. J Envir Qual 14:332–336.
Friedland, A. J., A. H. Johnson, and T. G. Siccama. 1984. Water Air Soil Pollut 21:161–170.
Galloway, J. N., and G. E. Likens. 1979. Limnol Oceanog 24:427–433.
Galloway, J. N., J. D. Thornton, S. A. Norton, H. L. Volchok, and R. A. N. McLean. 1982. Atmos Envir 16:1677–1700.
Glooschenko, W. A., and J. A. Capoblanco. 1982. Environ Sci Technol 16:187–188.
Gorham, E., S. J. Eisenreich, J. Ford, and M. V. Santelmann. 1985. *In* W. Stumm, ed. *Chemical processes in lakes,* 339–363. John Wiley and Sons, New York.
Groet, S. S. 1976. Oikos 27:445–456.
Hakanson, L. 1977. Can J Earth Sc 14:397–412.
Hanson, D. W., and S. A. Norton. 1982. Proc Internat Symp Hydromet, Denver, Colorado, American Water Research Association, 25–33.
Hanson, D. W., S. A. Norton, and J. S. Wiliams. 1982. Water Air Soil Pollut 18:227–239.
Heit, M., Y. Tan, C. Klusek, and J. C. Burke. 1981. Water Air Soil Pollut 15:441–464.
Hemond, H. F. 1980. Ecol Mon 50:507–526.
Herron, M. M., J. H. Cragin, and C. C. Langway, Jr. 1977. Geochim Cosmochim Acta 41:915–920.
Hogg, T. J., J. W. B. Stewart, and J. R. Bettany. 1975. J Environ Qual 7:440–447.
Husar, R. 1986. *In Acid deposition: Long-term trends,* 48–92. National Academy Press, Washington, D.C.
Jeffries, D. S. 1989. *In* S. A. Norton, S. E. Lindberg, and A. L. Page, eds. *Soils, aquatic processes, and lake acidification.* Springer-Verlag, New York.
Jeffries, D. S., and W. R. Snyder. 1981. Water Air Soil Poll 15:127–152.
Johansson, K. 1985. Verh Intern Verein Limnol 22:2359–2363.
Johnson, A. H., T. G. Siccama, and A. J. Friedland. 1982. J Envir Qual 11:577–580.
Johnson, D. L., and R. S. Braman. 1974. Environ Sci Technol 8:1003–1009.
Johnson, M. G. 1987. Can J Fish Aquat Sc 44:3–13.

Johnson, M. G., L. R. Culp, and S. E. George. 1986. Can J Fish Aquat Sc 43:754–762.

Johnston, S. E., S. A. Norton, C. T. Hess, R. B. Davis, and R. S. Anderson. 1982. *In* L. H. Keith, ed. *Energy and environmental chemistry: Acid rain,* 177–187. Ann Arbor Science, Ann Arbor, Mich.

Jonasson, I. R., and R. W. Boyle. 1972. Can Mining Metall Bull 65:32–39.

Kahl, J. S., and S. A. Norton. 1983. *Metal input and mobilization in two acid-stressed lake watersheds in Maine.* Final Report to U.S. Office of Water Research and Technology (Contract A-053). 70 p.

Kahl, J. S., S. A. Norton, and J. S. Williams. 1984. *In* O. P. Bricker, ed. *Geological aspects of acid deposition,* 23–35. Ann Arbor Science, Ann Arbor, Mich.

Kemp, A. L. W., J. D. H. Williams, R. L. Thomas, and M. L. Gregory. 1978. Water Air Soil Pollut 10:381–402.

Kim, J. P., and W. F. Fitzgerald. 1986. Science 231:1131–1133.

Krenkel, P. A. 1973. CRC Critical Rev Environ Cont 3:303–373.

LaZerte, B. D., R. D. Evans, and P. Grauds. 1989. Sc Tot Environ (in press).

Lindberg, S. E. 1987. *In* T. C. Hutchinson and K. M. Meema, eds. *Lead, mercury, cadmium, and arsenic in the environment,* 89–106. John Wiley and Son, Chichester.

Lindberg, S. E., and R. R. Turner. 1977. Nature 286:133–136.

Lindberg, S. E., and R. R. Turner. 1988. Water Air Soil Pollut 39:123–156.

Mayer, R., and S. E. Lindberg. 1985. *Heavy metals in the environment,* 351–355. CEP Publishers, Edinburgh.

McCarthy, J. H., Jr., J. L. Meuschke, W. H. Ficklin, and R. E. Leonard. 1970. *Mercury in the atmosphere.* U.S. Dept. Interior, Washington, D.C.

McCaffrey, R. J., and J. Thomson. 1980. Advances in Geophysics 22:165–236.

Mart, L. 1983. Tellus 35:131.

Meger, S. A. 1986. Water Air Soil Pollut 30:411–419.

Mierle, G. 1989. Environ Tox Chem (in review).

Murozumi, M., T. J. Chow, and C. C. Patterson. 1969. Geochim Cosmochim Acta 33:1247–1294.

Ng, A., and C. C. Patterson. 1981. Geochim Cosmochim Acta 45:2109–2121.

Norton, S. A. 1984. *Paleolimnological assessment of acidic deposition impacts on select acidic lakes.* Final Report to U.S. EPA (Contract APP-0119-1982). 127 p.

Norton, S. A. 1986a. Water Air Soil Pollut 30:331–345.

Norton, S. A. 1986b. *In Acid deposition: Long-term trends,* 369–434. National Academy Press, Washington, D.C.

Norton, S. A. 1987. *In* T. C. Hutchinson and K. M. Meema, eds. *Effects of atmospheric pollutants on forests, wetlands and agricultural ecosystems,* 561–576. Springer-Verlag, Berlin.

Norton, S. A. 1989. *Models to describe the geographic extent and time evolution of acidification and air pollution damage.* Finnish Water Authority, Helsinki. (in press).

Norton, S. A., R. W. Bienert, Jr., M. W. Binford, and J. S. Kahl. 1989. *In* D. F. Charles and D. B. Whitehead, eds. *Paleoecological investigations of recent lake acidification.* Junk Pub., Amsterdam, (in review).

Norton, S. A., R. B. Davis, and D. S. Anderson. 1982. *The distribution and extent of acid and metal precipitation in northern New England.* Final report to U.S. Fish and Wildlife Service (Contract 14-16-0009-79-040).

Norton, S. A., and J. S. Kahl. 1986. *In* P. M. Stokes, ed. *Pathways, cycling and transformation of lead in the environment,* 53–96. Royal Society of Canada, Toronto.

Norton, S. A., and J. S. Kahl. 1987a. In T. P. Boyle, ed. *New approaches to monitoring aquatic ecosystems*, 40–57. ASTM, Philadelphia.

Norton, S. A., and J. S. Kahl. 1987b. *Geochemical analysis of sediment cores, Wind River Mountains, Wyoming*. Final report to the U.S. Forest Service (Contract 53-84MB-5-0040). 49 p.

Norton, S. A., and J. S. Kahl. 1988. Atmospheric heavy metal pollution of lakes in the Sierra Nevada and Rocky Mountains, U.S.A. (abs). Ann Mtg Limnol Oceanogr, Denver, Colo. 60.

Nriagu, J. O. 1978. *In* J. O. Nriagu, ed. *The Biogeochemistry of lead in the environment*, 137–184. Elsevier, Amsterdam.

Nriagu, J. O. 1980. *Cadmium in the environment*. John Wiley and Sons, New York. 682 p.

Nriagu, J. O. 1986. *In* P. M. Stokes, ed. *Pathways, cycling and transformation of lead in the environment*, 17–36. The Royal Society of Canada, Toronto.

Nriagu, J. O., A. L. W. Kemp, H. K. T. Wong, and N. Harper. 1979. Geochim Cosmochim Acta 43:247–258.

Nriagu, J. O., H. K. T. Wong, and R. D. Coker. 1982. Environ Sci Technol 16:551–560.

Oldfield, F., P. G. Appleby, R. S. Cambray, J. D. Eakins, K. E. Barber, R. W. Battarbee, G. R. Pearson, and J. M. Williams. 1979a. Oikos 33:40–45.

Oldfield, F., A. Brown, and R. Thompson. 1979b. Quat Res 12:326–332.

Ouellet, M., and H. G. Jones. 1983. Can J Earth Sc 20:23–36.

Pacyna, J. M. 1989 *In* S. E. Lindberg, A. L. Page, and S. A. Norton, eds. *Sources, deposition, and canopy interactons*. Springer-Verlag, New York (in press).

Percy, K. E. 1983. Water Air Soil Pollut 19:341–349.

Pheiffer Madsen, P. 1981. Nature 293:127–129.

Puckett, K. J. 1986. *In* P. M. Stokes, ed. *Pathways, cycling and transformation of lead in the environment*, 225–256. The Royal Society of Canada, Toronto.

Rada, R. G., J. G. Wiener, M. R. Winfrey, and D. E. Powell. 1989. Arch Environ Cont Toxicol 18:175–181.

Rapin, F., A. Tessier, P. G. C. Campbell, and R. Carignan. 1986. Environ Sci Technol 20:836–840.

Rekolainen, S., M. Verta, and A. Liehu. 1986. Pub of the Water Res Inst, Nat Bd of Waters, Helsinki, Finland. 11–20.

Renberg, I. 1979. The National Swedish Environ Board, Rept PM 1151. Solna, Sweden. 318–324.

Royal Society of Canada. 1986. *Lead in the Canadian environment: Science and regulation*. Royal Society of Canada, Toronto. 374 p.

Ruhling, A., L. Rasmussen, K. Pilegaard, A. Makinen, and E. Steinnes. 1987. *Survey of atmospheric heavy metal deposition*. Nordisk Ministerrad, Copenhagen. 44 p.

Schell, W. R. 1987. Int J Coal Geol 8:225–237.

Schmitt, C. J., and D. R. Sasseville. 1978. *Northern Maine Mercury Investigattions*. (Rept submitted to) U.S. Army Corps of Engineers. Normandeau Associates, Inc, Bedford, NH.

Schroeder, W. H. 1982. Environ Sci Technol 6:394a–400a.

Schut, P. H., R. D. Evans, and W. A. Scheider. 1986. Water Air Soil Pollut 28:225–237.

Settle, D. M., and C. C. Patterson. 1980. Science 207:1167–1175.

Shirahata, H., R. W. Elias and C. C. Patterson. 1980. Geochim Cosmochim Acta 44:149–162.

Siccama, T. G., and W. H. Smith. 1978. Environ Sci Technol 12:593–594.

Siegel, S. M., and B. Z. Siegel. 1978. Water Air Soil Pollut 9:113–118.

SNV (Statens Naturevardsverk). 1987. *Monitor 1987: Tungmetaller,* Schmidts Boktryckeri AB, Helsingborg, Sweden.

Steinnes, E. 1988. *Soc Environ Tox Chem* (abs.), Arlington, Va.

Stephenson, M., and J. L. Mackie. 1988. Water Air Soil Pollut 38:121–136.

Swanson, K. A., and A. H. Johnson. 1980. Water Resourc Res 16:373–376.

Tessier, A., P. G. C. Campbell, and M. Bisson. 1979. Anal Chem 51:844–851.

Thomas, R. L. 1973. Can J Earth Sci 10:194–204.

Tolonen, K., and T. Jaakkola. 1983 Ann Bot Finnici 20:57–78.

Torrey, S. 1978. *Trace contaminants from coal.* Noyes Data Corporation, Park Ridge, N.J. 294 p.

Turner, R. S., A. H. Johnson, and D. Wang. 1985. J Envir Qual 14:305–314.

Varekamp, J. C., and P. R. Buseck. 1981. Nature 293:555–556.

Wahlen, M., and R. C. Thompson. 1980. Geochem Cosmochim Acta 44:333–339.

Weiss, H., K. Bertine, M. Koide, and E. D. Goldberg. 1975. Geochim Cosmochim Acta 39:1–10.

Weiss, H., M. Koide, and E. D. Goldberg. 1971. Science 174:692–694.

White, J. R., and C. T. Driscoll. 1985. Environ Sci Technol 19:1182–1187.

White, J. R., and C. P. Gubala. 1989. *In* D. F. Charles and D. R. Whitehead, eds. *Paleolimnological investigation of recent lake acidification.* Junk Pub., The Hague.

Williams, J. S. 1980. *The relative contribution of local and regional atmospheric pollutants to lake sediments in northern New England.* Unpublished M.S. Thesis, University of Maine, 59 p.

Wong, H. K. T., J. O. Nriagu, and R. D. Coker. 1984. Chem Geol 44:187–201.

Dry Deposition of Particles and Vapors

Cliff I. Davidson* and Yee-Lin Wu*

Abstract

This chapter reviews current knowledge of dry deposition of contaminants from the atmosphere. The first section introduces the topic and justifies its importance. In the second section, the dry deposition process is described in physical and mathematical terms. The physical process involves three steps: aerodynamic transport from the free atmosphere into relatively quiescent air near the surface, boundary layer transport across the quiescent region, and interactions with the surface. Mathematical relations for each step are expressed in terms of resistances to transport. The third section describes two methods by which predictions for dry deposition in the field have been developed. These include wind tunnel studies leading to development of empirical models and more detailed mathematical models based on theoretical formulations as well as field and laboratory data. In the fourth section, currently available methods for measuring dry deposition are discussed. The techniques include surface analysis methods, which involve measurement of the accumulation of contaminant on surfaces of interest, and atmospheric flux methods, which involve measurement of airborne contaminant concentrations used to infer the flux. Finally, dry deposition data from field and chamber experiments are summarized. Data are present for sulfur, nitrogen, and chloride species, as well as ozone, trace elements, and atmospheric particles. The data are compared with predicted dry deposition velocities using models presented earlier. The chapter concludes by acknowledging limitations in our understanding of dry deposition.

I. Introduction

Dry deposition of atmospheric contaminants is known to be an important removal mechanism, of comparable importance to wet deposition in many cases. For

*Departments of Civil Engineering and Engineering & Public Policy, Carnegie Mellon University, Pittsburgh, PA 15213, USA.

Definition of Symbols

A = exponential term in the model of Sehmel (1980)
Al_{air} = airborne concentration of aluminum (g/cm^3)
Al_{crust} = concentration of aluminum in the earth's crust (parts per million by mass)
a = empirical constant in the model of de la Mora and Friedlander (1982)
a' = empirical constant in the equation for R in the model of Slinn (1982)
a'' = empirical constant in the model of Shreffler (1978)
a_w = constant in the equation for f_w
B = sublayer Stanton number
b = vertical distance between the ground and the base of the foliage crown in the model of Bache (1979b) (cm)
b_e = constant in the equation for f_e
b_T = expression based on T_h, T_l, *and* T_o in the equation for f_T
b_w = constant in the equation for f_w
b' = empirical constant in the equation for r_s'
C = airborne concentration of contaminant (g/cm^3)
$\overline{C}$ = time-averaged airborne concentration of contaminant (g/cm^3)
C' = turbulent fluctuating component of airborne concentration of contaminant (g/cm^3)
C_o = airborne concentration of contaminant at z = 0 (g/cm^3)
C_r = airborne concentration of contaminant at reference height z_r (g/cm^3)
ΔC_i = airborne concentration of a trace metal in size range i (g/cm^3)
c = Cunningham slip correction factor
c_d = average drag coefficient for the vegetation at height z in the model of Slinn (1982)
c_v = portion of c_d arising from viscous drag in the model of Slinn (1982)
D = Brownian diffusivity of the contaminant (cm^2/s)
d = zero-plane displacement (cm)
d_c = cylinder diameter in the model of Davidson et al. (1982) (cm)
d_{c1} = diameter of "small" collectors in the canopy in the model of Slinn (1982) (cm)
d_{c2} = diameter of "large" collectors in the canopy in the model of Slinn (1982) (cm)
d_l = representative leaf dimension in the model of Shreffler (1978) (cm)
d_{min} = smallest dimension of a single leaf when projected on a plane normal to the airflow, in the model of Bache (1979b) (cm)
d_p = particle diameter (cm or μm)
d_p^+ = dimensionless particle diameter in the model of Sehmel (1980) = $d_p u_*/\nu$
d_s = particle stopping distance (cm or μm)
E = particle collection efficiency for all mechanisms without the influence of bounceoff in the model of Slinn (1982)
$E_{\text{diffusion}}$ = particle collection efficiency by diffusion without the influence of bounceoff in the model of Slinn (1982)
$E_{\text{interception}}$ = particle collection efficiency by interception without the influence of bounceoff in the model of Slinn (1982)
$E_{\text{impaction}}$ = particle collection efficiency by impaction without the influence of bounceoff in the model of Slinn (1982)
EF_{crust} = crustal enrichment factor
e = water vapor pressure (atm, or g/cm s^2)

$e_s(T)$ = saturated water vapor pressure at air temperature T (atm, or g/cm s^2)
F = flux or dry deposition rate of contaminant (g/cm^2s)
F_a = flux of air momentum (g/cm s^2)
F_H = sensible heat flux (calories/cm^2s)
F_W = flux of water vapor (g/cm^2s)
f = fraction of total interception by "small" collectors in the canopy in the model of Slinn (1982)
$f(z)$ = a function in the model of Bache (1979b)
f_e = correction factor for the effect of humidity in the equation for r_s'
f_R = a constant representing an approximation for $f(z)$ in the model of Bache (1979b)
f_T = correction factor for the effect of temperature in the equation for r_s'
f_w = correction factor for the effect of water stress in the equation for r_s'
f_x, f_z = structure coefficients for a vegetative canopy in the model of Bache (1979b)
g = gravitational acceleration (cm/s^2)
$g(z)$ = a function in the model of Bache (1979b)
g_R = a constant representing an approximation for $g(z)$ in the model of Bache (1979b)
h = height of the vegetative elements in a canopy (cm)
h_m = height of the center of the foliage crown in the model of Bache (1979b) (cm)
I = integral term in the model of Sehmel (1980)
I_p = photosynthetically active radiation in the equation for r_s' (watts/m^2)
K = eddy diffusivity of the contaminant (cm^2/s)
K_e = effective particle eddy diffusivity in the model of Sehmel (1980) (cm^2/s)
K_H = eddy diffusivity of heat (cm^2/s)
K_M = eddy viscosity of air (cm^2/s)
K_o = particle eddy diffusivity for $z < h$ in the model of Slinn (1982) (cm^2/s)
k = von Karman's constant = 0.4
L = Monin-Obukhov length (cm)
Le = Lewis number = κ/D
l = characteristic eddy size within the canopy in the model of Slinn (1982) (cm)
N = number of cylinders per unit area of ground in the model of Davidson et al. (1982)
n = empirical parameter in the equation for u, for $z < h$
Pr = Prandtl number = ν/κ
P_x = probability of particle capture by vegetative elements per unit distance in the horizontal direction in the model of Bache (1979b)
P_z = probability of particle capture by vegetative elements per unit distance in the vertical direction in the model of Bache (1979b)
R = reduction in collection efficiency caused by bounceoff in the model of Slinn (1982)
Re = Reynolds number = ud_c/ν
Re_* = surface Reynolds number based on friction velocity and roughness height = $u_* z_o/\nu$
r_a = aerodynamic resistance (s/cm)
r_b = boundary layer resistance (s/cm)
r_c = canopy resistance (or surface resistance) based on the flux to the horizontal area of the earth's surface (s/cm)
r_g = gravitational or sedimentation resistance (s/cm)
r_{soil} = resistance of the soil in the equation for r_c' (s/cm)

r_t = total resistance to transport (s/cm)
r'_c = canopy resistance (or surface resistance) based on the flux to the area of individual vegetative elements (s/cm)
r'_{cut} = resistance of the cuticle in the equation for r'_c (s/cm)
r'_m = resistance of the mesophyll in the equation for r'_c (s/cm)
r'_s = resistance of the stomata in the equation for r'_c (s/cm)
r'_{sm} = minimum value of r'_s (s/cm)
Sc = Schmidt number = ν/D
Sh = Sherwood number
St = Stokes number = $2\, d_s/d_c$
T = air temperature (°C or °K)
T_h and T_l = vegetation species-dependent high and low temperature extremes at which stomata no longer open (°C or °K)
T_o = temperature at which stomatal exchange is optimized (°C or °K)
u = air velocity in the direction of mean wind flow (cm/s)
$\bar{u}$ = time-averaged value of u (cm/s)
u_h = wind speed at height h (cm/s)
u_r = wind speed at reference height z_r (cm/s)
u_* = friction velocity (cm/s)
V_d = overall dry deposition velocity for a distribution of particles (cm/s)
v_a = deposition velocity of air momentum (cm/s)
v_d = dry deposition velocity (cm/s)
v_{dg} = deposition velocity for particles at the ground in the model of Slinn (1982) (cm/s)
v_{di} = dry deposition velocity corresponding to size range i of a distribution of particles (cm/s)
v_m = deposition velocity for momentum within the canopy in the model of Slinn (1982) (cm/s)
v_s = sedimentation velocity (cm/s)
w = vertical air velocity (cm/s)
$\overline{w}$ = time-averaged vertical air velocity (cm/s)
w' = turbulent fluctuating component of vertical air velocity (cm/s)
X_{air} = airborne concentration of element X (g/cm^3)
X_{crust} = concentration of element X in the earth's crust (parts per million by mass)
x = distance in the direction of airflow (cm)
z = height above the surface (cm)
z_o = roughness height or momentum sink (cm)
z_{oc} = contaminant sink (cm)
z_r = arbitrary reference height above the surface (cm)
z^+ = dimensionless height above the zero-plane displacement in the model of Sehmel (1980) = $(z-d)u_*/\nu$
α = surface area of vegetative elements per unit volume at height z (cm^{-1})
β = absorption coefficient for particle loss by the canopy per unit length of the flight path in the model of Bache (1979b)
γ = exponent of the Schmidt number in the proportionality for r_b
δ = thickness of boundary layer (cm)
ζ = dimensionless height = $\dfrac{z-d}{L}$ used in stability-dependent correction factors and in the model of Shreffler (1978)

ζ_h = dimensionless canopy height $= \frac{h-d}{L}$ in the model of Shreffler (1978)
ζ_{z_r} = dimensionless reference height $= \frac{z_r - d}{L}$ in the model of Shreffler (1978)
η = particle collection efficiency by all mechanisms in the model of Davidson et al. (1982)
$\eta_{\text{diffusion}}$ = particle collection efficiency by diffusion in the model of Davidson et al. (1982)
$\eta_{\text{impaction}}$ = particle collection efficiency by impaction in the model of Davidson et al. (1982)
$\eta_{\text{interception}}$ = particle collection efficiency by interception in the model of Davidson et al. (1982)
η' = average collection efficiency of particles within a vegetative canopy in the model of Slinn (1982)
$\theta = \tan^{-1}\frac{v_s}{u}$ = angle of particle trajectory under quiescent conditions
κ = thermal diffusivity (cm^2/s)
λ = leaf area index (cm^2 foliar area/cm^2 ground area)
μ = viscosity of air (g/cm s)
ν = kinematic viscosity of air (cm^2/s)
ξ = particle collection efficiency in the model of Slinn (1982) when it is assumed that $\xi_c = \xi_g = \xi$
ξ_c = particle/canopy element collection efficiency in the model of Slinn (1982)
ξ_g = particle collection efficiency at the ground in the model of Slinn (1982)
ρ_a = bulk density of air (g/cm^3)
ρ_p = bulk density of a particle (g/cm^3)
σ = a constant in the distribution of $\alpha(z)$ in the model of Bache (1979b)
σ_C = standard deviation of airborne contaminant concentration (g/cm^3)
σ_T = standard deviation of air temperature (°C or °K)
σ_v = standard deviation of the lateral (y-direction) windspeed (cm/s)
σ_W = standard deviation of water vapor concentration (g/cm^3)
σ_θ = standard deviation of the horizontal wind direction (degrees)
τ_o = shear stress at the surface ($g/cm\ s^2$)
ϕ_C = stability-dependent correction factor applied to eddy transport of contaminant mass
ϕ_H = stability-dependent correction factor applied to eddy transport of heat
ϕ_M = stability-dependent correction factor applied to eddy transport of air momentum
ψ = leaf water potential in the equation for f_w
ψ_C = stability-dependent correction factor in the equation for r_a for contaminant mass
ψ_M = stability-dependent correction factor in the equation for r_a for air momentum

example, Shannon (1981) has used budget studies of sulfur emissions in the eastern United States and Canada to estimate that dry removal, wet removal, and transport out of the region are of similar magnitude during summer months. Dry deposition is roughly half as large as the other terms during the winter. Lindberg and others (1986) have found that dry deposition represents more than half of the total annual input of SO_4^{2-}, NO_3^-, Ca, K, and acidity to an oak forest in Tennessee. Other studies (e.g., Galloway et al., 1984) suggest that dry deposition is less important than precipitation scavenging in the eastern United States, although still significant on an annual basis. Young and others (1987) estimate that dry and wet deposition of atmospheric acidity are roughly equal in magnitude in mountainous regions of the western United States; dry deposition may be dominant in arid western regions.

Despite its importance, our knowledge of dry deposition to natural surfaces is far from complete. For some chemical species, adequate methods exist for the purpose of estimating depletion from the atmosphere or accumulation onto surfaces. For other species, however, we are still unable to determine dry deposition reliably over the distance and time scales needed.

This chapter summarizes our current understanding of dry deposition of atmospheric contaminants. First, the process of dry deposition is discussed, including physical descriptions and mathematical relations. Wind tunnel studies and empirical models are then summarized. Techniques for measuring dry deposition in the field are described in the next section, including a discussion of the advantages and disadvantages of each method. Finally, published data for dry deposition are presented and compared with model results.

The focus of the chapter is on developments of the past decade. Vegetative canopies are of primary concern, although other surfaces are considered briefly. Most of the information summarized here is from material published after the 1979 Dry Deposition Workshop at Argonne National Laboratory (Hicks et al., 1980) and after the review papers of McMahon and Denison (1979), Sehmel (1980), and Hosker and Lindberg (1982). Information presented at the Dry Deposition Workshop in Harpers Ferry, West Virginia (Hicks et al., 1986a), as well as recent journal articles and reports, form the basis for much of the chapter.

II. The Process of Dry Deposition

Dry deposition may be broadly defined as the transport of particulate and gaseous contaminants from the atmosphere onto surfaces in the absence of precipitation. It is worthwhile to consider dry deposition as part of an overall atmosphere-surface exchange: Gases are sometimes reversibly adsorbed onto surfaces only to be reemitted, whereas particles may be deposited and subsequently resuspended. We often refer to *net* dry deposition as the resulting balance between downward and upward fluxes.

Developing an understanding of dry deposition requires familiarity with the various physical and chemical processes involved. In this section, we briefly explore these processes by first examining key factors influencing deposition. We then consider some of the details of the transport mechanisms involved. Finally, we examine mathematical models developed to describe the transport.

A. Key Factors Influencing Dry Deposition

The most important factors can be categorized into characteristics of the atmosphere, the nature of the surface, and properties of the depositing species. Atmospheric properties influence the rate at which contaminants are delivered to the surface. Especially important is the state of the atmosphere close to the ground.

The nature of the surface can have a marked effect on deposition. Of primary importance is the way in which the surface interacts with the atmosphere by exchanging momentum, heat, and water vapor. For example, the surface shape is important in influencing airflow patterns around it, whereas surface reflectivity and ability to retain moisture will affect temperature and humidity gradients that influence deposition. Also of importance are the physical and chemical properties of the surface that determine the ultimate fate of contaminants, once deposited. A smooth, relatively nonreactive surface may result in rapid bounceoff of particles and may not permit absorption or adsorption of certain vapors. A rougher, more reactive surface favors greater deposition rates. Natural surfaces such as vegetation are highly variable in their characteristics; fluxes are often strong functions of the status of the vegetation during the growth cycle.

Finally, properties of the depositing species will influence their transport to the surface and their ultimate fate after reaching the surface. Transport of gases through the atmosphere depends on their eddy and Brownian diffusivities. Transport of particles through the atmosphere depends on these diffusivities and on the rate of sedimentation. For gases, solubility and chemical reactivity may be dominant factors affecting uptake by the surface. For particles, the shape, size, and density may determine whether capture by surface roughness elements occurs. Table 5–1 summarizes properties of the atmosphere, the surface, and the contaminants that influence deposition.

It is clear that dry deposition is affected by a multiplicity of factors that often interact in complex ways. Now we will attempt to simplify the atmosphere-surface-contaminant system in order to illustrate how deposition takes place.

B. Transport Mechanisms: Physical Descriptions

The previous section indicated that interactions between the atmosphere and the surface can influence dry deposition of contaminant species. Let us examine these interactions to arrive at a physical description of the deposition process. First we will consider the structure of the atmospheric boundary layer above the surface onto which deposition is occurring. Then we will examine each step in the deposition process.

Table 5–1. Examples of properties of the atmosphere-surface-contaminant system influencing dry deposition.

Atmospheric properties	Surface properties	Depositing contaminant properties
Flow separation	Canopy structure	Gases:
Micrometeorological interactions with the surface:	Chemical/biological reactivity	Brownian and eddy diffusivities
Friction velocity	Electrostatic properties	Chemical reactivity
Roughness height	Geometry of roughness elements	Partial pressure in equilibrium with the surface
Zero-plane displacement	Leaf area index	
Relative humidity	pH effects	Particles:
Solar radiation	Penetration of canopy by contaminant	Brownian and eddy diffusivities
Stability class		Chemical reactivity
Temperature	Prior deposition loading	Density
Turbulence intensity	Terrain characteristics	Diameter
Wind speed	Thermal properties	Diffusiophoretic properties
	Wetness	Electrostatic properties
		Hygroscopicity
		Momentum
		Shape, size
		Solubility
		Thermal properties

Modified, with permission, from Sehmel (1980). Copyright © Pergamon Press plc.

1. Structure of the Atmospheric Boundary Layer

Any obstacle exposed to the ambient atmosphere will have a region of relatively quiescent air near its surface. This is a consequence of the *no slip* condition: The air an infinitesimal distance above a stationary surface will also be stationary. Because air is a viscous fluid (i.e., each air molecule interacts with those around it, causing internal friction when flowing), the air velocity increases gradually from zero with distance from the surface. The mainstream velocity is reached after a sufficient distance. The region of increasing air velocity is known as the *momentum boundary layer*.

Most surfaces exposed to the atmosphere consist of many irregularly shaped obstacles. Thus the overall momentum boundary layer has a complex structure with several interacting component boundary layers. There will be a boundary layer covering a distance of millimeters or smaller adjacent to each leaf, stem, or other element. This is known as the *quasi-laminar sublayer,* as the effects of turbulence are minimal close to the surface. There will also be a broader velocity gradient within and above the entire canopy and surrounding terrain. Known as the *surface layer,* this region includes the lowest several meters or tens of meters above the ground. The extent of the surface layer above any terrain is defined as the height over which the vertical turbulent fluxes of momentum and heat are roughly constant with respect to height. For many vegetated surfaces, this height is about 30 to 50 m. Finally, *the planetary boundary layer* represents the entire region where surface effects are important. This layer extends from the ground to heights of 300 to 500 m. Air motion in the planetary boundary layer is governed by shear stresses, large-scale horizontal pressure gradients, and Coriolis forces. Figure 5–1 illustrates development of the momentum boundary layer over a surface and shows the shapes of wind speed profiles in the planetary boundary layer above different types of surfaces.

When no heat is exchanged between layers of air and between air and the ground, the atmosphere is said to be *adiabatic*. Most of the time, however, there is heat exchange occurring. The resulting buoyant forces can either contribute to the turbulent energy of the flowing air or suppress the energy of turbulence. When air near the ground is colder than air aloft, such as at night, the buoyant forces suppress turbulence and the atmosphere is stable. When air near the ground is warmer than air at higher elevations, for example, due to solar heating of the surface during the day, the effect of buoyancy increases turbulence and the atmosphere is unstable. Dry deposition rates are usually greatest when the atmosphere is unstable, due to the rapid delivery of contaminants to the surface.

We normally consider the transport of particles and vapors to a surface to be composed of three steps. First, contaminants are carried through the lowest layers of the atmosphere and into the quasi-laminar sublayer. This is termed *aerodynamic* transport. During the second step, known as *boundary layer* transport, contaminants are carried across the sublayer to the surface. Finally, the particles or vapors interact with the surface. Each of these steps will now be discussed.

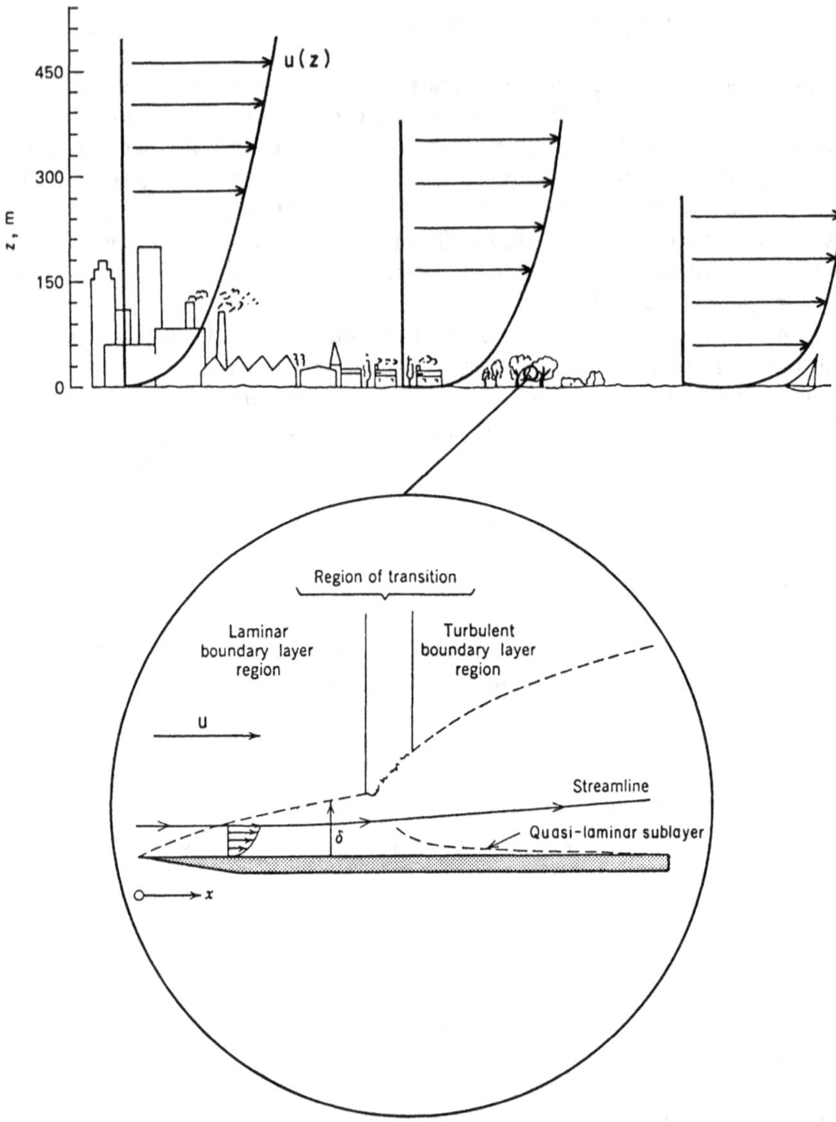

Figure 5–1. Development of a momentum boundary layer for air flowing over a smooth, flat surface. The quasi-laminar sublayer forms when the momentum boundary layer becomes turbulent at some distance downstream from the leading edge. Virtually all surfaces exposed to the ambient atmosphere have a quasi-laminar sublayer. Wind-speed profiles in the planetary boundary layer for different types of surfaces are also shown. (After Seinfeld, 1975 and Welty et al., 1976).

2. Aerodynamic Transport

Transport from the atmosphere to the sublayer can occur by eddy diffusion and sedimentation. The former mechanism refers to contaminant movement by turbulent wind eddies from regions of high concentration to regions of lower concentration. This movement reflects the stochastic nature of the wind: Turbulent motion is random, and hence there is a tendency for contaminants transported by the eddies to move away from regions of high concentration to form a more uniform system without strong concentration gradients. When the surface is a net sink for the contaminant, the concentration will always be low near the ground, and eddy diffusion will cause a continual flux of contaminant toward the surface. Both particles and vapors experience eddy diffusion.

Sedimentation is significant only for particles with diameters greater than about 1 μm. A particle accelerating downward under the influence of gravity will experience an aerodynamic drag force that increases as velocity increases. The drag force opposes gravity and retards the acceleration. After a short time period from the onset of particle settling (less than about 0.1 seconds for particles smaller than 100 μm in diameter), the magnitude of the drag force reaches that of the gravitational force and the acceleration decreases to zero. The particle is then falling at the sedimentation velocity, which is a function of the size, shape, and density of the particle.

3. Boundary Layer Transport

Transport across the quasi-laminar sublayer can occur by diffusion, interception, inertial motion, and sedimentation. These mechanisms are illustrated in Figure 5–2. In addition, electrostatic forces, thermophoresis, and diffusiophoresis may contribute to this transport.

Diffusion of contaminants results from the motion of air eddies as well as from Brownian motion. Air eddy motion in the sublayer is much weaker than in the free atmosphere. Nevertheless, several investigators have suggested that turbulent eddies may assist transport to within a few micrometers of a surface (e.g., Friedlander and Johnstone, 1957; Sehmel, 1973). Brownian diffusion transports particles and gases along a concentration gradient in a manner similar to that of eddy diffusion. However, the driving force is the random thermal energy of the molecules of air and molecules of contaminant species rather than turbulent energy. Brownian diffusion is important to gases and submicron particles and becomes significant only very close to the surface, where turbulent eddies are virtually nonexistent. Transport by Brownian and eddy diffusion is a function of atmospheric conditions, characteristics of the depositing contaminants, and the magnitude of contaminant concentration gradients.

Interception applies only to particles with diameters greater than a few tenths of a micrometer. The mechanism occurs when particles moving with the mean air motion pass sufficiently close to an obstacle to collide with it. Thus interception occurs only at places where the sublayer is smaller than the size of the particle, for

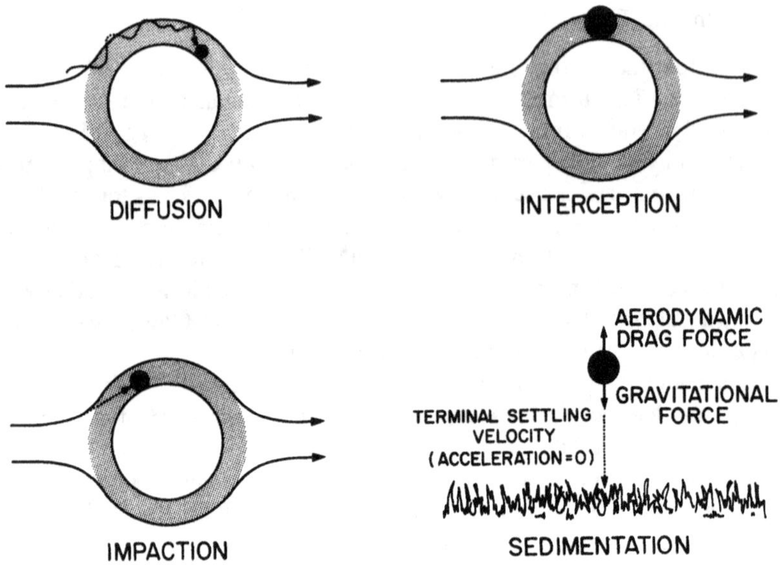

Figure 5–2. Mechanisms of particle deposition. In each case, the obstacle is represented as a circular cylinder (top view) surrounded by the quasi-laminar sublayer (shaded region). For simplicity, the sublayer is illustrated as uniform in thickness around the cylinder, although in reality the thickness can vary greatly with position.

example, at sharp corners or other places where there are rapid changes in the direction of airflow and where the sublayer may be narrow.

Inertial deposition similarly applies to particles with diameters greater than a few tenths of a micrometer. For obstacles that protrude into the airstream, inertial deposition by impaction may be important. This mechanism occurs when particles cannot follow rapid changes in direction of the mean airflow, and their inertia carries them across the sublayer and onto the surface. Like interception, impaction occurs where there are changes in the direction of airflow. Unlike interception, however, a particle subject to impaction leaves the air streamline and crosses the sublayer with inertial energy imparted from the mean airflow.

Inertial effects can also assist particle transport onto surfaces that do not protrude into the airstream. Known as *turbulent inertial deposition,* the mechanism applies to particles carried close to the surface by air eddies. The particles are carried across the relatively quiescent air of the sublayer by inertial energy imparted from the eddies. This mechanism differs from impaction in that the transport energy is derived from the component of the turbulent airflow which is normal to the direction of mean flow rather than from the mean flow.

Sedimentation through the quasi-laminar sublayer is significant for surfaces with components oriented horizontally. Although important mainly for supermicron particles, the mechanism can assist deposition of smaller particles once they are within a few micrometers of the surface.

Electrostatic forces, thermophoresis, and diffusiophoresis may also affect deposition to varying degrees. Chalmers (1949) reports that roughly 50% of condensation nuclei in the ambient atmosphere are uncharged; the remainder are equally divided between positively and negatively charged. For electric fields of a few volts per centimeter usually found at the earth's surface, Chamberlain (1960) has shown that electrostatic attraction is a minor force compared with observed field data for deposition of nuclei. Particles larger than typical nuclei will be even less affected by electrostatic forces due to the increasing importance of other mechanisms. *Thermophoresis* refers to motion along a temperature gradient, such as where warm air is in contact with colder surfaces. Chamberlain (1960) has used results of other investigators to show that thermophoretic forces are insignificant for temperature gradients commonly found over natural surfaces. Finally, *diffusiophoresis* refers to motion of a particle caused by nonuniformities in the suspending gas. For example, if the gas is composed of two types of molecules of different masses, then the particle will tend to move in the same direction as the heavier species. Diffusiophoresis also includes Stefan flow, where gas molecules flow toward a liquid surface during condensation, or away from the surface during evaporation. A particle suspended in a gas near a liquid surface will tend to move in the direction of Stefan flow (Hesketh, 1981). This effect may also influence deposition to a solid wet surface: Wind tunnel data (Sehmel, 1973) have shown that the deposition of submicron particles may be decreased somewhat by upward diffusion of water vapor from a wetted surface.

4. Interactions with the Surface

The final step in the deposition process occurs as particles or vapors reach the surface. Particles may simply adhere to the surface or may react chemically, producing irreversible changes in the deposited material. Vapors similarly may adsorb reversibly onto a surface or may undergo chemical reaction. Interactions with the surface may also involve resuspension of a fraction of the incoming particles or reemission of adsorbing vapors.

Figure 5–3 illustrates details of plant tissue that influence transfer of contaminants from the air. The size of stomatal openings, characteristics of the mesophyll, and stickiness of the cuticle will affect surface uptake. Other factors, such as the amount of moisture on the surface, the presence of previously deposited material, and plant exudates, may also be important.

The nature of interactions with the surface is dependent on the contaminant species. Gases such as SO_2, O_3, and NO_2 depositing on vegetation follow pathways similar to those of water vapor (Hicks et al., 1985). The presence of surface moisture and the chemical composition of this moisture may have a marked effect on deposition. For chemically reactive gases such as HNO_3, high deposition rates are expected for virtually any surface. Solid particles may experience significant bounceoff if the surface is smooth, whereas liquid particles are more likely to stick upon contact.

For a more detailed discussion of the various transport mechanisms discussed in

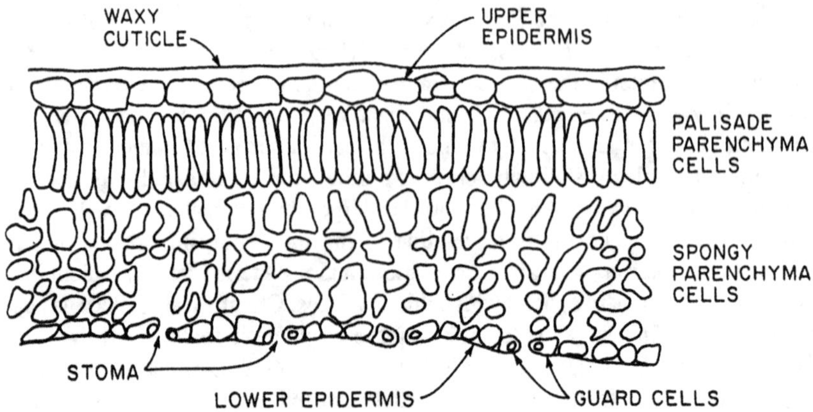

Figure 5–3. Characteristics of plant tissue that influence exchange of contaminants between vegetation and the atmosphere. (After O'Dell et al., 1977 and Hicks et al., 1985).

this section, the reader is referred to Fuchs (1964) and Friedlander (1977). For further discussion of reactions with surfaces, the reader is referred to the next chapter in this volume.

C. Transport Mechanisms: Mathematical Descriptions

This section presents mathematical relations that describe the transport processes defined above. First, we present the general concepts of *deposition velocity* and *resistance to transport,* building on the ideas of the previous section. Then we discuss mathematical relations for aerodynamic, boundary layer, and surface transport.

1. Deposition Velocity and Resistance to Transport

Most modeling efforts begin by defining the dry deposition velocity v_d (Chamberlain, 1953):

$$v_d(z) = -\frac{F}{C(z)} \tag{1}$$

where F is the flux of deposition rate per unit area (g/cm^2 s) and C is the airborne concentration of the contaminant (g/cm^3). The minus sign is needed because downward flux has a negative value but the deposition velocity is defined as a positive quantity. Because C is a function of height z above the surface, v_d is also a function of height. F is assumed to be constant over the appropriate range of heights.

The concentration gradient is often used to define a *concentration boundary layer,* analogous to the momentum boundary layer based on the air velocity gradient. Figure 5–4 shows plots of expected contaminant concentration $C(z)$ and

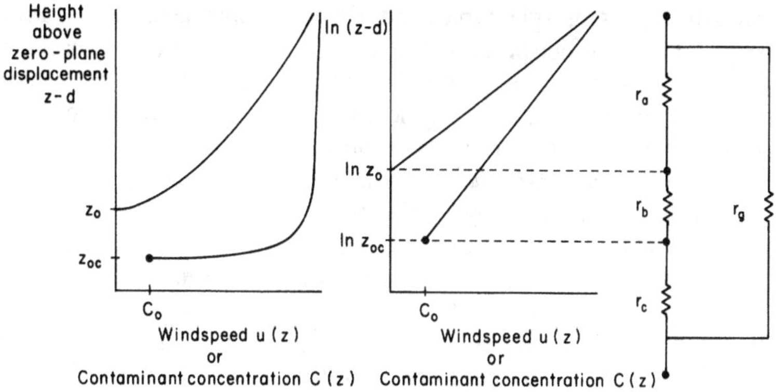

Figure 5–4. Wind speed and contaminant concentration as functions of height above the zero-plane displacement of a rough surface. The graphs are shown on linear and logarithmic z-axes. Hypothetical sinks for air momentum and contaminant concentration are denoted by $(z - d) = z_o$ and $(z - d) = z_{oc}$, respectively. A schematic diagram of resistance to transport is also shown, including the aerodynamic resistance r_a, boundary layer resistance r_b, canopy resistance r_c, and gravitational resistance r_g.

wind speed $u(z)$ above a rough surface such as a vegetative canopy. The z axis shows the height above the zero-plane displacement d. The displacement is merely a mathematical tool applicable to dense vegetation and other rough surfaces, where it is convenient to express concentration or wind speed with respect to $z = d$ rather than with respect to $z = 0$.

The shapes of the curves in Figure 5–4 can be explained by an analogy between contaminant (or momentum) transport and the flow of electrons in an electrical circuit. The concentration is analogous to voltage, whereas the deposition flux is represented as an electrical current. Ohm's law can then be used to determine a resistance to transport, analogous to an electrical resistance. The resistance to transport is merely the inverse of the deposition velocity. The value of the concentration gradient or velocity gradient at any height in Figure 5–4 depends on the resistance to contaminant transport or to momentum transport, respectively, at that height. Figure 5–4 shows that most of the resistance to momentum transport occurs well above the surface, as momentum is rapidly destroyed at the surface by friction. Hence, much of the velocity gradient appears over a wide range of heights. Unlike momentum transport, much of the resistance to contaminant transport is often near the surface. Contaminants are rapidly transported through the lowest layers of the atmosphere into the quasi-laminar sublayer, so the concentration is nearly constant with height above the sublayer. Because transport across the sublayer and reaction with the surface are relatively slow for many contaminants, a steep concentration gradient is often found adjacent to the surface. C_o represents the concentration adjacent to the surface; a surface that is a perfect sink for the contaminant will have $C_o = 0$.

The resistance to transport is typically divided into components representing the three steps in the deposition process. The first component applies to transport from any height z in the free atmosphere down to the quasi-laminar sublayer, termed the *aerodynamic resistance*, r_a. The second component, which applies to transport across the sublayer, is known as the *boundary layer resistance*, r_b. The last component, the *canopy resistance*, r_c, represents interactions with the surface. These series resistances must be placed in parallel with a *sedimentation resistance*, r_g, representing the influence of gravity. The resistances are illustrated in Figure 5–4. The concepts of deposition velocity and resistance to transport generally apply only to systems assumed to be in steady state, with fluxes that are constant with height. Using these concepts, we now explore the mathematical relations used to define the three steps in the dry deposition process.

2. Aerodynamic Transport

In order to understand contaminant transport from the atmosphere into the canopy, we first consider the flux of air momentum toward the surface:

$$v_a(z) = -\frac{F_a}{\rho_a u(z)} \tag{2}$$

where

v_a = deposition velocity of air momentum (cm/s)
F_a = flux of air momentum (g/cm s^2)
ρ_a = air density (g/cm^3), assumed constant with height

Note that $\rho_a u$ is essentially the concentration of air momentum (mass × velocity per unit volume), analogous to C in Equation 1. For simplicity, we assume an adiabatic atmosphere. The expression for momentum flux is given by:

$$F_a = -(\nu + K_M)\frac{d}{dz}(\rho_a u) \tag{3}$$

where

ν = kinematic viscosity of air (cm^{2}s)
K_M = kinematic eddy viscosity of air (cm^2/s) $= ku_*(z - d)$
k = von Karman's constant = 0.4
u_* = friction velocity (cm/s) $= (\tau_o/\rho_a)^{1/2}$
τ_o = shear stress at the surface (g/cm s^2) $= -F_a$ at $(z - d = 0)$

In general, the kinematic viscosity is much smaller than the kinematic eddy viscosity except very close to the surface. Neglecting ν and substituting the expression for K_M yields:

$$F_a = -ku_* \frac{d}{d\ ln\ (z - d)}(\rho_a u) \tag{4}$$

which can be integrated to give:

$$u(z) = \frac{u_*}{k} \ln \frac{z-d}{z_o} \tag{5}$$

Combining equations 4 and 2 shows that:

$$v_a(z) = \frac{u_*^2}{u(z)} \tag{6}$$

providing a simple expression for the deposition velocity of air momentum. Note that z_o is the height at which the wind speed profile extrapolated below the top of the canopy reaches zero, that is, the effective momentum sink. Equation 6 is merely an approximation for $v_a(z)$, as the wind-speed profile deviates somewhat from Equation 5 within the canopy ($z < h$).

For the flux of a contaminant transported by wind eddies, the analogue to Equation 3 is:

$$F = -(D + K) \frac{dC}{dz} \tag{7}$$

where D = Brownian diffusivity of the contaminant (cm^2/s) and K = eddy diffusivity of the contaminant (cm^2/s).

The Brownian diffusivity is much smaller than the eddy diffusivity except very close to the surface. Neglecting D and assuming $K = K_M$, Equation 7 may be integrated to obtain a solution for C:

$$C(z) - C_o = -\frac{F}{ku_*} \ln \frac{z-d}{z_{oc}} \tag{8}$$

In this equation, z_{oc} is a constant of integration representing the effective contaminant sink, where the concentration falls to C_o. Equations 7 and 8 apply to gases and small particles where sedimentation is negligible. For particles large enough to be influenced by gravity, these two equations become:

$$F = -(D + K) \frac{dC}{dz} - v_s C \tag{9}$$

$$C(z) = C_o \left(\frac{z_{oc}}{z-d}\right)^{v_s/ku_*} + \frac{F}{v_s} \left[1 - \left(\frac{z_{oc}}{z-d}\right)^{v_s/ku_*}\right] \tag{10}$$

where

$$v_s = \text{sedimentation velocity (cm/s)} = \frac{(\rho_p - \rho_a)\, g\, c\, d_p^2}{18\mu} \tag{11}$$

ρ_p = bulk density of the particle (g/cm^3)
g = gravitational acceleration (cm/s^2)
c = Cunningham slip correction factor
d_p = particle diameter (cm)
μ = viscosity of air (g/cm s) = $\rho_a \nu$

The assumptions of $D \ll K$ and $K = K_M$ have been applied to obtain Equation 10.

For small particles and gases where sedimentation is negligible, the above equations show that a straight line is obtained when $u(z)$ or $C(z)$ is plotted against $ln\,(z - d)$. This is a consequence of the eddy diffusivity being a linear function of height. The log-linear curves are illustrated on the right side of Figure 5–4. In practice, only those portions of the curves well above the top of the canopy will be log-linear. The minimum height of linearity will vary, depending on the properties of the particular system, but typical values are about twice the canopy height. The parameters z_o and z_{oc} are used merely as mathematical tools to fix the positions of the curves above the canopy and are without physical meaning.

We have shown that the equations describing the transport of air momentum and the equations describing the transport of atmospheric contaminants are of the same form, when the contaminants are not influenced by sedimentation. Hence, the deposition velocity due to aerodynamic transport given by Equation 6 can apply to both momentum and contaminants. Combining Equations 5 and 6 yields an expression for the aerodynamic resistance:

$$r_a(z) = \frac{1}{v_a(z)} = \frac{1}{ku_*} \, ln \, \frac{z - d}{z_o} \tag{12}$$

which also applies to both momentum and contaminants. For momentum, $v_a(_z)$ is the total deposition velocity because $r_b = r_c = 0$. For contaminants, however, $v_a(z)$ is only one part of the process: The boundary layer and surface resistances must also be considered. These resistances are discussed in the next section.

Note that the above expressions can be used only in an adiabatic atmosphere. Several investigators have developed modifications to these equations to account for the influence of buoyancy in a nonadiabatic atmosphere. For example, the expressions for the kinematic eddy viscosity for momentum and the contaminant eddy diffusivity become:

$$K_M = \frac{ku_* \, (z - d)}{\phi_M} \quad \text{and} \quad K = \frac{ku_* \, (z - d)}{\phi_C} \tag{13}$$

There is an analogous expression for the eddy diffusivity of heat K_H incorporating the function ϕ_H. The correction factors ϕ_M, ϕ_C, and ϕ_H are functions of $\zeta = (z - d)/L$, where L is the Monin-Obukhov length. L is positive in a stable atmosphere, negative in an unstable atmosphere, and approaches $+\infty$ or $-\infty$ when the atmosphere is neutral. L may be considered the height above ground where the production of turbulence by mechanical forces equals the production of turbulence by buoyant forces.

Appropriate expressions for the correction factors must be determined by experiment. Businger and others (1971) analyzed field data from a variety of stability conditions and suggested the expressions for ϕ_M and ϕ_H given in Table 5–2. Some authors have suggested that eddy transport of contaminant mass is similar to the eddy transport of heat (e.g., Galbally, 1971). Therefore, the expressions for ϕ_H in Table 5–2 are sometimes used identically for ϕ_c. According

Table 5-2. Expressions for stability-dependent correction factors for aerodynamic transport.

Reference	Parameter	Atmospheric condition		
		Stable	Neutral	Unstable
Businger et al. (1971)	ϕ_M	$1 + 4.7\ \zeta$	1	$[1 - 15\ \zeta]^{-1/4}$
Businger et al. (1971)	ϕ_H	$0.74 + 4.7\ \zeta$	0.74	$0.74\ [1 - 9\ \zeta]^{-1/2}$
Wesely and Hicks (1977)	ψ_M	$-5\ \zeta$	0	$\text{Exp}\ \{0.032 + 0.448\ ln\ (-\zeta) - 0.132\ [ln\ (-\zeta)^2\}$
Wesely and Hicks (1977)	ψ_C	$-5\ \zeta$	0	$\text{Exp}\ \{0.598 + 0.390\ ln\ (-\zeta) - 0.090\ [ln\ (-\zeta)^2\}$
Hicks et al. (1985)	r_a for contaminants	$\frac{4}{u\sigma_\theta^2}$	$\frac{4}{u\sigma_\theta^2}$	$\frac{9}{u\sigma_\theta^2}$

to this table, the equations developed by Businger and others (1971) show $\phi_H/\phi_M = 0.74$ for neutral conditions, rather than unity, as suggested by other authors (Lumley and Panofsky, 1964; Munn, 1966, Webb, 1970; Thom, 1975). Both alternatives have been used in eddy diffusion models found in the literature.

Rather than substituting the stability-corrected diffusivities into Equations 3 and 7, one can develop correction factors for nonadiabatic conditions to be applied after those equations are integrated:

$$r_a(z) \text{ for momentum} = \frac{1}{ku_*}\left[ln \frac{z-d}{z_o} - \psi_M\right]$$

$$r_a(z) \text{ for contaminants} = \frac{1}{ku_*}\left[ln \frac{z-d}{z_o} - \psi_C\right] \tag{14}$$

where ψ_M and ψ_C represent modifications applied directly to Equation 12. Wesely and Hicks (1977) give expressions for these functions, based on tabulations by Dyer and Hicks (1970). These expressions are shown in Table 5–2. Relations between the Pasquill stability categories (Turner, 1970) and values of ζ in the correction factors are discussed by Sheih and others (1979). Additional discussion of correcting r_a for nonadiabatic conditions is provided by Bache (1977), who considers difficulties in applying correction factors to tall canopies such as a pine forest.

As another way to correct for stability, Hicks and others (1985) suggest the use of σ_θ, the standard deviation of the horizontal wind direction. Based on Equation 12 and the relation $\sigma_\theta = \sigma_v/\bar{u}$, an expression for r_a can be written as:

$$r_a = \frac{\sigma_v^2}{u_*^2} \frac{1}{\bar{u}\,\sigma_\theta^2} \tag{15}$$

where σ_v is the standard deviation of the lateral (y-direction) wind speed and $\bar{u}$ is the time-averaged x-direction wind speed. The authors note that σ_v/u_* has a value of about 2 for stable and neutral conditions but increases asymptotically to about 3 in an unstable atmosphere, leading to the expressions shown in Table 5–2. The expression for r_a for an unstable atmosphere is used only if the net radiation is positive and if σ_θ exceeds about 10°.

3. Boundary Layer Transport

As discussed in section II, B, 3, a number of processes are responsible for carrying particles and vapors across the quasi-laminar sublayer. These include diffusion, interception, inertial motion, sedimentation, and influences of other minor mechanisms. The net effect of all of these mechanisms, when combined with aerodynamic transport, is represented by the contaminant sink z_{oc} introduced in Equation 8. Thus we can consider the region $(z - d) > z_o$ to be influenced by aerodynamic transport, whereas the region $z_{oc} < (z - d) < z_o$ to be influenced by boundary layer transport. Following Equation 12, we can define the boundary layer resistance as:

$$r_b = \frac{1}{ku_*} \ln \frac{z_o}{z_{oc}} \tag{16}$$

We can also define the sublayer Stanton number B, which relates the transport of contaminant mass to the transport of air momentum (after Owen and Thompson, 1963; Chamberlain, 1966):

$$B^{-1} = \frac{1}{k} \ln \frac{z_o}{z_{oc}} \tag{17}$$

so that $r_b = 1/Bu_*$. For cases where the canopy resistance r_c is zero and where sedimentation is negligible ($r_g \rightarrow \infty$), the total resistance to transport r_t can be expressed as the sum of aerodynamic and boundary layer resistances:

$$\begin{aligned} r_t(z) &= \frac{1}{v_d(z)} = \frac{1}{ku_*} \ln \frac{z-d}{z_{oc}} \\ &= r_a(z) + r_b = \frac{1}{v_a(z)} + \frac{1}{Bu_*} = \frac{1}{ku_*} \ln \frac{z-d}{z_o} + \frac{1}{ku_*} \ln \frac{z_o}{z_{oc}} \end{aligned} \tag{18}$$

It is generally difficult to determine z_{oc} directly, and thus Equations 16 through 18 are not often used to estimate boundary layer resistance in the field. Rather, one must consider the specific transport mechanisms occurring in the atmosphere-surface system. As an example, Hicks and others (1985) propose a simple expression for r_b for gases and small particles transported to vegetative canopies by diffusion. To examine the influence of the diffusion coefficient, they call attention to models suggesting that r_b is proportional to Sc^{γ}, where Sc = Schmidt number = ν/D. Values of γ are generally in the range of 0.5 to 0.8, with a commonly accepted value of $\frac{2}{3}$ (based on models such as that of Brutsaert, 1975). Then they consider the influence of surface roughness by examining the parameter $k\, u_*\, r_b$ versus the Reynolds number Re_*, where $Re_* = u_* z_o/\nu$. Note that $k\, u_*\, r_b$ is equivalent to $\ln (z_o/z_{oc})$, according to Equation 16. Despite model predictions of monotonically increasing values of $k\, u_*\, r_b$ with increasing roughness as in Equation 16, field results indicate that $k\, u_*\, r_b$ is approximately constant as Re_* varies, having a limiting value of about 2 for vegetation. Results of such field studies for sensible heat transport are shown in Figure 5–5. Based on this figure, Hicks and others (1985) suggest the following expression for boundary layer resistance:

$$r_b \approx 2 \left(\frac{1}{ku_*}\right) \left(Sc/Pr\right)^{2/3} \tag{19}$$

where Pr = Prandtl number = ν/κ, and κ is the thermal diffusivity (cm^2/s). The Prandtl number correction is included to account for the difference between heat and momentum transport, whereas the Schmidt number accounts for the difference

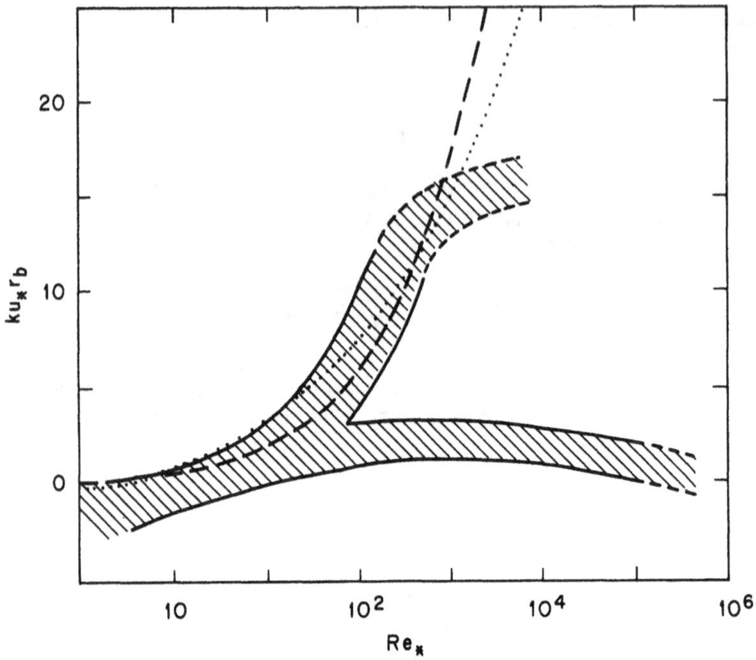

Figure 5–5. Values of the parameter $ku_* r_b$ as a function of the surface Reynolds number taken from Hicks et al. (1985). The dashed curve represents the model of Owen and Thompson (1963); the dotted curve shows the model of Brutsaert (1975). The shaded areas indicate field data for sensible heat transport. The upper branch represents data obtained for bluff roughness elements, and the lower branch represents data obtained for vegetation and fibrous roughness elements.

between momentum and mass transport. Some investigators have used the Lewis number $Le = \kappa/D$ in place of Sc/Pr.

Other expressions have been developed for transport by diffusion, interception, inertial motion, and sedimentation across the boundary layer. The expressions have been developed as parts of models describing the combined effects of several mechanisms. We will present these models, which include mathematical relations for the individual mechanisms, in section III.

4. Interactions with the Surface

The presence of a nonzero concentration at the surface, denoted by C_o in Equation 8, indicates a nonzero canopy resistance r_c. The expression for canopy resistance is merely:

$$r_c = -\frac{C_o}{F} \tag{20}$$

It is often difficult to determine a representative value of C_o for a complex surface such as a vegetative canopy, so Equation 20 cannot readily be used to calculate the canopy resistance. Rather, we must quantify those processes that determine the rates of contaminant uptake in order to estimate r_c. In cases where the total resistance to transport can be measured, r_c can be estimated as the residual resistance when r_a and r_b are subtracted from the total. Let us now consider how details of the surface influence the canopy resistance.

Figure 5–3 shows that contaminants can interact with a leaf through the stomata or through the cuticle. We therefore consider the stomata and cuticle resistances to be in parallel for a single leaf. In the "big leaf" model (Hicks et al., 1985), this concept is extended to an entire canopy: The overall resistance of the stomata in a canopy is assumed to be in parallel with the overall cuticle resistance. Figure 5–6 illustrates several types of component resistances that may be included in this model.

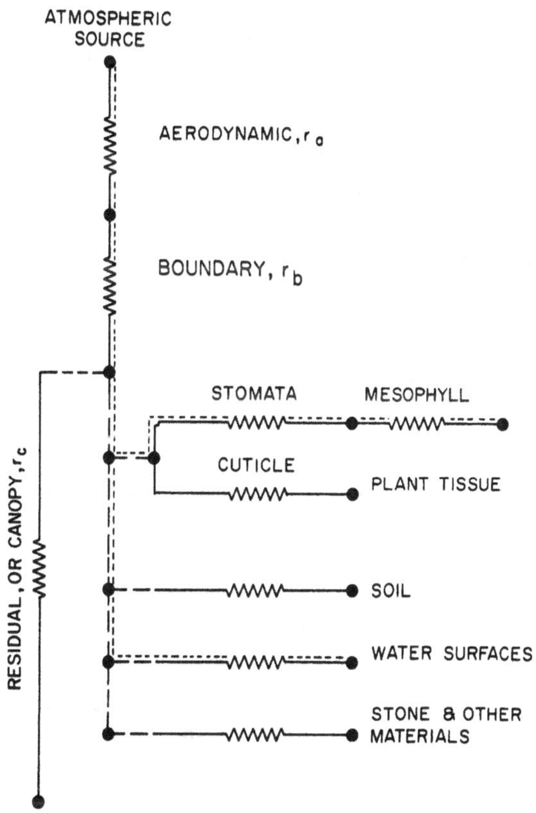

Figure 5–6. Resistance diagram for the transport of contaminants to various types of surfaces, taken from Hicks et al. (1985). The relative magnitudes of the resistances vary for different contaminants. For example, the dotted line illustrates a probable route for SO_2 deposition. Highly reactive gases such as HNO_3 would have all components of r_c near zero.

It is necessary to distinguish between two types of resistances when discussing r_c. As we have seen, resistance to transport is expressed as the airborne concentration divided by the flux, where the flux is the mass of contaminant depositing per unit area per unit time. When expressed without a prime, the resistance incorporates a flux based on the horizontal area of the earth's surface. When expressed with a prime, the resistance incorporates a flux based on the surface area of a vegetative element, for instance, a particular leaf. The relation between the two definitions is given by:

$$r_c = \frac{r_c'}{\lambda} \tag{21}$$

where λ is the leaf area index for a particular canopy expressed as cm^2 foliar area/cm^2 ground area.

For gases depositing on vegetative canopy, Figure 5–6 shows that the overall canopy resistance is related to various component resistances according to:

$$r_c'^{-1} = (r_s' \frac{Sc}{Pr} + r_m')^{-1} + r_{cut}'^{-1} + r_{soil}^{-1} \tag{22}$$

where

r_s' = resistance of the stomata (s/cm)
r_m' = resistance of the mesophyll (s/cm)
r_{cut}' = resistance of the cuticle (s/cm)
r_{soil} = resistance of the soil (s/cm)

The stomatal resistance is expressed as:

$$r_s' = \frac{r_{sm}' \, [1 + (b'/I_p)]}{f_e f_w f_T} \tag{23}$$

where

r_{sm}' = minimum value of r_s', which varies with plant species (s/cm)
b' = empirical constant
I_p = photosynthetically active radiation (watts/m^2)
f_e = correction factor for the effect of humidity
f_w = correction factor for the effect of water stress
f_T = correction factor for the effect of temperature

Expressions for the correction factors have been developed by Hicks and others (1985) based on the literature survey of Jarvis and others (1976). The expressions may be summarized as:

$$f_e = 1 - b_e \, [e_s \, (T) - e] \tag{24}$$

where

b_e = a constant indicative of the response of a given plant species
$e_s(T)$ = saturated water vapor pressure at air temperature T (atm or g/cm s^2)

e = actual water vapor pressure (atm or g/cm s^2)
= $e_s(T) \times$ relative humidity

$f_w = a_w\psi + b_w$ *if* ψ is less than the threshold for water stress
= 1 if ψ is above the threshold for water stress

where a_w, b_w = constants.

$$\psi = \text{leaf water potential} \tag{25}$$

$$f_T = \left(\frac{T - T_l}{T_o - T_l} \times \frac{T_h - T}{T_h - T_o}\right)^{b_T} \tag{26}$$

where

T = air temperature (°C or °K)
T_h and T_l = species-dependent high and low temperature extremes at which stomata no longer open
T_o = temperature at which stomatal exchange is optimized

$$b_T = \frac{T_h - T_o}{T_h - T_l}$$

The expression for f_w is based on the "discontinuous switch" model of Fisher and others (1981), in which the stomatal resistance is relatively independent of ψ until ψ drops below the threshold value. Values of T_h, T_l, *and* T_o are given by Hicks and others (1985) for a number of plant species; typically b_T is about 0.4. Overall, values for f_e, f_w, and f_T lie in the range 0 to 1.

We cannot yet predict r'_m and r'_{cut} quantitatively from canopy and meteorological data. Hosker and Lindberg (1982) report that the transfer of gases through the cuticle is generally much smaller than transfer through the stomata and is often neglected. They estimate typical values of r'_{cut} for water vapor diffusion through leaves as 30 to 200 s/cm compared with r'_s in the range 1 to 20 s/cm. The mesophyll resistance depends on the solubility of the contaminant gas, chemical reactions in the cavity, and reactions occurring on the surfaces of the mesophyll tissue (Meyers and Yuen, 1987). This resistance is believed to be small, based on chamber studies (Taylor et al., 1982; Taylor and Tingey, 1983) and micrometeorological studies (Wesely and Hicks, 1977). Meyers and Yuen (1987) have assumed that $r'_m = 0$ for SO_2 and O_3 transport onto vegetation as part of the CORE deposition program (Hales et al., 1987). Additional information on cuticle and mesophyll resistances is provided by Hicks and others (1985) and Baldocchi and others (1987).

For particles, our understanding of r_c is poor. Many models assume perfect retention of particles coming in contact with surfaces despite evidence that bounceoff occurs for certain types of particles and surfaces. For example, Esmen and others (1978) and Aylor and Ferrandino (1985) have provided evidence that bounceoff is well correlated with the kinetic energy of the depositing particle. Slinn (1982) has attempted to account for bounceoff in modeling dry deposition to natural vegetation; a reduction in the vegetative element collection efficiency by bounceoff is assumed to be a function of the Stokes number, the ratio of inertial forces on the particle to viscous forces in the air. This formulation is discussed further in section III, B, 2.

We now examine a way in which separate expressions for r_a, r_b, and r_c have been combined to arrive at an overall description of dry deposition.

5. Combining Expressions for Aerodynamic, Boundary Layer, and Surface Transport

Hicks and others (1985) have combined equations for each of the three transport steps into a general computer model applicable to a number of vegetative canopies. The model has been developed for several gases as well as for fine particles.

For aerodynamic resistance, the model uses Equation 15 with u and σ_θ as inputs. Explicit determination of the friction velocity is not necessary due to the assumed value of σ_v/u_* of about 2 to 3. Atmospheric stability is determined by measurement of net radiation and σ_θ.

For the boundary layer resistance, the model incorporates Equation 19. Note that this is a modified version of Equation 16 with the simplifying assumption $ln(z_o/z_{oc}) = 2$ and with a correction factor for the different diffusivities of momentum and mass. Equation 19 requires u_* as an input, obtained from the approximation $u_* = [u(z)/r_a(z)]^{½}$based on Equations 6 and 12, with r_a determined from Equation 15.

Finally, the canopy resistance for gases is computed from Equations 21 and 22. Many of the details of estimating the component resistances in Equation 22 have been discussed in the preceding section. For particles, Hicks and others (1985) suggest that current models incorporating details of the surface are inadequate. They therefore propose the empirical relationship $v_d = (r_t)^{-1} = 0.003\ \sigma_\theta \bar{u}$ as a rough approximation, based on the data of Wesely and others (1983a) for fine particle deposition over a deciduous forest.

It is useful to note some limitations of the model. Hicks and others (1985) point out that the extension of the "big leaf" model to an entire canopy is valid only for vegetation with leaf area index λ near unity. For canopies with larger values of λ, a layer-by-layer computation of net overall resistance may be necessary because stomatal resistance is a strong function of radiation that is attenuated and diffused by the canopy.

A similar problem is discussed by Bache (1986b), who notes that separating canopy and boundary layer resistances is acceptable only for large canopy resistances. For cases where r_c is small, it may not be possible to separate r_c from r_b, leading to a breakdown in the parallel resistance concept of Figure 5–6.

We now consider predictions of dry deposition by mathematical models based on laboratory data, field data, and theory that have been designed for specific surface and atmospheric conditions. The models use the concepts discussed above for aerodynamic, boundary layer, and surface transport steps.

III. Wind Tunnel Studies and Semiempirical Models

This section considers two categories of research for predicting dry deposition. First, we examine wind tunnel studies that have provided data for use in developing deposition models. These studies have explored the influence of wind speed, turbulence intensity, surface roughness, and contaminant properties on dry deposition under controlled conditions. Second, we explore studies that have taken information from the literature to develop semiempirical models for specific vegetative canopy and atmospheric conditions. Results of wind tunnel experiments, field experiments, and theoretical investigations are used to formulate these detailed models. Because of space considerations, we summarize only a limited number of studies of each type.

A. Wind Tunnel Studies

Several studies of contaminant deposition in wind tunnels have been reported in the literature. Most of the investigations have involved surfaces with relatively small roughness heights and a wide range of wind speeds. Various types of monodisperse particles and gaseous species have been used. Table 5–3, taken from Schack and others (1985), summarizes several of these studies.

Two investigators, A. C. Chamberlain and G. A. Sehmel, have been responsible for most of the work listed in Table 5–3. Some of the detailed experiments conducted by these investigators will now be discussed.

1. Studies of Chamberlain (1966, 1967)

In one of the earliest sets of experiments, Chamberlain examined the transport of gases and particles to various surfaces in a wind tunnel. The results of the initial experiments, which examined radioactive iodine vapor depositing on artificial surfaces, were reported by Chamberlain in 1953. More detailed work published in the following decade involved ^{212}Pb vapor and several types of particles depositing onto a number of surfaces.

The experiments with ^{212}Pb utilized 7-cm-high Italian rye grass, much shorter lawn grass, artificial grass made from PVC strips, toweling, and roughened glass (Chamberlain, 1966). The surfaces were used in a wind tunnel that provided friction velocities in the range 7.9 to 200 cm/s. The airborne concentrations and deposition fluxes of ^{212}Pb were measured in each experiment, yielding estimates of deposition velocities. Because the wind tunnel data suggested that all of the surfaces were perfect sinks ($r_c = 0$), the results of these experiments yielded information specifically on transport across the quasi-laminar sublayer.

The experimental findings were used to estimate the reciprocal sublayer Stanton number B^{-1} (defined by Equation 17) for several surfaces and windspeeds. As indicated by Equations 17 and 18, $1/Bu_*$ is defined as the boundary layer resistance r_b, which is added to r_a to yield the total resistance to mass transport. It is convenient to consider the value of B^{-1} in comparison with the value of $u(z_r)/u_*$, where z_r is an arbitrary reference height above the surface. B^{-1} is the dimensionless resistance to mass transport across the quasi-laminar sublayer, whereas $u(z_r)/u_*$ is the dimensionless resistance to momentum transport between the reference height and the sublayer.

Results of the wind tunnel tests showed that values of B^{-1} were slightly greater than values of $u(5\text{ cm})/u_*$ for real and artificial grass by factors of $\sim$1 to 3. This

Table 5-3. Studies of particle and gas deposition in wind tunnels.

Number	Surface	Aerosol	Reference	d_p (μm)	u_* (cm/s)	z_o (cm)	$v_d - v_s$ (cm/s)
(1)	Sticky artificial grass	Aitken nuclei, PSL,[a] TCP,[b] ragweed, *Lycopodium*, H_2O, ^{212}Pb	Chamberlain (1966, 1967)	2.66×10^{-4}–32	15.8–170	1.0	0.007–34
(2)	Rye grass	Aitken nuclei, PSL, TCP, oleic acid, ragweed, *Lycopodium*, ^{212}Pb	Chamberlain (1966, 1967); Clough (1975)	5.62×10^{-4}–32	13.5–183	0.6–1.0	0.011–7.3
(3)	0.45–1.6 cm gravel	Uranine	Sehmel et al. (1974)	0.03–28	22–133	0.13–0.18	0.01–9.5
(4)	Artificial grass	Uranine	Sehmel (1972); Sehmel et al. (1973)	0.03–14	19–144	0.12–0.4	0.07–8.4
(5)	Water	Uranine	Sehmel and Sutter (1974)	0.25–28	44	0.02	0.005–32
(6)	Water	Uranine	Sehmel and Sutter (1974)	0.25–28	117	0.10	0.02–33
(7)	Moss	Aitken nuclei, PSL, Zn–CdS,[c] *Lycopodium*	Clough (1975)	0.08–32	37–87	0.37	0.06–7.7

(8)	Various spheres, cylinders, wave-forms	H_2O, ^{212}Pb	Chamberlain (1968)	2.66×10^{-4} 5.62×10^{-4}	12–222	0.02–0.60	0.38–8.5
(9)	Short rye grass	^{212}Pb	Chamberlain (1966)	5.62×10^{-4}	25–176	0.21–0.25	1.7–9.1
(10)	Barley, wheat, soil	*Lycopodium* spores	Chamberlain and Chadwick (1972)	32	31–190	0.30–1.2	1.3–19.2
(11)	Toweling	*Lycopodium*, H_2O, ^{212}Pb	Chamberlain (1967)	2.66×10^{-4}–32	10.6–146	0.045	0.65–10.2
(12)	Rough glass	*Lycopodium*, camphor, ^{212}Pb	Chamberlain (1966, 1967); Owen and Thomson (1963)	5.62×10^{4}–32	7.9–339	0.02–0.023	0.22–14.0
(13)	Water	PSL, misc., H_2O, CO_2, ^{212}Pb	Moller and Schumann (1970)	2.66×10^{-4}–1.2	40	0.002	0.01–3.9
(14)	3.8–5.1 cm gravel	Uranine	Sehmel et al. (1974); Sehmel and Hodgson (1977)	0.25–6	15,107	0.6, 0.3	0.05–33

After Schack et al. (1985).

[a]Polystyrene latex.

[b]Tricresyl phosphate.

[c]Zinc–cadmium sulfide.

suggests that there is slightly more resistance to transport across the quasi-laminar sublayer than across the lowest 5 cm above the surface. When applied to greater values of z_r, however, the results suggest that B^{-1} will be much smaller than $u(z_r)/u_*$. Thus, the deposition of ^{212}Pb from heights of a few meters above the surface is rate limited by eddy diffusional transport through the lowest layers of the atmosphere, not by transport across the sublayer.

For toweling and rough glass, values of B^{-1} were smaller than the corresponding values of $u(5\text{ cm})/u_*$. Resistance to transport across the sublayer was apparently smaller than across the lowest 5 cm in the wind tunnel for these surfaces.

In other experiments Chamberlain (1967) used radioactively tagged particles to measure deposition velocities to various surfaces in the wind tunnel. The particles included *Lycopodium* spores (d_p = 32 μm), ragweed pollen (d_p = 19 μm), polystyrene spheres (d_p = 5 μm), tricresyl phosphate droplets (d_p = 1 μm and 2 μm), and Aitken nuclei ($d_p \sim 0.08$ μm). The surfaces considered were real and artificial grass, toweling, and rough glass. The tunnel was operated at a number of wind speeds to give friction velocities in the range 18 to 183 cm/s. Data from these experiments were used to estimate z_{oc} in Equation 10, assuming a perfect sink surface ($C_o = 0$) and no zero-plane displacement ($d = 0$). The value of z_{oc} was then compared with the momentum sink z_o and with the particle stopping distance:

$$d_s = \frac{v_s u}{g} \tag{27}$$

In this equation, d_s represents the horizontal distance a particle will travel when injected with velocity u into stationary air, such as that approximated by the quasi-laminar sublayer.

Figure 5–7 provides an example of Chamberlain's findings. The open circles show deposition velocities of *Lycopodium* spores onto sticky artificial grass, where deposition is expected to be efficient. Curve A represents the deposition velocity calculated as $-F/C$, based on a rearrangement of Equation 10 with C_o = 0, d = 0, z = 7.5 cm, and $z_{oc} = z_o$. Reasonable agreement between the experimental data and the theoretical curve suggests that the effective sink for particles is at the same height as that for momentum. At the upper end of the curve, $v_s \ll u_*$ and Equation 10 reduces to Equation 8; because $z_{oc} = z_o$, the upper end of curve A approaches $v_d = u_*^2/u$ as given by Equation 6.

The Xs in Figure 5–7 show deposition velocities of *Lycopodium* spores onto real grass. The much smaller values of v_d reflect poorer retention of the spores that reach the surface. Chamberlain presents additional data showing that the stickiness of the artifical grass is probably responsible for the greater surface retention of large particles such as *Lycopodium* and ragweed spores, but v_d for 0.08, 1, and 5 μm particles is greater for real grass than for artificial grass due to the importance of hairs and other irregularities in trapping small particles.

In experiments with *Lycopodium* spores depositing on sticky rough glass, Chamberlain found good agreement between the data and Equation 10 with z_{oc} = d_s. For friction velocities above 20 cm/s, true of most of the data, the stopping

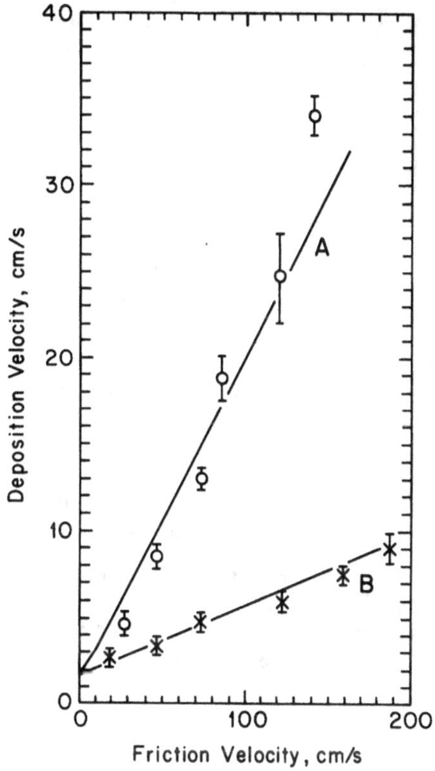

Figure 5–7. Deposition velocity versus friction velocity for *Lycopodium* spores in a wind tunnel, redrawn from Chamberlain (1967). The open circles represent data for sticky artificial grass; the *X*s show data for real grass.

distance was greater than z_o and the deposition velocities for the spores were greater than those for air momentum. These results are attributed to the high resistance to momentum transport due to the relatively smooth surface (rough glass); the spores were able to traverse the relatively stagnant sublayer by virtue of their inertia. Furthermore, the sticky coating on the surface helped maintain relatively high deposition velocities for the spores by minimizing bounceoff.

Figure 5–8 shows deposition velocities for the various particle sizes used in the wind tunnel experiments. These data were obtained with a real grass surface and a friction velocity of 70 cm/s. A curve representing sedimentation from Equation 11 is also shown.

The shape of the curve, representing a fit to the experimental data, can be explained on the basis of the various transport mechanisms. Very small particles are influenced by Brownian diffusion; hence the deposition velocity decreases as particle size increases. Interception and impaction become important for diameters greater than about 0.1 to 1 μm. Deposition due to these mechanisms increases with

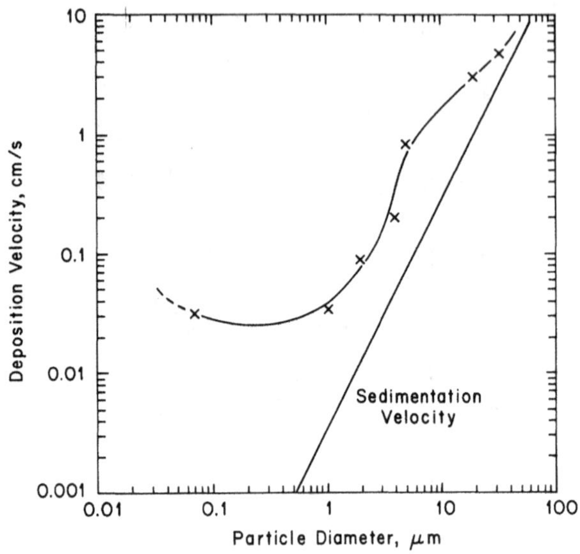

Figure 5–8. Deposition velocity versus particle diameter for several types of particles in a wind tunnel, redrawn from Chamberlain (1967). Results of the wind tunnel experiments are shown as *X*s; the curve is drawn as a reasonable fit to the data. The sedimentation equation is also graphed. The data were obtained at a friction velocity of 70 cm/s over a real grass surface.

increasing diameter. Finally, very large particles are influenced primarily by sedimentation. Most of the subsequent wind tunnel studies and mathematical models have confirmed the general shape of this curve, first determined experimentally by Chamberlain. Several other findings are reported in these early studies, including the results of field tests involving ^{212}Pb and *Lycopodium* spores.

In additional work, Wells and Chamberlain (1967) studied particle deposition to a vertical smooth brass surface and to the same surface covered with filter paper having fibers about 100 μm long. Although both surfaces were aerodynamically smooth and had similar momentum transport characteristics, particle deposition rates onto the filter paper were much greater than those onto the brass. The differences were attributed to the role of the fibers in capturing and preventing bounceoff of the particles.

2. Studies of Sehmel (1972, 1973, 1980)

After Chamberlain's work with radioactive gases and particles, Sehmel conducted a series of wind tunnel experiments to provide estimates of particle deposition velocity for application to field conditions. The experiments involved monodisperse uranine particles produced with a spinning disc generator. Particle diameters ranged from 0.03 to 29 μm, used with a number of surfaces: smooth brass, artificial grass, water, and two sizes of gravel. Friction velocities ranged from 11 to 144 cm/s.

The data were used as inputs to an empirical model based on Equation 9. However, Sehmel's solution to the equation differs from that described earlier in this chapter. The solution described earlier, presented as Equation 10, assumes that the particle eddy diffusivity K is equal to $k\,u_*\,(z-d)$ for all heights down to the particle sink $(z-d) = z_{oc}$; that assumption also applies to Chamberlain's analysis. In contrast, Sehmel's model assumes that the particle eddy diffusivity is equal to $k\,u_*\,(z-d)$ only for heights greater than 1 cm above the zero-plane displacement. In the lowest 1 cm, the particle eddy diffusivity assumes a more complex functional form with empirical coefficients derived from the wind tunnel data. This effective particle eddy diffusivity, denoted by K_e, is thus a function of diffusion, interception, inertial transport, and any other mechanisms (excluding sedimentation) influencing transport onto the surface. The expressions for K_e are derived so as to apply from 1 cm down to $(z-d) = d_p/2$, corresponding to the height of the center of the particle above the zero-plane displacement. Sehmel has formulated this model for adiabatic as well as nonadiabatic conditions; in the latter case, the expression for K_e for $(z-d) > 1$ cm includes the effects of buoyancy using the corrections of Businger and others (1971) in Table 5–2.

Sehmel's final solution of Equation 9 in integral form, when rearranged to provide the deposition velocity, is:

$$v_d = \frac{v_s}{1 - A^{-1}} \tag{28}$$

where

$$A = \exp\left(-\frac{v_s I}{u_*}\right)$$

$$I = \int_{\frac{d_p^+}{2}}^{z_r^+} \frac{\nu\, dz^+}{K_e + D}$$

$z^+ = (z-d)u_*/\nu =$ dimensionless height above the zero-plane displacement
$d_p^+ = d_p\, u_*/\nu =$ dimensionless particle diameter

Note that v_d is defined at the reference height z_r, which represents the upper limit of the integral I.

After establishing the appropriate empirical functions for K_e on the basis of the wind tunnel tests, Sehmel used the model to obtain graphs of deposition velocity versus particle diameter. Families of curves have been obtained for various friction velocities, roughness heights, atmospheric stabilities, and particle densities. Separate curves are reported for reference heights of 1 cm, 10 cm, 1 m, and 10 m above the surface.

Examples of these curves are shown in Figures 5–9 and 5–10, taken from

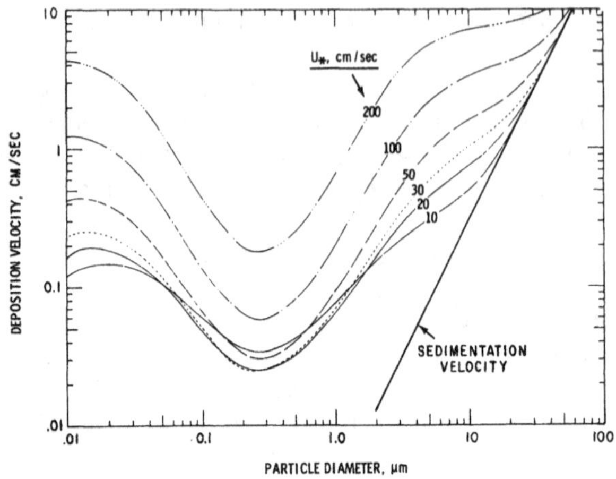

Figure 5–9. Deposition velocity versus particle diameter for several friction velocities. The curves represent model predictions on wind tunnel data, taken from Sehmel and Hodgson (1978). All of these curves correspond to a particle density of 1 g/cm^3, a roughness height of 3 cm, and an adiabatic atmosphere. The reference height is 1 m.

Sehmel and Hodgson (1978). Figure 5–9 shows the influence of friction velocity on v_d, whereas Figure 5–10 illustrates the effect of atmospheric stability and reference height at which v_d is measured. Note that the shapes of these curves resemble that of Figure 5–8, as expected.

The graphs illustrate a number of interesting features. Figure 5–9 shows generally increasing deposition velocities with increasing u_*, due to more rapid delivery of particles to the surface and greater efficiencies of capture by the roughness elements. For very small particles, the deposition velocities are relatively invariant with particle size as v_d is rate limited by delivery to the surface via wind eddies.

Figure 5–10 indicates smaller deposition velocities for greater reference heights. This is attributed to the increased aerodynamic resistance to transport from high above the surface. Note that $C(z)$ increases with height according to Figure 5–4, and hence v_d, which is inversely proportional to C, should decrease. Figure 5–10 also shows that v_d is slightly greater for an unstable atmosphere with $L = -10$ m than for stable conditions with $L = 10$ m. Other figures provided by Sehmel and Hodgson (1978) and Sehmel (1980) show that v_d increases as z_o increases, and as ρ_p increases. For additional details, the reader is referred to the original papers.

The wind tunnel studies described above have been useful for exploring the influence of various parameters on dry deposition. They have provided direct measurements of dry deposition velocities that often have been applied to estimate fluxes in the field. The principal weakness with such extrapolations, however, is that field conditions are likely to differ from those in the laboratory. For example, the maximum size of turbulent eddies in a wind tunnel are comparable to the height

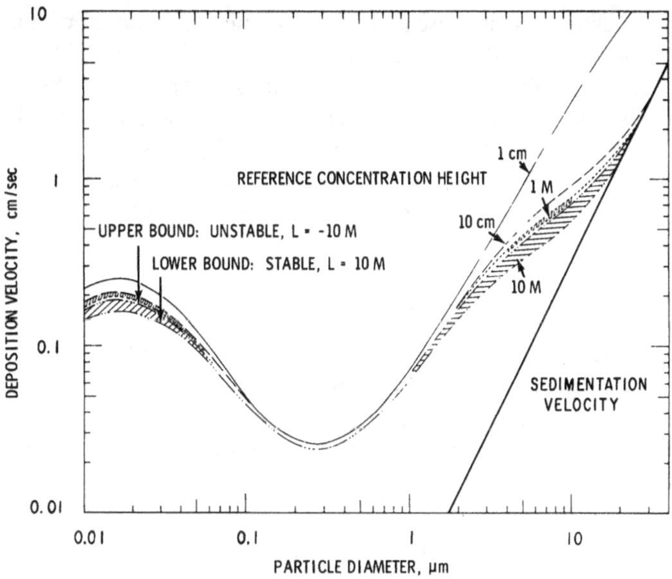

Figure 5–10. Deposition velocity versus particle diameter for different reference heights and atmospheric stabilities, taken from Sehmel and Hodgson (1978). *L* corresponds to the Monin-Obukhov length. The curves are based on a particle density of 1 g/cm^3, a roughness height of 3 cm, and a friction velocity of 20 cm/s.

of the tunnel, while turbulent eddies in the field can be larger by orders of magnitude. The structure of vegetative canopies and other surfaces exposed to the atmosphere is quite different from that of the surfaces used in wind tunnel experiments. Profiles of contaminant concentration in the field also may differ from profiles in the laboratory due to the spacing of sources and variations in types of surfaces found outdoors. Some of these differences may not cause serious problems, such as where deposition from the free atmosphere is rate limited by transport immediately adjacent to the surface. Nevertheless, the inability to simulate field conditions in the laboratory remains a weakness in extrapolating wind tunnel data to the field.

To overcome this problem, several investigators have developed models specific to certain types of surfaces and atmospheric conditions. These models have incorporated details of the micrometeorology and surface structure in many instances, using theoretical formulations and laboratory data to explain dry deposition in the field. Four models of this type developed for natural vegetation will now be discussed.

B. Semiempirical Models for Vegetative Canopies

The models summarized below pertain to a variety of vegetation types. First, the model of Davidson and others (1982), incorporating the detailed structure of five species of natural grass, is presented. This formulation has been developed for particles in the 0.01- to 100-μm-size range. The model of Slinn (1982) for a variety of vegetation types is then discussed, with results for a *Eucalyptus* forest presented as an example. The same particle size range applies. A model developed by Bache (1979a, 1979b, 1984) for large particles depositing by impaction and sedimentation is then presented and applied to a conifer forest. Finally, the model of Shreffler (1978) for SO_2 deposition onto a deciduous forest canopy and onto grass is summarized.

1. Model of Davidson and Others (1982)

For use in modeling deposition of particles, Davidson and others have suggested that some natural grass canopies can be approximated as a collection of cylinders of various heights, diameters, and orientations. The overall flux from the atmosphere to the surface is related to the efficiency of collection of an individual cylinder:

$$F = -Nd_c \int_{z_{oc}}^{h} \eta(z)\, u(z)\, C(z)\, dz \tag{29}$$

where

N = number of cylinders of diameter d_c and height h per unit area of ground

η = efficiency of collection of a cylinder

$$= \frac{\text{particles/second depositing on cylinder}}{\text{particles/second passing through the area } d_c h}$$

$$u = \text{wind speed within the canopy} = u(h) \exp\left[-n\left(1 - \frac{z}{h}\right)\right] \tag{30}$$

n = empirical parameter in the wind-speed profile usually in the range $2 < n < 5$

Equation 29 assumes all cylinders are vertical and have identical diameter and height, although the concept can be extended to include distributions of plant sizes. The efficiency η depends on the specific mechanism involved. For very small particles, Brownian diffusion controls transport across the sublayer; the appropriate expression for η is derived by analogy to heat flow for cylinders (Holman, 1972):

$$\eta_{\text{diffusion}} = \frac{D\pi\, Sh}{d_c u} \tag{31}$$

where

Sh = Sherwood number = $0.683\, Re^{0.466}\, Sc^{1/3}$
Re = Reynolds number = $u\, d_c/\nu$.

This form of the Sherwood number applies to the interval $40 < RE < 4{,}000$.

For particles with diameters greater than a few tenths of a μm, interception and inertial impaction are important. The efficiency for interception is given by:

$$\eta_{\text{interception}} = 2\,\frac{d_p}{d_c} \tag{32}$$

and the efficiency for impaction is:

$$\eta_{\text{impaction}} = \frac{St^3}{St^3 + 0.753\ St^2 + 2.796\ St - 0.202} \tag{33}$$

$$St = \text{Stokes number} = \frac{(\rho_p - \rho_a)\ u\ c\ d_p^{\ 2}}{9\ \mu\ d_c} = 2\,\frac{d_s}{d_c} \tag{34}$$

Equation 32 is derived from theoretical considerations (Fuchs, 1964), whereas Equation 33 is an empirical fit to laboratory data (Davidson and Friedlander, 1978). Both expressions apply to potential flow. The overall efficiency for use in Equation 29 is:

$$\eta = 1 - [(1 - \eta_{\text{diffusion}})(1 - \eta_{\text{interception}})(1 - \eta_{\text{impaction}})] \tag{35}$$

Once the expression for η is known, Equations 9 and 29 are solved simultaneously for the unknowns F and z_{oc}. This is accomplished by using separate expressions for K within and above the canopy, for substitution into Equation 9:

$$K = ku_*(z - d) \qquad \text{for} \quad z > h \tag{36}$$

$$= ku_*h\ exp[-n\,(1 - \frac{z}{h})] \quad \text{for} \quad z < h \tag{37}$$

Note that the second expression for K is based on Equation 30 for wind speed within the canopy. The boundary condition $C(z - d = z_{oc}) = 0$ has been assumed, implying perfect retention of particles that contact the surface ($r_c = 0$).

This model has been applied to five specific canopies in rural Pennsylvania, each dominated by a different species of wild grass. Measurements have been made of the distribution of plant sizes in each canopy, as well as wind profiles above and within the vegetation for use as model inputs. The characteristics of the vegetation are shown in Figure 5–11; parameters in the wind speed profiles are given in Table 5–4.

Running the model for a variety of particle sizes yields the curves shown in Figure 5–12. Values of v_d are similar to those in Figures 5–8 through 5–10, although differences in u_*, z_o, z_r, and surface characteristics make detailed comparison difficult. It is of interest that v_d for *Agrostis hyemalis* is greater than that of *Phleum pratense* for d_p >0.1 μm despite the smaller leaf area index of the former species. This reflects the higher efficiency of impaction on the *Agrostis* canopy due to fine hairs and hence larger Stokes numbers. The importance of considering details of the canopy structure when modeling dry deposition for certain applications is apparent.

Despite the presence of the hairs on *Agrostis*, the deposition velocities over a

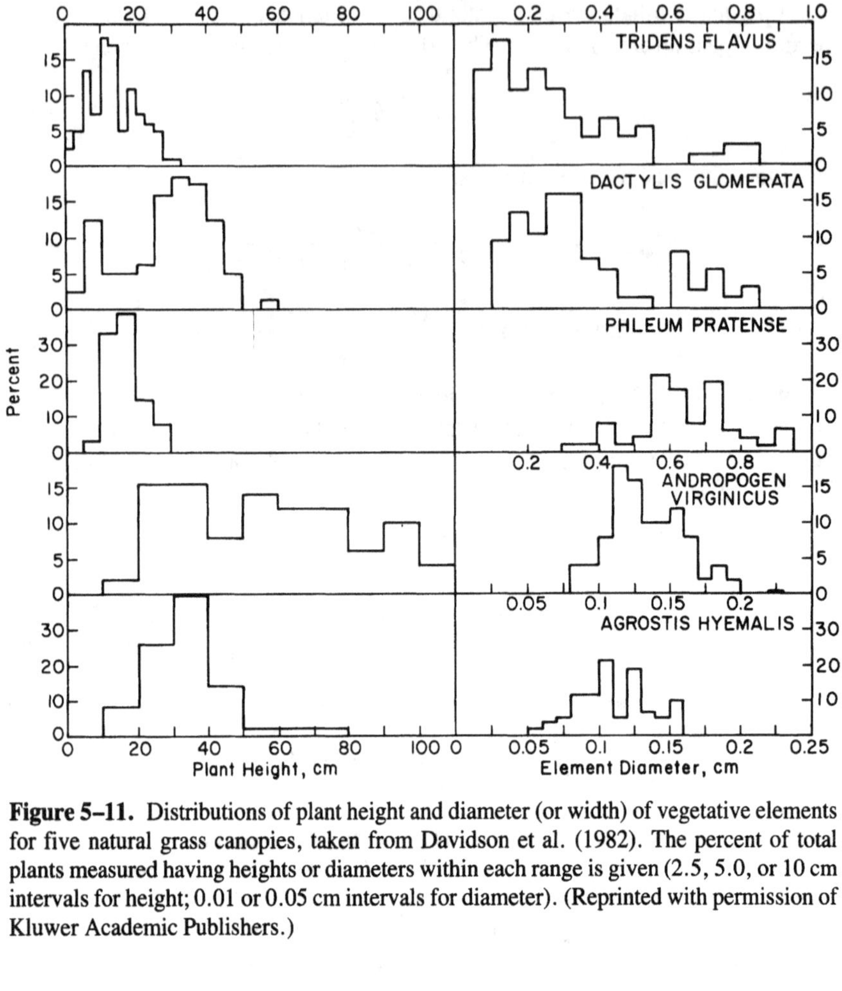

Figure 5–11. Distributions of plant height and diameter (or width) of vegetative elements for five natural grass canopies, taken from Davidson et al. (1982). The percent of total plants measured having heights or diameters within each range is given (2.5, 5.0, or 10 cm intervals for height; 0.01 or 0.05 cm intervals for diameter). (Reprinted with permission of Kluwer Academic Publishers.)

Table 5-4. Wind profile data used as inputs for the model of Davidson et al. (1982) used to construct Figure 5-12.

	Number of profiles measured	u (200 cm) (cm/s)	u_* (cm/s)	d (cm)	z_o (cm)	n
Tridens flavus	6	270	34.5	0	8.1	1.25
Dactylis glomerata	14	580	63.8	14	5.0	1.75
Phleum pratense	8	420	42.1	0	4.0	2.5
Andropogen virginicus	7	240	32.3	14	9.0	2.0
Agrostis hyemalis	3	260	40.4	13	14	1.0

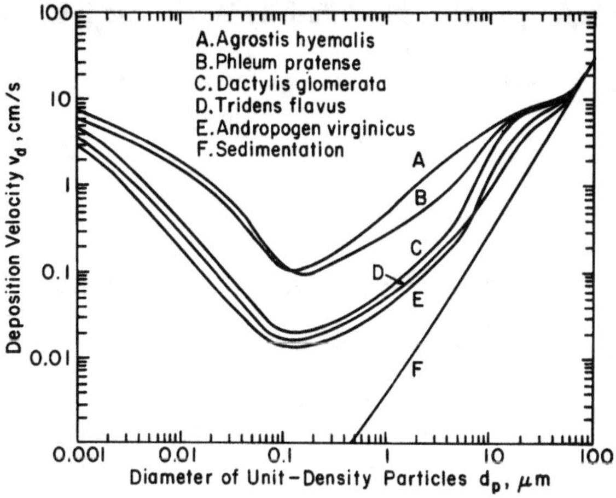

Figure 5–12. Deposition velocity versus particle diameter based on the model of Davidson et al. (1982), using data from Figure 5–11 and Table 5–4 as inputs. The reference height is 2 m. (Reprinted with permission of Kluwer Academic Publishers.)

wide range of particle sizes for each of the five canopies are small. This reflects the generally poor collection efficiencies of the vegetation elements. Such low efficiencies are probably rate limiting in the overall dry deposition process: Delivery of particles to the canopy by the atmosphere is much faster than removal at the surface. This is the situation illustrated in Figure 5–4, which shows that $r_a \ll r_b + r_c$.

2. Model of Slinn (1982)

Slinn uses a similar framework for developing a particle deposition model but with several variations. He begins with a relation analogous to Equation 29:

$$\frac{d}{dz}\left(K\frac{dC}{dz}\right) = \alpha(z)\ \eta'(z)\ u(z)\ C(z) \tag{38}$$

where

α = surface area of vegetation elements per unit volume at height z (cm^{-1})

$\int_0^h \alpha(z)\, dz$ = cumulative leaf area index = λ

η' = average collection efficiency of particles within the canopy = $c_d(z)$ XI_c

c_d = average drag coefficient for the vegetation at height z

ξ_c = particle/canopy element collection efficiency

The expression for particle eddy diffusivity in this case is given by:

$$K = ku_* (z - d) = ku_* (z - h) + K_o \quad \text{for} \quad z > h$$
$$= K_o \quad \text{for} \quad z < h$$

where

$K_o = ku_* l$
$l = h - d =$ characteristic eddy size within the canopy (cm)

The boundary condition for Equation 38 is:

$$\lim_{z \to 0} [K_o \frac{dC}{dz} = v_{dg} C] \tag{39}$$

where

$v_{dg} =$ deposition velocity for particles at the ground (cm/s) $= v_m \xi_g$
$v_m =$ deposition velocity for momentum in the canopy (cm/s) (40)
$\xi_g =$ particle collection efficiency at the ground

Note that v_m is similar to v_a in Equation 2 but refers specifically to transport for $z < h$. We can repeat the steps leading to v_a given by Equation 6, using the wind speed profile within rather than above the canopy. With the wind speed given by Equation 30 instead of equation 5, the result is:

$$v_m = \frac{nK_o}{h} \tag{41}$$

It is assumed that $\xi_c = \xi_g = \xi$. Furthermore, the product $\alpha\ c_d u$ is assumed constant with height in the canopy. When Equations 38 and 39 are solved with these assumptions, expressions for C above and within the canopy are obtained. Transport above the canopy is similar to that discussed in the previous section, yielding a logarithmic profile resembling Figure 5–4. Transport within the canopy depends on the values of n and ξ.

Families of curves of C/C_r versus z/h are shown in Figure 5–13. Here C_r represents the concentration at an arbitrary reference height z_r above the canopy. It is assumed that $u_r = 2\ u_h$ (where $u_r =$ wind speed at height z_r and $u_h =$ wind speed at height h). The curves for $\xi = 1$ correspond to momentum, for which removal by the canopy is very efficient. The curves for smaller values of ξ reflect the less efficient deposition of particles.

Slinn proposes additional equations to calculate ξ for a particular canopy. He begins with the relation:

$$\xi = ER \tag{42}$$

where

E = average, particle-size-dependent collection efficiency for the canopy, without the influence of bounceoff
R = reduction in collection efficiency caused by bounceoff = $\exp(-a' \sqrt{St})$
a' = an empirical constant

Expressions for E are analogous to Equations 31 through 33 and Equation 35:

$$E_{\text{diffusion}} = \frac{c_v}{c_d} Sc^{-2/3} \tag{43}$$

$$E_{\text{interception}} = \frac{c_v}{c_d} [f \frac{d_p}{d_p + d_{cl}} + (1 - f) \frac{d_p}{d_p + d_{c2}}] \tag{44}$$

$$E_{\text{impaction}} = \frac{St^2}{1 + St^2} \tag{45}$$

$$E = 1 - [(1 - E_{\text{diffusion}})(1 - E_{\text{interception}})(1 - E_{\text{impaction}})] \tag{46}$$

where

c_v = portion of c_d arising from viscous drag (as opposed to form drag)
f = fraction of total interception by "small" collectors in the canopy
d_{c1} = diameter of "small" collectors
d_{c2} = diameter of "large" collectors

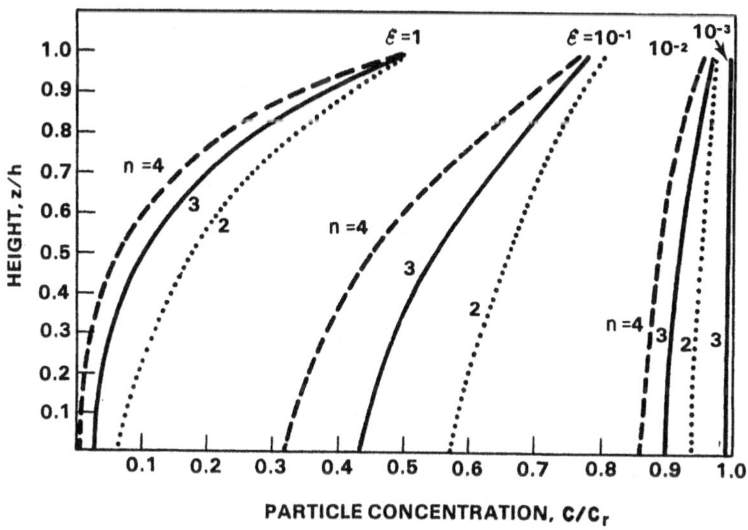

Figure 5-13. Predicted profiles of airborne particle concentration within vegetative canopies, taken from Slinn (1982). The differences among the curves represent different values of particle collection efficiency XI and wind speed profile parameter n.

Equation 43 predicts a deposition velocity that varies as $D^{2/3}$, in agreement with the dependence implied by Equations 19 and 31. Equation 44 is based on the approximation of two distinct sizes of vegetative elements in the canopy. The Stokes number in Equation 45 is similar to that defined by Equation 34 but is based on u_* and d_{c2} rather than u and d_c.

Equations 38 through 46 are used to determine the concentration profile $C(z)$, which determines the flux and deposition velocity, excluding the effects of gravity. The resulting deposition velocity is then added to v_s to obtain the final value of v_d.

Slinn has used this model to estimate dry deposition as a function of particle size for several surfaces. These include artificial and natural grass used in the wind tunnel experiments of Chamberlain (1966, 1967), a *Eucalyptus* forest following canopy and wind characteristics of Leuning and Attiwill (1978), and canopies of grass and wheat with characteristics reported by Plate (1971). As an example, Figure 5–14 shows deposition velocities for the *Eucalyptus* forest. Parameter values used by Slinn to construct the figure are given in Table 5–5.

The curves are similar in shape to those shown earlier. One difference between Figure 5–14 and the other models can be seen in the curve for u_r = 10 m/s, where v_d decreases slightly for $d_p \gtrsim 10$ μm due to the influence of bounceoff.

As illustrated in the previous model, the deposition velocities in Figure 5–14 are small over a wide range of particle sizes. Slinn points out that inefficient particle collection by the vegetation elements appears to be rate limiting for the canopies considered in his model and probably for most canopies occurring in natural settings. He notes that even tall, dense vegetation may not have efficient removal because deeper canopies require more time for particles to penetrate them. It is

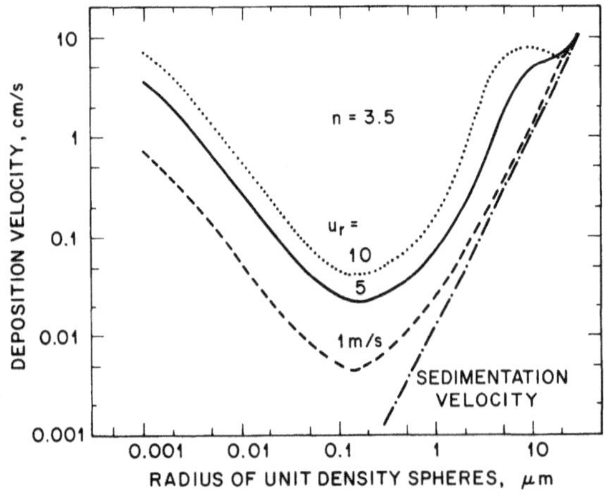

Figure 5–14. Deposition velocity versus particle diameter using model predictions for a *Eucalyptus* forest, taken from Slinn (1982). Values of input parameters used to generate these curves are given in Table 5–5.

Table 5-5. Values of input parameters for the model of Slinn (1982) used to construct Figure 5-14.

Parameter	Value
h	27.4 m
d/h	0.79
z_o/h	0.068
u_*/u_h	0.295
n	3.5
u_r/u_h	2
c_v/c_d	1/3
d_{c1}	10 μm
d_{c2}	1,000 μm
f	0.01
a'	2

thus likely that the shapes of the profiles in Figure 5–4, comparing momentum transport and contaminant transport, are representative of most types of vegetation when there is at least some wind.

3. Model of Bache (1979a, 1979b)

Bache has considered dry deposition of particles with diameters of 5 to 300 μm onto vegetative canopies. These particles are large enough to be deposited primarily by impaction and sedimentation. The analysis begins with the equation of mass conservation, analogous to Equations 29 and 38:

$$u \frac{\partial C}{\partial x} - v_s \frac{\partial C}{\partial z} - \frac{\partial}{\partial z}\left(K \frac{\partial C}{\partial z}\right)$$
$$= -\beta \left[u\, C \cos\theta + \left(v_s C + K \frac{\partial C}{\partial z}\right) \sin\theta\right] \tag{47}$$

where

x = distance in the direction of airflow (cm)
β = absorption coefficient for particle loss by the canopy per unit length of the flight path
$= P_x \cos\theta + P_z \sin\theta$
$P_x = \alpha f_x \eta_{\text{impaction}}$ = probability of capture per unit distance in the horizontal direction
$P_z = \alpha f_z$ = probability of capture per unit distance in the vertical direction
f_x, f_z = structure coefficients for the canopy in the horizontal and vertical directions, respectively ($f_x = 0$ for infinitesimally thin horizontal leaf, $f_z = 0$ for infinitesimally thin vertical leaf)
$\theta = \tan^{-1} \frac{v_s}{u}$ = angle of particle trajectory under quiescent conditions

The expression for efficiency of impaction in this case is given by:

$$\eta_{impaction} = \frac{St^2}{(St + 0.6)^2} \tag{48}$$

In this equation, the Stokes number is defined similar to that expressed by Equation 34 with one important difference: The characteristic length scale is taken as the smallest dimension of a single leaf when projected on a plane normal to the flow, denoted by d_{min}, rather than an equivalent cylinder diameter d_c. Equation 48 represents a rough fit to the data of May and Clifford (1967).

Bache applies this scheme to three vertical regimes in a forest. The regime $z > h$ is characterized by $\beta = 0$ as no particle capture occurs above the canopy. The wind-speed profile is assumed to be logarithmic (Equation 5), and the eddy diffusivity is assumed to be equal to $ku_* (z - d)$ (Equation 36). This is similar to the problem posed in section II, C, 2: The resulting concentration profile is logarithmic if $v_s = 0$ and obeys a power law if $v_s > 0$ (Equations 8 and 10).

The second regime is $b < z < h$, where b is the base of the foliage crown. All of the leaves and/or needles of the trees are assumed to exist in this regime. The expressions for wind speed and eddy diffusivity are given by Equations 30 and 37, respectively. The vertical distribution of foliage in the crown is assumed to obey a Gaussian relationship, such as that characteristic of a pine forest (Belot et al., 1976):

$$\alpha (z) = \frac{\lambda}{\sigma \sqrt{2\pi}} \exp \left[- \frac{(z - h_m)^2}{2 \sigma^2} \right] \tag{49}$$

where

$\sigma = 0.18\ (h - b) =$ a constant associated with the distribution (cm)
$h_m = (h + b)/2 =$ height of the center of the foliage crown (cm)

In the third regime, it is assumed that the wind speed is $u(z = b)$ and eddy diffusivity is $K(z = b)$ for all heights $0 < z < b$. Furthermore, β is assumed to be zero in this regime because the tree trunks and understory growth are assumed to have neglibible surface area compared with that of the foliage crown. However, Bache notes that if significant particle loss occurs in this regime, it can be specified with a vertical flux term:

$$F = (v_d C)_{z = b} \tag{50}$$

applied to the lower boundary of the second regime.

To obtain an analytical solution for this system, the concentration is assumed constant with downwind distance x so that the first term in Equation 47 is zero. This yields a differential equation expressed only in terms of z:

$$\frac{d^2C}{dz^2} + f(z) \frac{dC}{dz} + g(z)\ C = 0 \tag{51}$$

where

$$f(z) = \frac{v_s}{K} + \frac{1}{K}\frac{dK}{dz} - \beta \sin \theta \tag{52}$$

$$g(z) = -\frac{\beta}{K}(u^2 + v_s^2)^{1/2} \tag{53}$$

Bache then assumes that $f(z)$ and $g(z)$ can be replaced by coefficients f_R and g_R independent of z, which represent the functions evaluated at a height where the product $g(z)C(z)$ equals its depth-average value in the foliage crown. A sensitivity analysis has suggested that g is often the most significant parameter in determining v_g(h). Bache solves the resulting equation subject to the boundary conditions:

$$\left(\frac{dC}{dz}\right)_{z=h} = \frac{v_d(h) - v_s}{K(h)} C(h) \tag{54}$$

$$\left(\frac{dC}{dz}\right)_{z=b} = \frac{v_d(b) - v_s}{K(b)} C(b) \tag{55}$$

Figure 5–15 shows deposition velocity as a function of friction velocity for several particle sizes. The figure includes two solutions to Equation 51: the analytic result incorporating f_R and g_R and a numerical solution to the equation when $f(z)$ and $g(z)$ are retained. These results are based on representative values of input parameters for a Scots pine canopy: $h = 15$ m, $d = 10.5$ m, $b = 7.5$ m, and $\lambda = 11$. If a Stokes number defined by Equation 34 is used to compute $\eta_{\text{impaction}}$ in this model (by Equation 48), the equivalent value of d_{min} used by Bache is 4 mm. Furthermore, Bache chooses $f_x = f_z = 0.2$ after Belot and Gauthier (1975).

The curves indicate overall greater deposition velocities for larger particle sizes. For 100 μm and 300 μm particles, $v_d(h)$ is relatively insensitive to u_* due to the dominance of sedimentation. For 5 to 30 μm particles, however, impaction is important and $v_d(h)$ increases with increasing u_*.

Bache also points out several details in the comparison of analytic and numerical solutions. For all particle sizes considered, use of a single representative height for determining f_R and g_R provides a reasonable estimate of the deposition velocity. For $d_p \gtrsim 30$ μm, however, the result is sensitive to the height chosen due to the vertical foliage distribution. The solution is much less sensitive to choice of height for smaller particle sizes.

The above analytical framework has been modified by Bache (1984) for numerical solution using an initial value approach. The new scheme provides better accuracy and greater flexibility than the earlier technique. Later research by this author suggests that Equation 37 for eddy diffusivity K is valid only for a uniform canopy, rather for a canopy with Gaussian vertical distribution of foliage (Equation 49); a new analysis for the Gaussian distribution has been proposed (Bache, 1986a).

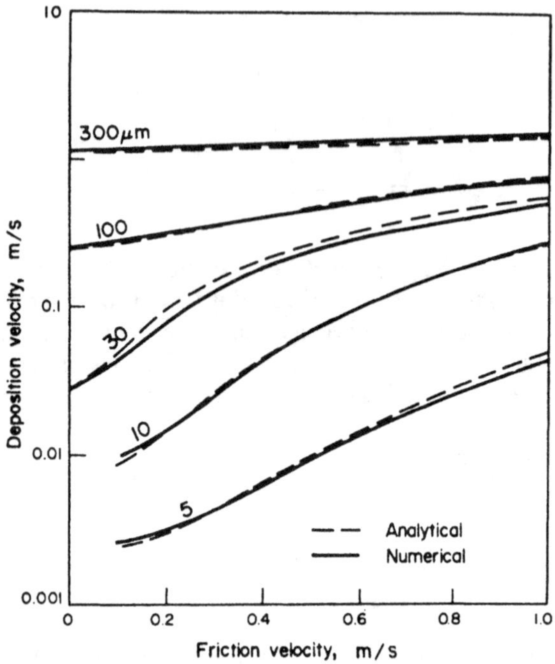

Figure 5–15. Deposition velocity of particles versus friction velocity using model predictions for a Scots pine canopy, taken from Bache (1979b). The particle density is 1 g/cm^3; values of other parameters are given in the text.

4. Model of Shreffler (1978)

Shreffler has considered the transport of SO_2 to vegetative canopies. The analysis begins with a mass conservation equation of the same form as Equation 51. In this case, however, the functions $f(z)$ and $g(z)$ are given by:

$$f(z) = \frac{1}{K}\frac{dK}{dz} \tag{56}$$

$$g(z) = \frac{a'' \alpha}{K\,[r_a + r_b + r_s]} \tag{57}$$

where a'' = an empirical constant.

The aerodynamic and boundary layer resistances are taken as a sum defined by:

$$r_a + r_b = 1.4\,[\frac{d_l}{\nu\, u(z)}]^{1/2}\, Sc^{1/3} \tag{58}$$

where d_l = representative leaf dimension (cm).

Note that Equations 56 and 57 are similar to Equations 52 and 53 but incorporate Brownian diffusion as the mechanism of transport across the boundary layer rather than impaction and sedimentation. Transport by eddy diffusion above the boundary layer is assumed in both models. Equation 58 is adopted from Thom (1968) and Chamberlain (1974) for values of d_l in the range 4 to 11 cm. Shreffler cautions that this relation may not be valid for values outside this range.

The boundary conditions for this system are given by:

$$C(z_r) = C(h) + \frac{h-d}{\phi_H(\zeta_h)}\left[ln\,\frac{z_r-d}{h-d} - \int_{\zeta_h}^{\zeta_{z_r}} \frac{1-\phi_H(\zeta)}{\zeta}\,d\zeta\right]\left(\frac{dC}{dz}\right)_{z=h} \quad (59)$$

$$\left(\frac{dC}{dz}\right)_{z=0} = 0 \quad (60)$$

where

$\zeta = \dfrac{z-d}{L}$ = dimensionless height above the zero-plane displacement

$\zeta_h = \dfrac{h-d}{L}$ = dimensionless canopy height

$\zeta_{z_r} = \dfrac{z_t-d}{L}$ = dimensionless reference height

Note that expressions for ϕ_H are given in Table 5–2. Shreffler has solved this system using a stability correction form of Equation 5 for wind speed above the canopy and Equation 30 for wind speed within the canopy. For eddy diffusivity, he uses Equations 13 and 37 for conditions above and within the canopy, respectively.

The model has been run for two cases. First, a deciduous forest canopy with sparse grass undergrowth has been taken from Rauner (1976), who provides the leaf area density as a function of height. The second canopy represents any low-set vegetation, including a field crop, and is similar to the distribution of sparse grass undergrowth but with a leaf area density five times as large. Values of the parameters used by Shreffler are given in Table 5–6.

Figure 5–16 shows the resulting deposition velocity of SO_2 as a function of friction velocity for both canopies. Curves are shown for a surface resistance (i.e., canopy resistance) of 0 s/cm, implying a perfect sink, and 1 s/cm, approximating a lower bound for stomatal resistance in many cases (Saugier, 1976; Rauner, 1976). Also shown for comparison is a corresponding curve for the ocean surface after Hicks and Liss (1976).

Shreffler points out that the curves for $r_c = 0$ s/cm, including the ocean surface, show a roughly linear increase of v_d with u_*. If a surface resistance is introduced, the increase in deposition velocity will be limited as u_* increases. The effective shielding of low-level foliage by the upper canopy can also be seen in these curves: Deposition velocities for the vegetative canopies exceed those for the ocean by amounts much smaller than the leaf area index of each canopy.

Several other authors have attempted to model dry deposition of gases and particles onto vegetative canopies, focusing on specific situations. For example,

Table 5-6. Values of input parameters for the model of Shreffler (1978) used to construct Figure 5-16.

	Parameter	Value
Forest	h	16 m
	λ	6.6
	n (wind speed)	1.0
	n (eddy diffusivity)	2.0
Low-set vegetation	h	0.6 m
	λ	5.0
	n (wind speed)	2.5
	n (eddy diffusivity)	3.0
Both canopies	d/h	0.75
	$z_o/(h-d)$	0.36
	z_r-d	10 m
	d_l	2 cm
	a''	1
	L	∞ (neutral atmosphere)
	D_{SO_2}	0.122 cm^2/s

Lassey (1982, 1983) has considered particle transport through vegetation as analogous to chromatographic transport through an ion-exchange column. Unlike the previous models, which have focused on microscopic contaminant-vegetation interactions, Lassey's formulation includes overall average values of key parameters to determine the fate of depositing aerosol. Legg and Price (1980) have examined the role of sedimentation in particle deposition onto vegetation. Results of their modeling effort suggest that turbulent transport into a canopy followed by sedimentation onto the lower leaves of the vegetation is an important process for aerosols with $d_p \lesssim 30 \mu m$. This applies to wet vegetation in low and moderate wind speeds, where impaction is inefficient, and for dry vegetation in all wind speeds where bounceoff is significant.

Other investigators have developed models for better extrapolation of available wind tunnel data (e.g., Table 5–3) to a variety of applications, such as predicting deposition in the field. For example, de la Mora and Friedlander (1982) have correlated deposition data from Chamberlain (1966, 1967) with filtration theory for blades of artificial grass. The theory is based on transport by diffusion and interception at high Reynolds numbers. Using dimensionless groups, they found a satisfactory correlation over nine orders of magnitude of the deposition variable, extending well into the region of substantial inertial effects. The results imply that interception is a key mechanism for particle sizes of $\simeq 1 \mu m$ to 20 to 30 μm. The authors suggest that particle deposition velocities in the interception range can be expressed as $v_s + a\, u_*^{3/2} d_p^2$, where a is an empirical constant characteristic of each type of surface.

Models have also been developed for dry deposition to other natural surfaces. For example, Slinn and Slinn (1980; 1981) have used a two-layer model to

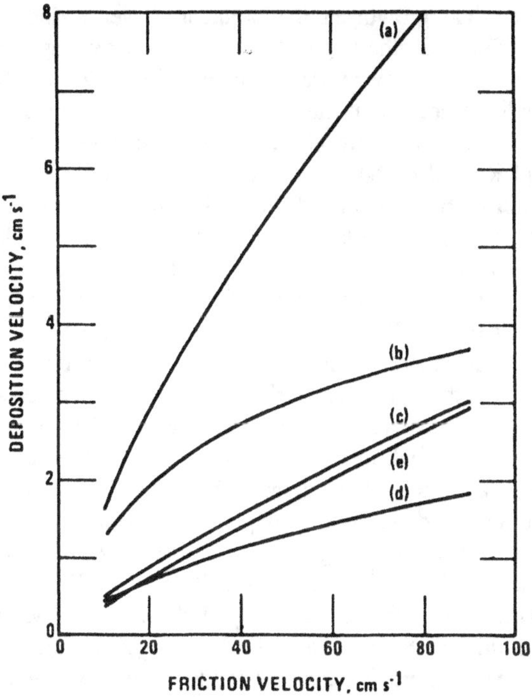

Figure 5-16. Deposition velocity of SO_2 versus friction velocity using model predictions for two vegetative canopies, taken from Shreffler (1978). The input parameters are given in Table 5-6. Curve (a) corresponds to a deciduous forest canopy with $r_c = 0$; curve (b) is similar but with $r_c = 1$ cm/s. Curve (c) corresponds to low-set vegetation with $r_c = 0$; curve (d) is similar but with $r_c = 1$ cm/s. Curve (e) is for an ocean surface.

describe the deposition of particles to water. The first layer involves particle transport by eddy diffusion and sedimentation from well above the surface and is termed the *constant flux layer* because F is invariant with height. The second layer involves transport through the boundary layer by Brownian diffusion, inertial impaction, and other mechanisms. The model includes the effect of slip at the water surface by setting the dependence of the flux on particle diffusivity equal to $D^{1/2}$, rather than $D^{2/3}$, as used in the canopy models. The effect of hygroscopic growth due to high humidity in the boundary layer is also included. The model assumes a perfect sink surface. A variation of this model that accounts for a nonzero surface resistance has been offered by Fairall and Larsen (1984). These authors use the concept of an effective surface source strength to account for the production of seaspray and atmosphere-surface interactions close to the water. Williams (1982) has developed a model for dry deposition to water incorporating a similar two-layer concept. He accounts for the breaking of waves by considering separate expressions for smooth and rough areas of the water surface. The effect of hygroscopic particle growth in the boundary layer is included. As another example, Ibrahim and others (1983) have used a two-layer model to predict particle deposition to a snow surface. The model includes the effects of interception onto small ice needles at the surface, as well as estimates of transport by diffusion and impaction. Hygroscopic growth is accounted for.

McRae and Russell (1984) have used a model that links source emissions, atmospheric transport, physical and chemical transformations, and dry deposition to examine how NO_x emissions influence acidic deposition. The dry deposition part of their effort begins with the aerodynamic transport described by Equation 7. This equation is modified for nonadiabatic conditions, using the correlations of Businger and others (1971) in Table 5–2. Transport through the boundary layer is described by Equation 19. An upper limit for v_d is obtained by assuming a perfect sink surface; the authors then adjust this deposition velocity downward by citing published experimental data that provide estimates of the surface resistance. The final equations for deposition are used as one component in the overall transport model. McRae and Russell have used this model with input data from the California South Coast Air Basin to predict airborne concentrations and dry deposition of nitrogen-containing species for a smoggy day during summer 1974. Included in the model are NO, NO_2, PAN, HONO, $HONO_2$, NH_4NO_3, NH_3, and other species that react with nitrogen compounds. When predictions are compared with measured airborne concentrations for NH_4NO_3 and O_3, reasonable agreement is obtained. The model predicts that 35% of the nitrogen emitted into the air of the South Coast Air Basin is removed by dry deposition during one day. The authors conclude that dry deposition can remove virtually all nitrogen oxides and their acid oxidation products within a few days of travel from the source.

Clearly a wide variety of models exist for predicting dry deposition to natural and artificial surfaces. Several of the models for particles are summarized in Figure 5–17, showing deposition velocity as a function of particle diameter. The large differences in predicted deposition velocities among the models are due to differences in input data and in model assumptions. Note that none of the models has been verified adequately in the field over the range of parameter values commonly encountered. For this reason, many individuals prefer to use field data when estimating dry deposition rates rather than model predictions. The next section summarizes techniques for measuring dry deposition of particles and gases in the field. For each method, we include a description of the technique, the spatial and temporal scales of the measurement, and the relative advantages and disadvantages

IV. Measurement Methods

Over the years, several techniques have been developed for measuring dry deposition. These can be divided conveniently into two categories. The first category, *surface analysis methods,* include all types of measurements that examine contaminant accumulation on surfaces of interest. The second category,

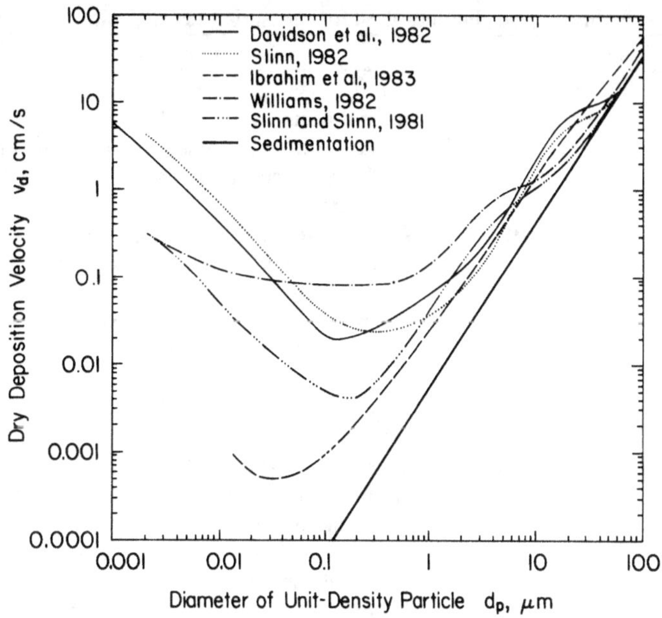

Figure 5–17. Dry deposition velocity as a function of particle diameter, for the models of Davidson et al. (1982) for a canopy of *Dectylus glomerata* (orchard grass) (canopy height = 57 cm, friction velocity = 64 cm/s, roughness height = 5 cm, zero-plane displacement = 14 cm); Slinn (1982) for a *Eucalyptus* forest (canopy height = 27.4 m, friction velocity = 75 cm/s, roughness height = 1.86 m, zero-plane displacement = 21.6 m); Ibrahim et al. (1983) for snow (reference height = 10 cm, mean wind speed = 2.5 m/s, relative humidity = 90%), Williams (1982) for water (reference height = 10 m, mean wind speed = 10 m/s, relative humidity = 99%); and Slinn and Slinn (1981) for water (reference height = 10 m, mean wind speed = 5 m/s, relative humidity = 99%). The curve for sedimentation velocity is also shown.

atmospheric flux methods, involves measurement of contaminants in the atmosphere from which one may estimate the flux. The NAPAP Dry Deposition Workshop in March 1986 included discussions of measurement methods in both categories, the results of which are presented in Table 5–7. This section provides some details of these methods, taken largely from the NAPAP Workshop Proceedings (Hicks et al., 1986a).

A. Surface Analysis Methods

Many natural and man-made surfaces that serve as sinks for airborne contaminants can be analyzed for the total mass accumulation for species of interest. The dry deposition flux F is then calculated from knowledge of the surface area and the exposure time. Measurements are made over spatial scales ranging from the size of individual leaves to the dimensions of a watershed. Variations in exposure time range from hours to months or even years.

Table 5-7. Summary of techniques used for measuring dry deposition in the field.

Method	Description
Surface Analysis Methods	
Foliar extraction	Material accumulated by dry deposition is removed from individual leaves by washing, and species concentrations in wash fluid are measured
Throughfall and stemflow	Wet deposition estimated from precipitation measurements above the canopy is subtracted from deposition estimated from throughfall plus stemflow measured within the canopy to yield the net flux of a contaminant; net flux is separated statistically into components representing dry deposition and canopy exchange
Cloud droplet collection	Deposition of water on the canopy is estimated from TF + SF during cloud events; species concentration in cloudwater above the canopy is multiplied by water deposition to yield the contaminant deposition rate
Watershed mass balance	Flux of a species into a watershed by precipitation is subtracted from the total export from the watershed in streams and groundwater, yielding the net dry deposition input
Isotopic tracers	Ratios of certain radionuclides in soil or other media are used to estimate the fractions of contaminant aerosol originating in the atmosphere and the fractions derived from other sources
Snow sampling	Contaminant concentrations in fresh snow are subtracted from concentrations in older surface snow onto which dry deposition has occurred to yield the net dry deposition flux
Aerodynamically designed surrogate surfaces	Surfaces with well-defined airflow characteristics, which are inert or have other known surface resistances, are exposed to the ambient atmosphere over specified time periods to allow accumulation of the contaminants of interest
Other surrogate surfaces	A wide variety of artificial surfaces that do not have well-defined airflow characteristics have been exposed to the ambient atmosphere to allow accumulation of the contaminants of interest
Surface analysis methods for materials effects	Several of the preceding methods developed for deposition in ecosystems have analogues in materials effects research, involving estimates of contaminant accumulation on man-made surfaces

Table 5-7. (*continued*)

Method	Description
Atmospheric Flux Methods	
Tower-based eddy correlation	The turbulent fluctuating component of the vertical wind velocity w' and of the airborne contaminant concentration C' are measured simultaneously above the canopy; the turbulent deposition flux is calculated as the time average of the product $w'C'$
Vertical gradients	The airborne contaminant concentration is measured at two or more heights above the canopy, and the flux is inferred as $-K(dC/dz)$ where K is the contaminant eddy diffusivity
Eddy correlation from aircraft	Eddy correlation is used during aircraft flights to determine horizontal and vertical variations in the mean contaminant concentration and the flux above the terrain
Aerometric mass balance	Airborne contaminant concentrations are measured on the boundaries of a specified volume of air; the difference between inflows and outflows, when used with information on sources and chemical reactions in the air volume, is used to determine the dry deposition
Multiple artificial tracers	A nondepositing tracer is released simultaneously with a contaminant; decreases in tracer concentration with downwind distance are used to estimate the dispersion, which can be subtracted out of the contaminant data to estimate loss by deposition
Eddy accumulation	One integrating sampler measures the contaminant concentration in downdrafts, and another measures the concentration in updrafts; the difference in the two measurements is used to estimate the net downward flux
Variance	Variances in contaminant concentration σ_C^2 and in the temperature σ_T^2 are measured simultaneously with the heat flux F_H, and the contaminant flux is inferred as $F_H \sigma_C / \sigma_T$

After Hicks et al. (1986a).

1. Foliar Extraction

In this method, atmospheric material accumulated on surfaces of leaves or other plant parts is separated from contaminants internal to the plant by selective extraction procedures, normally washing the leaves with water. The technique applies to individual leaves or groups of leaves with spatial scales of cm^2 to m^2; typical exposure times are tens to hundreds of hours.

The main advantage of the method is that a direct analysis of deposited material on surfaces of interest is obtained. Washing procedures have been developed and tested for a number of chemical species. The method can be calibrated by dosing selected leaves with known amounts of contaminants. Spatial variability in dry deposition can be assessed by washing plants at several locations within a canopy.

The principal disadvantage is that foliar extraction requires large numbers of samples to obtain reliable data. This is because of variations in fluxes caused by differences in leaf orientations, shadowing effects, and larger scale inhomogeneities in the canopy. Access to leaves in the upper parts of forest canopies may be difficult, requiring towers. It may also be difficult to distinguish between different forms of a single chemical species, such as deposited gaseous and particulate sulfur. Finally, measurements must be made in the absence of precipitation, fog, or dew drip.

2. Throughfall and Stemflow

This method relies on precipitation to remove contaminants previously dry deposited on the foliage: Concentrations in throughfall and stemflow (TF + SF) within the canopy will be greater than concentrations in precipitation above the canopy due to this natural washoff. However, the difference between the two concentrations, termed *net* TF + SF, is affected by exchange of the contaminant within the canopy as well as by dry deposition. For example, leaching of the contaminant from internal plant tissues during precipitation is a major canopy exchange process that may complicate data interpretation. To separate the effects of canopy exchange and dry deposition, statistical regressions involving data from several storms are sometimes used. Previous work has shown that the net TF + SF from separate storms can be regressed with precipitation volume and length of the dry period prior to the storm. The coefficient of the dry period term can then be used as a measure of the dry deposition flux.

The method can be applied to individual trees or to much greater areas, such as large forested regions. For chemical species and vegetation types where canopy exchange is negligible, the net TF + SF provides a direct measure of dry deposition fluxes for the period prior to the storm. For more complex systems, several storms (typically 20) are needed; total averaging times for such systems range from $\simeq$ 2 months to several years.

The major advantage of throughfall and stemflow analysis is that dry deposition fluxes integrated over entire canopies and for lengthy time periods are obtained. The method is relatively inexpensive and easy to implement. Furthermore, the analysis can be applied to nonhomogeneous canopies and complex terrain. As with foliar extraction, each set of samples can provide data for several chemical species of interest. Recent work has shown that foliar leaching by rain is negligible for sulfate deposited on some hardwood trees; hence, net TF + SF provides direct estimates of dry deposition for these cases.

The main disadvantage is that not all dry-deposited material is removed by precipitation. For example much of the deposited SO_2, NO_2, HNO_3, and certain ionic species (H^+, NH_4^+, and some trace metals) may be irreversibly absorbed by the canopy. Furthermore, these and other contaminants may be deposited on parts of the canopy that are not washed during precipitation. Exudation of internal plant material during dry periods may also cause complications, along with surface residues remaining from previous storms. The assumptions in the statistical analysis have not been tested under controlled conditions. Finally, large numbers of storm events are needed, and thus a substantial time period may exist between initiation of the measurements and estimation of the dry deposition rate.

3. Cloud Droplet Collection

For some canopies, such as those in mountainous terrain, impaction and interception of cloud droplets containing contaminants may be an important deposition mechanism. Concentrations of a contaminant in the cloudwater can be measured above the canopy, whereas the total volume of cloudwater reaching the canopy can be estimated by TF + SF measurement. The product of these two quantities yields the total contaminant input rate by dry deposition of cloud droplets.

The method can be applied to a small spatial scale, such as that of an individual tree, or to a large forested area. The exposure time is the length of the cloud event, which may be as short as a few hours or as long as several days.

The main advantage of the method is that a direct estimate of contaminant deposition from cloudwater is obtained relatively easily. A wide variety of chemical species can be easily measured. The technique is applicable to mountainous terrain and nonhomogeneous canopies. In the latter case, however, a considerable amount of sampling may be needed to account for spatial variability in TF + SF.

Another problem is that it may be difficult to measure contaminant concentrations in the cloudwater, particularly during mixed rain-cloud events. The method is not useful during subfreezing weather. In addition, corrections for evaporation may require collection of meteorological data under severe conditions.

4. Watershed Mass Balance

In this method, it is assumed that the total input of a contaminant to a watershed is the sum of inputs by wet deposition and dry deposition and that the total input equals the total export by stream and groundwater flow when internal sources and sinks are accounted for. The dry deposition input can thus be determined by measurement of precipitation input, total export, and contaminant sources and sinks within the watershed.

The mass balance is applicable to vegetated areas of $\simeq$5 to 10 hectares, although areas as small as 10 m^2 may be feasible if edge effects are controlled. Temporal scales are relatively long due to the residence time of water in the area of interest, typically several months to a year.

The advantage of the watershed mass balance is that dry deposition estimates over large spatial scales and long time periods are obtained. The terrain may include a variety of surface types. Long-term use of a watershed may provide reliable estimates of trends in dry deposition for only a modest effort following the initial calibration.

The main problem with the technique is that there are stringent criteria for selection of the particular watershed for measurement. The area must by hydrologically well characterized and must have known internal sources and sinks. In addition, the initial setup of equipment may be expensive and time-consuming. Information on chemical speciation is not obtained.

5. Isotopic Tracers

This method allows one to distinguish between amounts of a contaminant derived from wet plus dry deposition from the atmosphere and amounts derived from other sources, such as that naturally occurring in the soil. For example, the ratio $^{87}Sr/^{86}Sr$ in coarse aerosols is different from that in bedrock at some sites. Thus the value of this ratio in throughfall can be used to estimate the fraction of coarse aerosol originating outside the ecosystem and the fraction resulting from local resuspension. In some cases, distinctions may be possible between wet and dry deposition. As an example, the ratio $^{137}Cs/^{210}Pb$ in precipitation is more than double the corresponding ratio in submicron aerosols in surface air. Thus values of the ratio can be related to the fraction of submicron aerosol input by dry as opposed to wet deposition.

The method can be applied to large scale spatial scales by collecting samples over wide areas. The time scales provided by the technique depend on the radionuclides used. $^{87}Sr/^{86}Sr$ in throughfall can be used for individual storm events or for decades of accumulation when measured in soil; $^{137}Cs/^{210}Pb$ data from soil samples provide deposition inputs integrated over decades.

The principal advantage of the method is that the radioisotopes provide definitive identification of atmospheric material. However, information is generally obtained only for categories of airborne material, such as coarse aerosol or submicron aerosol, rather than for specific chemical species.

Another use of isotopic ratios for deposition estimates involves natural radionuclides whose half-lives are shorter than or of the same magnitude as time periods of dry deposition. For example, ^{214}Pb, ^{212}Pb, ^{35}S, and ^{3}Be are all associated with submicron aerosol and have relatively short half-lives. These species may therefore be used to provide direct measurement of dry deposition. As with long-lived radionuclides, the method provides information only on generic categories of airborne material.

6. Snow Sampling

This method relies on snow as a natural surface onto which dry deposition can be easily measured after sufficient accumulation of a contaminant. The technique can be used at many points throughout sizable areas to provide data over large spatial scales. The minimum time scale is typically at least 1 to 2 days to allow sufficient deposition. However, use of snowpacks may provide data for wet plus dry deposition integrated over months or years.

The main advantage of snow sampling is that deposition on a natural surface of interest is directly measured with relative ease. The method can be used to examine long-term integrated data in cold regions with sufficient snowfall. A major disadvantage, however, is that the data may be difficult to interpret due to problems of sublimation, contaminant migration within the snow, and contamination by soil dust and biological debris. The method cannot distinguish between gaseous and particulate forms of certain contaminants, such as sulfur species or nitrogen species. Furthermore, concentrations of some contaminants may be very small, requiring preconcentrations of the snow samples.

7. Aerodynamically Designed Surrogate Surfaces

For this method, a passive artificial collector with known airflow characteristics is positioned in an area with well-defined atmospheric flow patterns. There are several possible shapes that might be used. One example is a symmetric airfoil made of Teflon-coated aluminum, as illustrated in Figure 5–18. The device is designed to have a laminar or transition boundary layer over its upper surface where a deposition plate is positioned. Plates may be made of various materials: A Teflon coating will provide an inert surface for measuring particle deposition, whereas a nylon coating will provide a perfect sink surface for HNO_3. The device measures dry deposition on surfaces of $\simeq 100\ cm^2$, but several such devices can be set up to assess variability in deposition over large areas. Minimum acceptable time periods for exposure range from a few hours in polluted areas to several days in remote regions.

The main advantage of the method is that several of the parameters influencing dry deposition, particularly r_b and r_c, can be accounted for. Hence the mechanisms responsible for dry deposition can be explored. Wind tunnel tests have shown that sedimentation is the principal mechanism influencing particle transport, so the method is most useful for measuring fluxes of large particles. The device may also be useful for measuring deposition of certain gases whose surface resistances are known. Because the method is inexpensive and allows a considerable degree of control, the device may be suitable for routine monitoring. This is particularly true for identification of trends in deposition over long time periods.

The main disadvantage is that use of the devices in the field may be difficult at certain locations: Protection from rain, soil resuspension, and biological debris is necessary. Site selection may be a problem in nonuniform terrain. It may also be difficult to use data from the devices to provide information on deposition to surfaces of interest, such as vegetative canopies.

8. Other Surrogate Surfaces

Over the past several years, various types of surrogate surfaces have been used to measure dry deposition in the field. These include glass microslides, Teflon plates, petri dishes, dustfall buckets, different types of filter paper, and other

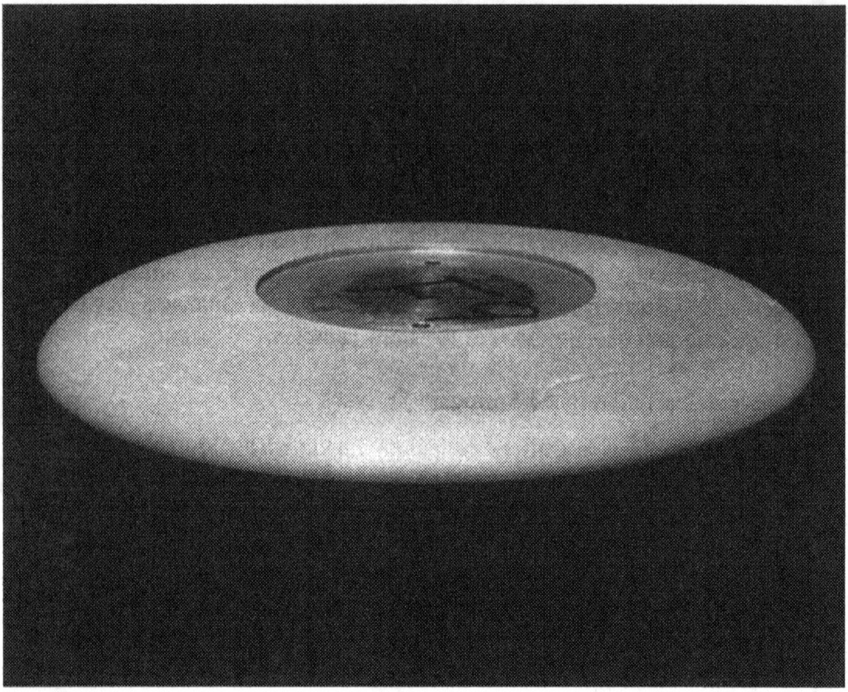

Figure 5–18. Symmetric airfoil for use in measuring dry deposition of large (supermicron) particles and other species.

devices. Table 5–8 summarizes characteristics of many of the surrogate surfaces reported in the literature; some of these are illustrated in Figures 5–19 through 5–23.

The main advantage of these surfaces is that they are inexpensive and easy to use. For this reason, surrogate surfaces have found widespread use in routine monitoring networks. For example, dustfall buckets have been used for many years by the National Atmospheric Deposition Program.

The major problem with the method is that deposition to such surrogate surfaces is difficult to relate to surfaces of interest. The deposition flux of some contaminants to buckets, for example, may be very sensitive to wind speed; the way in which deposition varies with wind speed may be quite different for the buckets than for vegetation or other surfaces of interest. Complications caused by soil resuspension and biological debris may be severe. Deposition of gaseous species on previously deposited particles may change surface resistance characteristics with time during exposure. Because of the difficulties in interpreting data from surrogate surfaces with poorly defined airflow characteristics, many monitoring networks have discontinued their use.

9. Surface Analysis Methods for Materials Effects

There has been considerable interest in acidic deposition damage to buildings, statues, monuments, and other structures. Until recently, however, measurements of dry deposition to these surfaces were extremely limited.

Many of the methods discussed above for measuring deposition in natural settings have analogues to measuring deposition to the materials composing man-made structures. For example, washing deposited contaminants from surfaces of structures may be performed in a manner similar to foliar extraction; collection of runoff from structures simultaneously with collection of incident precipitation is analogous to the throughfall and stemflow method. A mass balance may be performed for an urban area as for a watershed. Similar to studies in ecosystems, analysis of radionuclides accumulated on structures can provide information on the fraction of deposited contaminant originating in the atmosphere. Finally, small samples of materials commonly used in buildings and other structures can be placed in aerodynamically designed holders to study boundary layer and surface resistances.

In general, most of the surface analysis techniques discussed in this section offer the advantage of direct measurement at relatively low cost and the opportunity to analyze the samples for a wide variety of chemical species of interest. However, analysis of contaminant accumulated on vegetation, structures, or other surfaces may require large numbers of samples, and the resulting data may be difficult to interpret. Furthermore, the information obtained often pertains to spatial scales that are too small and time scales that are too long to be useful inputs to regional air-quality models. Because of problems with all of the surface analysis methods, techniques have been developed to estimate dry deposition by measuring concentrations of contaminants in the atmosphere. These techniques are now discussed.

B. Atmospheric Flux Methods

The methods in this category generally rely on known properties of the atmosphere to infer contaminant fluxes based on airborne concentration data. For most of the methods, micrometeorological data are obtained simultaneously with the concentration measurements to define atmospheric conditions. Unlike surface analysis techniques that involve relatively long field exposures followed by chemical analysis in the laboratory, many of the atmospheric flux methods employ fast-response sensors to detect contaminants in the field and provide estimates of deposition over short time periods.

1. Tower-Based Eddy Correlation

Because the atmosphere is virtually always in turbulent flow, any property of the atmosphere at a given instant of time can be represented as the sum of two components. These are the time-average value of the property and the deviation from the time-average value caused by random turbulent fluctuations. The latter component may be positive or negative and by definition has an average value of

Table 5-8. Characteristics of surrogate surfaces reported in the literature.

Surface type	Dimensions	Deposition surface material	Height (m) above ground unless otherwise indicated	Remarks	Reference
Microscope slides	2.5 cm × 7.6 cm	Glass	0.0	Glass slides coated with silicon grease	Raynor (1976)
Flat plate	9.4 cm dia.	FEP Teflon	1.5 m above ground level; 1.5 m above building roof	0.3-cm rim on stainless steel holder, top and bottom deposition measurements (Figure 5-19)	Davidson (1977); Davidson and Friedlander (1978); Davidson et al. (1985a)
Flat plate	17.6 cm dia.	FEP Teflon	1.5	1.0-cm rim	Elias and Davidson (1980); Davidson and Elias (1982)
Flat plate	13.3 cm dia.	FEP Teflon	0.4, 1.5	Mounted flush in aluminum holders; some experiments used fixed rainshield 30 cm or 45 cm above plate (Figure 5-18)	Dolske and Gatz (1982); Davidson et al. (1985b)
Flat plate	76 cm × 76 cm	Teflon-laminated aluminum sheet	1.0	Rimless	Smith and Friedman (1982); Dolske and Gatz (1982)
Flat plate	300 cm^2 (round)	Polyethylene	10, 20	Rimless	McDonald et al. (1982)
Filter paper	25 cm × 20 cm	Cellulose (Whatman 541)	1.0	Used a fixed 1 m × 1 m rainshield 12 cm above deposition surface, filter paper mounted on Perspex frame (Figure 5-20)	Cawse (1974, 1975, 1976); Peirson et al. (1973); Ibrahim et al. (1983)
Filter paper	25 cm × 20 cm	Cellulose	1.0	Used automatic rainshield (Figure 5-21)	Pattenden et al. (1982)

Filter paper	14.2 cm dia.	Nylon membrane (Nylasorb, Membrana Corp.	17	Mounted on a 15.0-cm dia., Teflon-coated stainless steel plate	Japar et al. (1985)
Filter paper	14.2 cm dia.	PTFE Teflon membrane (Zefluor, Membrana Corp.)	17	Mounted on a 15.0-cm dia., Teflon-coated stainless steel plate	Japar et al. (1985)
Petri dish	9.5 cm dia. 1.3 cm deep	Polyethylene	—	Located in foliage, no rainshield	Lindberg and Harriss (1981)
Petri dish	9.5 cm dia. 1.3 cm deep	Polycarbonate	15–19	Automatic rainshield located in foliage (Figure 5–23)	Lindberg and Lovett (1985)
Pluviometer (rain collector)	400 cm^2 area	—	—	Inverted for dry deposition measurements	Servant (1976)
Sangamo Precipitation Collector	18-cm-dia., 10-cm-deep cup	—	1.5	Cup suspended in a bucket	Ibrahim et al. (1983)
O.M.E. Collector Funnel	50 cm × 50 cm at top	Teflon-coated collection surface	1.0	—	Ibrahim et al. (1983)
Funnel	26 cm dia. at top, 2.0 cm dia. at bottom, 20-cm height	Polyethylene	0.8	—	Ibrahim et al. (1983)
Aerochem-Metrics and HASL type dustfall buckets	25 cm dia., 28 cm deep	Linear polyethylene	1.5	Automatic rainshield (Figure 5–22)	Dolske and Gatz (1982); Semonin et al. (1984); Cadle and Dasch (1985); Dasch (1985a, 1985b); Feely et al. (1985)

After Gamble and Davidson (1986), with permission.

Figure 5–19. Flat Teflon plate, 9.4 cm diameter, used by Davidson et al. (1985a). Copyright © American Chemical Society. Reproduced by permission.

zero. Two properties of interest in the eddy correlation method are the vertical wind velocity $w(t)$ and the contaminant concentration $C(t)$. When expanded into components, we can write $w(t) = \overline{w} + w'$ and $C(t) = \overline{C} + C'$, where the overbar represents the time average of the variable and the prime denotes the turbulent fluctuating component. Note that $\overline{w} = 0$ over level terrain.

Measurement of dry deposition by the eddy correlation method involves the flux as represented by the time-average value of the product of w' and C': $F = \overline{w'C'}$. The fluctuating component of the vertical wind is measured with a fast-response airflow sensor such as a hot film anemometer. The fluctuating component of the concentration is measured with a sensor for the contaminant of interest, which provides data in real time.

The flux measured by eddy correlation pertains to a spatial scale of several hundred meters upwind, with a zone typically $\pm 15°$ from the average wind direction. Wind and contaminant sensors must have response times of about 1 second or faster, and fluxes are usually computed for 15- to 60-minute averages.

The main advantage of the method is that fluxes can be measured with reasonably high accuracy and good time resolution when conditions permit. Fast-response sensors are available for several important species, including O_3,

Figure 5–20. Deposition measurement apparatus with Whatman 541 filter paper used by Cawse (1974) and others.

Figure 5–21. Automatic wet and dry deposition collector with Whatman 541 filter paper used by Pattenden et al. (1982).

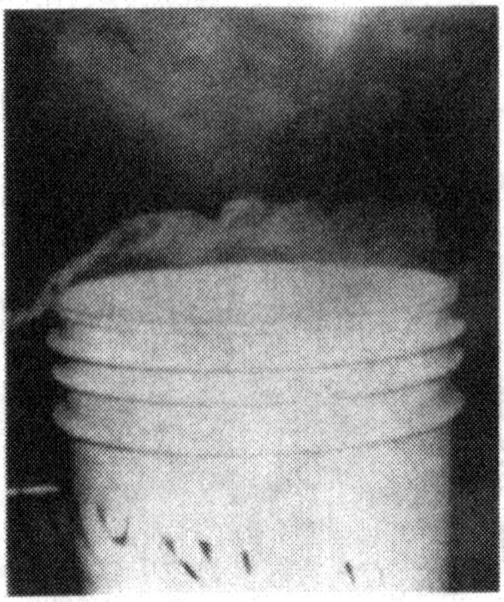

Figure 5–22. Aerochem-Metrics dustfall bucket during wind tunnel flow visualization experiments, used in the field by several investigators (e.g., Dolske and Gatz, 1982; Semonin et al., 1984; Cadle and Dasch, 1985; Feely et al., 1985).

SO_4^{2-}, SO_2, NO, and NO_2. The method has been tested by measuring fluxes of heat and water vapor using eddy correlation and comparing the results with heat and water vapor fluxes measured independently using surface energy budgets and weighing lysimeters.

The principal disadvantage is that the technique is useful only when there is uniform vegetation and adequate fetch. In addition, the sensors are expensive and subject to noise and detection limit problems. Sensors with adequate time response are not yet available for many species of interest.

2. Vertical Gradients

For species where fast-response sensors are not available, measurement of the gradient in concentration above the surface can provide estimates of the flux. The gradient can be determined by measuring the concentration at two or more heights using real-time continuous monitors or using filters for later chemical analysis. The eddy diffusivity K of the contaminant must also be estimated. Usually, it is assumed that the eddy diffusivities of contaminant mass and sensible heat are equal, where the latter is determined by measuring the sensible heat flux. The contaminant flux is then determined as $F = -K\,(dC/dz)$.

The applicable spatial scales are similar to those of eddy correlation. Data averaged over time periods on the order of 1 hour have been used to estimate fluxes.

The principal advantage of the gradient approach is that fast-response sensors are not needed as with eddy correlation. Interpretation of the data is straightforward, provided satisfactory gradients are obtained. The major problem with the

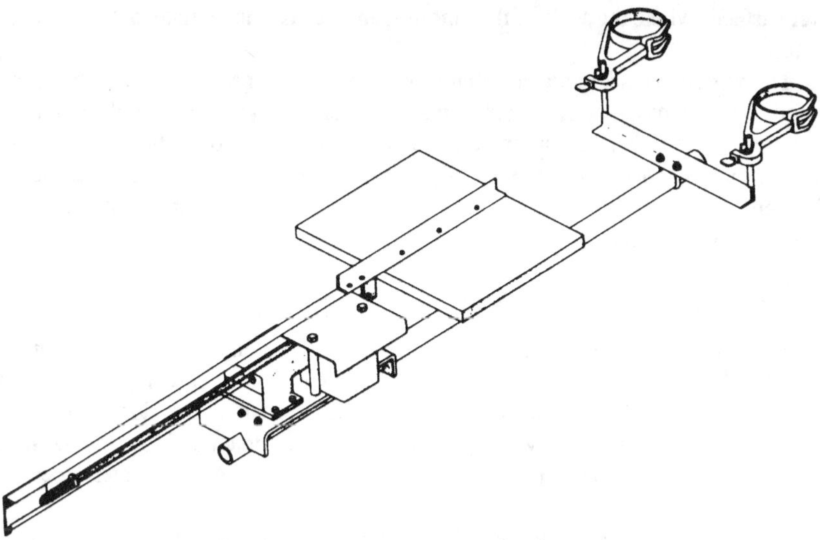

Figure 5–23. Petri dish holder with automatic rainshield, used by Lindberg and Lovett (1985).

method is that the contaminant data must be extremely precise, because small differences in concentration must be used to determine the gradient; the method works best for species with low surface resistance where the gradient is most pronounced. The technique is useful for short vegetation in uniform canopies, but the analogy between eddy diffusivities of contaminant mass and sensible heat breaks down over deep canopies.

3. Eddy Correlation from Aircraft

In this method, eddy correlation is applied to large areas by mounting the sensors on board aircraft. The fluxes are measured at different heights by the aircraft and then extrapolated to the surface to give dry deposition at ground level. The technique can also be used to obtain airborne concentration and flux estimates for constructing small-scale aerometric mass balances for the contaminant in a defined region (see next section).

The method applies to spatial scales of at least tens of square kilometers due to flight path considerations. Typical time scales are on the order of several minutes to a few hours averaging time.

The main advantage of the technique is that conventional eddy correlation can be extended to considerable horizontal and vertical distances and can be used to identify variations in the flux over these distances. The method permits eddy correlation to be used over oceans, swamps, and other surfaces not amenable to the use of towers. Furthermore, the method can be applied to heterogeneous terrain for surfaces where the scale of the inhomogeneities is smaller than that of the flight path.

The principal disadvantage is that sensors with response times faster than those used in tower-based eddy correlation are needed, due to the speed of the aircraft relative to the wind. Measurements are necessarily limited to short time periods corresponding to the duration of flights. Because the aircraft must maintain a reasonable height above the surface, the method is often used in conjunction with tower-based sensors.

4. Aerometric Mass Balance Studies

In this technique, a balance is defined between the amount of a contaminant produced or flowing into a region and the amount lost or flowing out. Aircraft sampling is usually conducted along the perimeter of the region of interest, defining inflows and outflows. Sources of the contaminant and loss by chemical reaction are quantified. The loss by dry deposition is then determined by difference.

The method applies to regions at least 100 km on a side so that the dry deposition loss is measurable. Time scales are determined by the transit times across the region by wind, typically a few hours or more.

The principal advantage of the method is that dry deposition estimates over large regions or even continents can be obtained. The technique is most usable for species with atmospheric reaction rates that are slow compared with the transit times across the region.

The main problem is that a considerable effort is required to obtain the necessary input data for a mass balance. Extensive aircraft sampling with sensitive instruments results in high expenses for implementation of the method.

5. Multiple Artificial Tracers

Airborne concentrations of a contaminant will decrease with distance downwind of a source due to dry deposition, dispersion of the plume, and atmospheric chemical reactions. This technique focuses on contaminants for which the last process is either negligible or can be accounted for. The method involves release of a nondepositing, nonreacting tracer simultaneously with the contaminant of interest. The decrease in tracer concentration (and hence contaminant concentration) due to dispersion is obtained, and thus the decrease in contaminant concentration attributed to dry deposition can be determined.

The method is useful over distances as small as 100 m or as great as 100 km, depending on the species of interest and experimental design. Time scales from 1 hour to 1 day are applicable.

The tracer method offers the advantage of studying specific details of the dry deposition process and can provide reasonably accurate deposition data when loss rates are high. However, the method is more suitable for intensive field studies than for routine dry deposition monitoring. The technique cannot be used easily over heterogeneous terrain.

6. Eddy Accumulation

This method relies on the difference in contaminant concentration in updrafts and in downdrafts to infer the flux. Two integrating samplers, such as filters, are set up; one operates only when the vertical wind velocity is directed downward, and the other operates when the wind is directed upward. This is achieved using two independent sampling systems controlled by a fast-response vertical anemometer. After sufficient material has been collected, the filters are chemically analyzed.

The spatial scales for the method are similar to those of eddy correlation and gradient techniques. The minimum time resolution is determined by the amount of sampling needed for high-precision chemical analysis. The method has been used with 1-week sampling times.

Eddy accumulation permits direct measurement of the flux over reasonably long averaging times for species that do not have fast-response sensors available. Analyses can be conducted for several chemical species of interest using a single pair of filters. The main problem is that accurate flow controllers and precise chemical analyses are necessary. As with other micrometeorological methods, there must be adequate fetch and uniform canopy characteristics.

7. Variance

This method incorporates fluctuations in contaminant concentration and fluctuations in temperature or humidity to estimate contaminant fluxes. When the sensible heat flux F_H is known, the contaminant flux may be inferred as $F = F_H\sigma_C/\sigma_T$, where σ_C is the standard deviation of the contaminant concentration and σ_T is the standard deviation of the temperature. When the water vapor flux is known, the contaminant flux may be inferred as $F = F_W\sigma_C/\sigma_W$, where σ_W is the standard deviation of the water vapor concentration. These relations make use of the fact that turbulent fluctuations in temperature, humidity, and contaminant mass concentration are highly correlated in the atmosphere. Several other variance approaches have also been devised.

The spatial scales are similar to those of the other micrometeorological methods. Typical averaging periods are about 1 hour.

The main advantage of the variance method is that estimates of the flux are obtained without the need for fast-response wind sensors, as needed in eddy correlation and eddy accumulation. Commercial sensors are available for several species of interest, including SO_2, SO_4^{2-}, NO_2, and O_3. However, these contaminant sensors must be fast-response and are therefore expensive and difficult to maintain. Another disadvantage is that fluctuations in concentration may be due to instrument noise as well as atmospheric turbulence and lead to overestimation of the contaminant flux. The usual fetch and uniform canopy criteria for micrometeorological methods apply.

In general, atmospheric flux methods offer certain advantages over surface analysis techniques in that the flux estimates pertain to short periods of time; in some cases, direct measurement of the flux is possible. However, all of the methods in this category involve measurement of airborne contaminant concentrations, requiring sensors and/or sampling systems that may be expensive and not always reliable. Furthermore, many of the methods suffer from data interpretation problems. Because none of the methods in either category can be used successfully in all situations, it is generally acknowledged that a combination of techniques may be needed to maximize chances of arriving at reliable dry deposition estimates in the field.

We now consider dry deposition data reported for a variety of contaminants, using the measurement methods outlined above.

V. Dry Deposition Data and Comparison with Model Predictions

A large number of dry deposition experiments have been reported in the literature. Review papers by McMahon and Denison (1979), Sehmel (1980), and others have summarized data obtained prior to the late 1970s. In this section, we review experimental studies published in the interval up to early 1987. First we list dry deposition velocity data obtained in field and laboratory experiments for several contaminant species; then we use some of the models discussed in section III, B to predict particle deposition to various surfaces and briefly compare the predictions with field data.

A. Dry Deposition Data

Measurement of dry deposition in the field and in laboratories over the past 10 years has involved primarily six categories of contaminants. These include sulfur species, nitrogen species, chloride species, ozone, trace elements, and atmospheric particles. A very limited amount of data has also been reported for other species such as CO_2, CO, water vapor, and radioisotopes, although data for these latter species will not be discussed here.

1. Sulfur Species

Dry deposition data for sulfur may be conveniently divided into three categories: gaseous SO_2, particulate SO_4^{2-}, and other sulfur-containing vapors. Examples of data for each category are given in Tables 5–9, 5–10, and 5–11, respectively.

The data for SO_2 show deposition velocities ranging from near zero to 3.4 cm/s. The wide variation in reported deposition velocities is due to differences in rates of aerodynamic transport to the surface as well as variations in surface characteristics. Of particular importance is the stomatal resistance expressed by Equation 23. Measurements involving daytime sampling over vegetation are expected to have higher v_d values due to open stomata, whereas those at night are expected to have lower values. In addition, differences in measurement methods may be significant. It is important to note that values of v_d reported in the field studies include the effects of aerodynamic as well as boundary layer and surface resistance, whereas

Table 5-9. Dry deposition data for sulfur dioxide.

References	v_d (cm/s)	Number of samples averaged	Type of measurement	Time of exposure	Date
Barrie and Walmsley (1978)	0.25 ± 0.20 (stable)		Snow sampling		Mar. 3-9, 1976
Cadle et al. (1985)	0.057 ± 0.025 (cold)	8	Snow sampling	7 days	Winter, 1982-84
	0.15 ± 0.13 (warm)	8			
Dasch and Cadle (1986)	0.082 ± 0.062	9	Snow sampling	2 weeks	Dec. 18, 1983-Apr. 6, 1984
	0.12 ± 0.11	6	Cutoff polyethylene bucket with snow/water	A few days	
	0.69 ± 0.37	5	Cutoff polyethylene bucket with water	A few days	
Davies and Mitchell (1983)	0.04-3.4		Gradient over grass $h = 4$-20 cm	5-10 hr	June 1977-Feb. 1978
	0.74 ± 0.38 (summer, day, dry)				
	1.3 ± 0.43 (summer, day, wet)				
	0.45 ± 0.39 (summer, night, dry)				
	1.0 ± 0.94 (summer, night, wet)				
	0.54 ± 0.38 (winter, day, dry)				
	0.95 ± 0.51 (winter, day, wet)				
	0.42 ± 0.26 (winter, night, dry)				
	0.41 ± 0.27 (winter, night, wet)				
Davis and Wright (1985)	<0.1-5		Gradient over grass $z = 0.75, 1.5, 5.25, 5.8$ m	30 min	June 1982
Dolske and Gatz (1985)	2.1		Gradient over grass	30 min	June 1982
Fowler (1978)	0.1-1.5		Gradient over wheat $z < 2$ m	1-2 hr	

Table 5-9. (*Continued*)

References	v_d (cm/s)	Number of samples averaged	Type of measurement	Time of exposure	Date
Fowler and Cape (1983)	0.35	73	Eddy correlation over Scots pine	20 min	Oct–Nov. 1981
	(0.15–0.96)		$z = 11$ m		
	0.2–1.0 (day, dry)		$h = 9$ m		
	0.1–0.6 (day, wet)				
	0.01–0.2 (night, dry)				
	0.1–0.4 (night, wet)				
Granat and Johansson (1983)	<0.09		Chamber with spruce or pine		
	0.1		Chamber with snow		
Hicks et al. (1983)	1.0 ± 0.26		Eddy correlation over wheat	25 min	June 1982
Japar et al. (1985)	0.52 ± 0.16		Nylon filter	5 days	Aug. 8–25, 1983
Johansson et al. (1983)	0.5 (summer)				
	0.1 (winter)		Chamber with coniferous tree		
John et al. (1985)	0.27 ± 0.04		Foliar extraction of *Ligustrum japonicum* leaves	4 days	Aug. 9–12, 1983
	0.25 ± 0.04		Foliar extraction of *Ligustrum ovalifolium* leaves		
	<0.03		Artificial leaves in ambient atmosphere $z = 50$ cm		

	<0.10		Foliar extraction of *Ligustrum japonicum* leaves		Aug. 21–25, 1983
	<0.04		Foliar extraction of *Ligustrum ovalifolium* leaves		
	<0.05		Artificial leaves in ambient atmosphere $z = 2$ m		
Lorenz and Murphy (1985)	≃0.72 ± 0.65		Gradient over loblolly pine $z = 9.7, 10.7, 12.7,$ 15.8 m above forest floor; $h = 8.5$ m	1 hr	June 1982– May 1983
Milne et al. (1979)	0.008 ± 0.23		Chamber with soil and grass	20 min	May 1983
	0.11 ± 0.58		Chamber with grass		
Mulawa et al. (1986)	0.15 ± 0.11 (dew)		Teflon plate $z = 1$ m	17 hr	June 1981, July 1983
Platt (1978)	0.7–2.1		Gradient $z = 30, 80, 120$ m	1–2 hr	May 1975– Sept. 1976
Sprugel and Miller (1979)	0.5–0.9	19	Open-air fumigation with growing soybean	≃4.1 hr	July 20– Aug. 27, 1978
			Chamber with:		
Taylor et al. (1983)	0.32		*P. vulgaris*		
	0.42		*G. max*		
	0.27		*L. esculentum*		
	1×10^{-4}		Teflon		
Wesely et al. (1983b, 1985)	< 0.5		Eddy correlation over lush grass	30 min	Sept. 1981, June 1982

Table 5-10. Dry deposition data for sulfate.

References	v_d (cm/s)	MMD (μm)	Number of samples averaged	Type of measurement	Time of exposure	Date
Bytnerowicz et al. (1987)	0.15		7	Foliar extraction of *Ceanothus crassifolius* leaves	2 weeks	May–Oct. 1985
	0.50		7	Nylon filter over *Ceanothus crassifolius* canopy		
	0.11		7	Polycarbonate petri dish over *Ceanothus crassifolius* canopy		
Dasch and Cadle (1985)	0.63 ± 0.48 0.29 (summer) 0.76 (fall) 0.71 (winter) 0.83 (spring)			Polyethylene bucket	5 days	June 15, 1981–July 24, 1982
Davidson et al. (1985b)	0.22 ± 0.11 (0.17–0.42)	0.5–1.0	5	Teflon plate over grass	2 days	June 1982
	0.35 ± 0.17 (0.18–0.61)	0.5–1.0	5	Petri dish over grass	2 days	
Davidson et al. (1985c)	0.03 ± 0.01		1	Snow sampling	3 days	May 10–13, 1983
Dolske and Gatz (1982)	0.19 ± 0.01		14	Teflon plate over grass $z = 40$ cm	2–3 days	Sept. 1981
	0.36 ± 0.19		17	Polycarbonate petri dish over grass $z = 40$ cm		

	0.91–0.35		15	Polyethylene bucket over grass $z = 1.5$ m		
Dolske and Sievering (1979)	0.5	0.1–2.0		Diabatic drag coefficient over water		
Doran and Droppo (1983)	<0.4			Gradient over grass $z = 0.75, 1.5, 3.0, 6.0$ m	1.5–2 hr	June 1982
Droppo (1980)	0.1 (winter)			Gradient over vegetation		
Everett et al. (1979)	1.2 ± 0.3[a]		15	Gradient over grass	2 hr	July 6–13, 1976
	1.7 ± 0.8[a]		12	$z = 11.5$ and 34.5 m		
	1.4 ± 0.4 (overall)[a]		27			
Feely et al. (1985)	0.45 ± 0.31 (0.1–1.2)		18	Polyethylene bucket over grass	1 day	June 1982
Garland (1978)	0.03–0.56	0.4–1		Gradient over grass		
Gravenhorst et al. (1983)	0.7–1.5 (summer) 0.3–0.9 (winter)			Throughfall in *Fagus silvatica* canopy		
	1.0–2.0 (summer) 0.8–1.8 (winter)			Throughfall in *Picea abies* canopy		
Hicks and Wesely (1982)	<0 (night)[a] 0.7 (day)[a] (0.41–1.44) 0.5–2.0 (day)[b]			Eddy correlation over loblolly pine $z = 23$ m		July 1977
Hicks et al. (1983)	0.35 ± 0.13[a] 1.2 ± 0.26[b]			Eddy correlation over wheat	25 min	Sept. 1979
Hicks et al. (1986b)	≃0.4 (day)			Eddy correlation		
Hofken and Gravenhorst (1982)	1.1 (summer)			Throughfall in *Fagus silvatica* canopy		
Hofken et al. (1983)	1.1 ± 0.4	0.63		Throughfall in beech canopy		1980

Table 5-10. *(Continued)*

References	v_d (cm/s)	MMD (μm)	Number of samples averaged	Type of measurement	Time of exposure	Date
	1.5 ± 0.5	0.62		Throughfall in spruce canopy $h = 25$–30 m		
Ibrahim et al. (1983)	0.039 (stable)	0.7	4	Snow sampling		Feb. 26, 1980–Mar. 14, 1981
	0.096 (unstable)	0.7	4			
	0.056 (unstable)	7	8			
	0.012 (stable)	0.7	3	Sangamo collector at 1.5 m		
	0.031 (unstable)	7	3			
	3.0 (unstable)	7	5			
	0.005 (stable)	0.7	3	Harwell collector with Whatman 41 filter at 1.5 m		
	0.007 (unstable)	0.7	3			
	0.104 (unstable)	7	5			
	0.018 (stable)	0.7	2	O.M.E. collector with Teflon-coated funnel at 1.0 m		
	0.008 (unstable)	0.7	2			
	0.54 (unstable)	7	5			
	0.014 (stable)	0.7	2	Polyethylene funnel at 0.8 m		
	0.007 (unstable)	0.7	3			
Japar et al. (1985)	0.053			Teflon-coated aluminum plate	5 days	Aug. 8–25, 1983
Johannes and Altwicker (1983)	0.87 (0.15–2.43)			Polyethylene bucket	7 days	
John et al. (1985)	<0.1	80% >2.5		Artificial leaves in ambient atmosphere $z = 50$ cm	4 days	Aug. 9–12, 1983

	<0.14	90% >2.5		Foliar extraction of *Ligustrum japonicum* leaves	4 days	Aug. 21–25, 1983
	<0.06	90% >2.5		Foliar extraction of *Ligustrum ovalifolium* leaves		
	<0.05	90% >2.5		Artificial leaves in ambient atmosphere $z = 2$ m		
Lindberg and Harriss (1981)	0.09	0.6	1	Polyethylene petri dish in oak canopy	7 days	May 9–16, 1977
	0.13	0.6	1	Foliar extraction of oak leaves		
	0.13 ± 0.04	0.6	4	Polyethylene petri dish in oak canopy	4–7 days	May–July, 1977
Lindberg and Lovett (1985)	0.13 ± 0.02		15	Polycarbonate petri dish in oak canopy $z = 15.2, 18.0, 18.3, 19.8$ m	80–340 hr	July 1981–Oct. 1982
Lovett and Lindberg (1984)	0.60 (growing season) 0.38 (dormant season)			Throughfall in *Quercus prinus* canopy		
Mulawa et al. (1986)	0.10 ± 0.09			Teflon surface $z = 1$ m	17 hr	June 1981–July 1983
Neumann and den Hartog (1985)	0.33 (unstable, day)[b] 0.21 (stable, day)[b]			Eddy correlation over grass		June 1982
Shanley (1989)	0.18–1.0			Polyethylene petri dish $z = 2$ m	7 days	Sep.–Nov. 1983
	0.09–1.2			Foliar extraction of spruce leaves		

Table 5-10. *(Continued)*

References	V_d (cm/s)	MMD (μm)	Number of samples averaged	Type of measurement	Time of exposure	Date
Sievering et al. (1979)	0.2 ± 0.16 (stable) (0–0.51)			Aerometric mass balance over water	5 days	May 18–20, 1977
Sievering (1986)	1.0 ± 0.5 (stable, night)	0.3–0.7		Gradient over wheat $z = 5.5$, 15.8 m		June 1982
	2.9 ± 1.1 (unstable, day)	0.3–0.7				
Smith and Friedman (1982)	0.04–0.29			Teflon sheet		
Vandenberg and Knoerr (1985)	0.14 ± 0.03			Cellulose-glass filter paper	Average 86 hr (24–143 hr)	Jan. 1981–Apr. 1982
	0.10 ± 0.03			Polycarbonate membrane		
	0.03 ± 0.01			Polystyrene petri dish		
	0.31 ± 0.09			Polystyrene petri dish with cellulose filter		
	0.11 ± 0.03			Polyethylene bucket (inside surface)		
	0.01 ± 0.003			Polyethylene bucket (outside surface)		
	0.03 ± 0.01			Teflon-coated steel, upfacing		
	0.01 ± 0.003			Teflon-coated steel, downfacing		
	0.03 ± 0.01			Solid Teflon, upfacing		
	0.02 ± 0.01			Solid Teflon, downfacing		

			Eddy correlation (z = 5–10 m) over:
Wesely et al. (1983b; 1985)	0.18 ± 0.02 (unstable)	12	Short mixed pasture
	0.18 ± 0.02 (stable)	6	Grass, drought
	0.26 ± 0.11 (unstable)	8	Grass, well watered
	0.53 ± 0.18 (strongly unstable)	12	Lush grass, well watered
	0.41 ± 0.05 (moderately unstable)	50	Lush grass, well watered
	0.10 ± 0.07 (stable)	10	Lush grass, well watered
	0.76 ± 0.08 (unstable)	18	Pine forest, drought
	−0.05 ± 0.24 (unstable)	14	Deciduous forest, winter
Wesely and Hicks (1979)	0.5		Eddy correlation over loblolly pine canopy

[a]Particulate sulfur.

[b]Total sulfur.

Table 5-11. Dry deposition data for gaseous sulfur compounds.

References	Species	v_d (cm/s)	Type of measurement	Time of exposure
Cope and Spedding (1982)	H_2S	0.05–0.17 (summer)	Chamber with *Pinus radiata*	30 min
		0.03–0.12 (winter)	Chamber with *Pinus radiata*	
		0.11–0.17	Chamber with *Lolium perenne*	30 min
Kluczewski et al. (1985)	COS	3.1×10^{-4}	Chamber with soil	
		$\sim 5.7 \times 10^{-4}$		
			Chamber with:	$\simeq$
Taylor et al. (1983)	H_2S	0.20	*P. vulgaris*	
		0.36	*G. max*	
		0.17	*L. esculentum*	
		6×10^{-5}	Teflon	
	COS	0.14	*P. vulgaris*	
		0.31	*G. max*	
		0.04	*L. esculentum*	
		2×10^{-5}	Teflon	
	CH_3SH	0.06	*P. vulgaris*	
		0.23	*G. max*	
		0.06	*L. esculentum*	
		5×10^{-5}	Teflon	
	CS_2	0.06	*P. vulgaris*	
		0.12	*G. max*	
		0.02	*L. esculentum*	
		1×10^{-5}	Teflon	

the values reported in chamber studies include only the last two resistances plus any transport resistance specific to the chamber.

Several studies in Table 5–9 have used micrometeorological techniques over natural vegetation in the field (eddy correlation and gradient) and report an average deposition velocity or a range of values for each set of measurements. If a separate average v_d is computed for each range of values, and these are combined with the average deposition velocities reported in the table, there are a total of 16 average v_d values in this category. The overall average and standard deviation of these values is 0.95 ± 0.62 cm/s. Similarly, there are four values corresponding to field measurements of v_d for snow, with an overall average and standard deviation of 0.13 ± 0.09 cm/s.

Table 5–10 summarizes deposition velocities for SO_4^{2-}. The table shows values of the mass median aerodynamic diameter (MMD) for the size distributions of SO_4^{2-}-containing particles, if reported in the original references. Most of the data in the table refer to particulate SO_4^{2-}. Values marked with an [a] include all forms of particulate sulfur, such as SO_4^{2-} and SO_3^{2-}. Values marked with a [b] include all forms of gaseous and particulate sulfur, such as SO_2, H_2S, SO_4^{2-}, SO_3^{2-}, and other species. Variations in the way each group reported their findings may result in additional differences. For example, Bytnerowicz and others (1987) reported v_d for SO_4^{2-} onto a nylon filter, whereas Japar and others (1985) considered the influence of both SO_2 and SO_4^{2-} deposition onto the same type of surface.

The deposition velocities for SO_4^{2-} listed in Table 5–10 have been combined with data from earlier published studies and graphed in Figure 5–24. Note that the values reported in the figure and table span three orders of magnitude.

The data may be categorized by measurement method, and overall average deposition velocities can be computed following the technique applied to Table 5–9. Only field data for ambient SO_4^{2-} given in Table 5–10 have been used. Results of these calculations are: micrometeorological methods 0.55 ± 0.65 cm/s (N = 20), surrogate surface exposures 0.26 ± 0.25 cm/s (N = 26), foliar extraction 0.23 ± 0.24 cm/s (N = 5), and throughfall 1.0 ± 0.41 cm/s (N = 9); N refers to the number of values averaged. Upper limit values listed in the table were included along with the other values in the computations. Only values for particulate SO_4^{2-} were used, however; data with [a] and [b] footnotes were excluded. Also excluded were data from downward-facing surrogate surfaces and the value of v_d for the outside of a dustfall bucket.

It is interesting to compare the average values for each measurement category. The surrogate surface and foliar extraction values are in agreement, although the large standard deviations indicate that there is great variability within each category. One would not necessarily expect close agreement: Lindberg and Lovett (1985) observed differences in v_d between simultaneous petri dish and foliar extraction sampling and attributed the results to differences in particle collection efficiencies, biological uptake of dry-deposited material on leaves, and differences in SO_2 uptake. Comparing the surrogate surface/foliar extraction and micrometeorological average values may be somewhat tenuous. Several investigators have shown that dry-deposited particulate material on surrogate surfaces

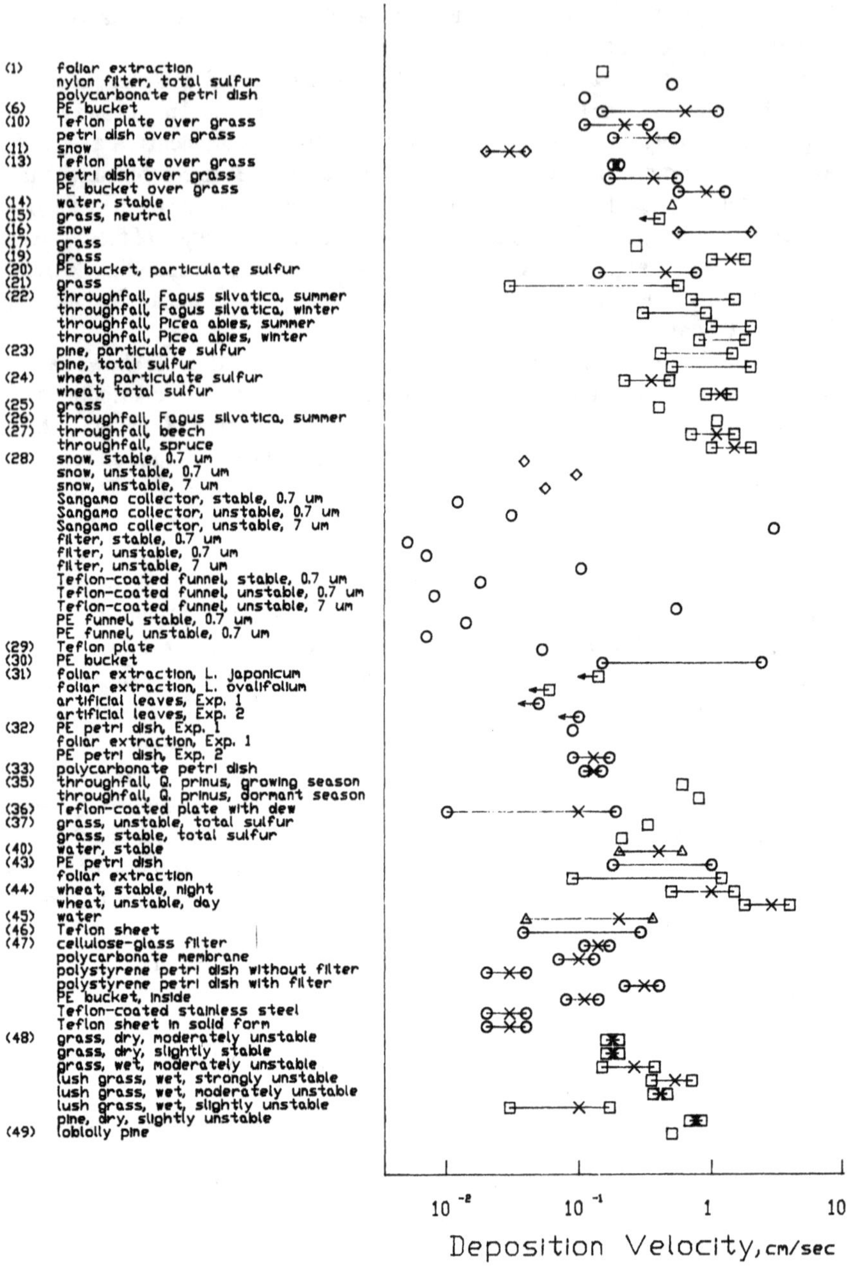
(1) foliar extraction
nylon filter, total sulfur
polycarbonate petri dish
(6) PE bucket
(10) Teflon plate over grass
petri dish over grass
(11) snow
(13) Teflon plate over grass
petri dish over grass
PE bucket over grass
(14) water, stable
(15) grass, neutral
(16) snow
(17) grass
(19) grass
(20) PE bucket, particulate sulfur
(21) grass
(22) throughfall, Fagus silvatica, summer
throughfall, Fagus silvatica, winter
throughfall, Picea abies, summer
throughfall, Picea abies, winter
(23) pine, particulate sulfur
pine, total sulfur
(24) wheat, particulate sulfur
wheat, total sulfur
(25) grass
(26) throughfall, Fagus silvatica, summer
(27) throughfall, beech
throughfall, spruce
(28) snow, stable, 0.7 um
snow, unstable, 0.7 um
snow, unstable, 7 um
Sangamo collector, stable, 0.7 um
Sangamo collector, unstable, 0.7 um
Sangamo collector, unstable, 7 um
filter, stable, 0.7 um
filter, unstable, 0.7 um
filter, unstable, 7 um
Teflon-coated funnel, stable, 0.7 um
Teflon-coated funnel, unstable, 0.7 um
Teflon-coated funnel, unstable, 7 um
PE funnel, stable, 0.7 um
PE funnel, unstable, 0.7 um
(29) Teflon plate
(30) PE bucket
(31) foliar extraction, L. japonicum
foliar extraction, L. ovalifolium
artificial leaves, Exp. 1
artificial leaves, Exp. 2
(32) PE petri dish, Exp. 1
foliar extraction, Exp. 1
PE petri dish, Exp. 2
(33) polycarbonate petri dish
(35) throughfall, Q. prinus, growing season
throughfall, Q. prinus, dormant season
(36) Teflon-coated plate with dew
(37) grass, unstable, total sulfur
grass, stable, total sulfur
(40) water, stable
(43) PE petri dish
foliar extraction
(44) wheat, stable, night
wheat, unstable, day
(45) water
(46) Teflon sheet
(47) cellulose-glass filter
polycarbonate membrane
polystyrene petri dish without filter
polystyrene petri dish with filter
PE bucket, inside
Teflon-coated stainless steel
Teflon sheet in solid form
(48) grass, dry, moderately unstable
grass, dry, slightly stable
grass, wet, moderately unstable
lush grass, wet, strongly unstable
lush grass, wet, moderately unstable
lush grass, wet, slightly unstable
pine, dry, slightly unstable
(49) loblolly pine
10^-2
10^-1
1
10
Deposition Velocity, cm/sec

and natural vegetation is dominated by the small fraction of large airborne particles, as these particles have the greatest deposition velocities (e.g., Davidson et al., 1982; Garland, 1983; Coe and Lindberg, 1987). Thus the surrogate surface data in Table 5–10 probably reflect deposition from the upper end of the SO_4^{2-} size distribution onto the collectors. The eddy correlation and gradient data, however, pertain to submicron particle fluxes from the atmosphere onto the vegetative

Figure 5–24. Measured dry deposition velocities for SO_4^{2-} reported in the literature. Vegetative canopies are shown as squares, surrogate surfaces as circles, water surfaces as triangles, and snow surfaces as diamonds. Where available for each study, the data show the mean value (indicated as an *X*) and the values one standard deviation above and below the mean (indicated with the surface-specific symbols listed above). A horizontal bar shown without an *X* indicates that no mean value was given in the original reference; the length of the bar indicates the range of values. Data shown as a single point refer to individual measurements, or to an average value where no standard deviation was given. Single points with arrows refer to upper limit values. The data presented include all studies given in Table 5–10 plus data reported in the literature prior to the late 1970s. Each study includes a reference number from the list below.

Ref.	Author(s)	Ref.	Author(s)
1	Bytnerowicz et al. (1987)	26	Hofken and Gravenhorst (1982)
2	Cawse (1974)	27	Hofken et al. (1983)
3	Cawse (1975)	28	Ibrahim et al. (1983)
4	Cawse (1976)	29	Japar et al. (1985)
5	Cawse (1977)	30	Johannes and Altwicker (1983)
6	Dasch and Cadle (1985)	31	John et al. (1985)
7	Davidson (1977)	32	Lindberg and Harriss (1981)
8	Davidson and Elias (1982)	33	Lindberg and Lovett (1985)
9	Davidson et al. (1985a)	34	Little and Wiffen (1977)
10	Davidson et al. (1985b)	35	Lovett and Lindberg (1984)
11	Davidson et al. (1985c)	36	Mulawa et al. (1986)
12	Dedeurwaerder et al. (1983)	37	Neumann and den Hartog (1985)
13	Dolske and Gatz (1982)	38	Pattenden et al. (1982)
14	Dolske and Sievering (1979)	39	Peirson et al. (1973)
15	Doran and Droppo (1983)	40	Prahm et al. (1976)
16	Dovland and Eliassen (1976)	41	Rohbock (1982)
17	Droppo (1980)	42	Servant (1974)
18	El-Shobokshy (1985)	43	Shanley (1988)
19	Everett et al. (1979)	44	Sievering (1986)
20	Feely et al. (1985)	45	Sievering et al. (1979)
21	Garland (1978)	46	Smith and Friedman (1982)
22	Gravenhorst et al. (1983)	47	Vandenberg and Knoerr (1985)
23	Hicks and Wesely (1982)	48	Wesely et al. (1983b; 1985)
24	Hicks et al. (1983)	49	Wesely and Hicks (1979)
25	Hicks et al. (1986)	50	Young (1978)

canopies. Because of the different surface characteristics and the different particle sizes responsible for deposition, one would not necessarily expect similar values of v_d. The average value for throughfall is greater than for any of the other categories; this may reflect deposition of SO_2 in addition to SO_4^{2-} onto the canopies, as well as differences in canopy characteristics and time periods of sampling.

Table 5–11 gives deposition velocities for three chamber studies involving gaseous sulfur compounds. The values range from near zero for several species depositing onto Teflon to 0.36 cm/s for H_2S depositing onto vegetation.

2. Nitrogen Species

Table 5–12 summarizes deposition velocities for NO_2, NO_x, NO_3^-, HNO_3, NH_3 and NH_4^+ in field and chamber experiments. Note the large values of v_d for HNO_3 in several of the field studies, reflecting values of r_s near zero for most surfaces. Deposition velocities for NO_3^- particles appear to be somewhat larger than corresponding values for SO_4^{2-}; this may reflect the larger particle sizes associated with NO_3^- (Voldner et al., 1986; Milford and Davidson, 1987).

3. Chloride Species

Table 5–13 summarizes a limited number of studies that have examined fluxes of chloride from the ambient atmosphere. These results show deposition velocities in the range 1.0 to 5.1 cm/s for chloride-containing particles and a value of 0.73 cm/s for HCl deposition to dew. Overall, the values of v_d for particulate chloride are fairly large, probably reflecting the association of chloride with large particles. Note that the highest values are reported in winter when road salt is expected to be a major contributor of chloride aerosol. Dasch and Cadle (1985) have noted maximum chloride fluxes in winter due to road salt.

4. Ozone

Several studies have used micrometeorological techniques to estimate O_3 fluxes to vegetative canopies. Some examples are shown in Table 5–14. Deposition velocities are in the range 0 to 1.5 cm/s; the overall average of 15 values in this table is 0.39 ± 0.21 cm/s, computed in the same manner as for SO_2 and SO_4^{2-}. Note that the deposition velocities determined by the gradient method in the Doran and Droppo studies (Doran and Droppo, 1983; Droppo and Doran, 1983; Droppo, 1985) have been computed using both K_M and K_H; the latter is in better agreement with measurement by eddy correlation obtained simultaneously, and hence v_d based on momentum has not been included in the average. Values for the nighttime hours are often smaller than those during the day, due to increased aerodynamic resistance (stable atmosphere conditions) and increased surface resistance (closed stomata) at night. However, Wesely and others (1982) point out that O_3 is much more easily removed by soil and plants with closed stomata than NO_2, suggesting that the differences in v_d between day and night are expected to be less for O_3 than for NO_2. This is apparent when the data for Wesely and others in Tables 5–12 and 5–14 are compared.

Table 5-12. Dry deposition data for nitrogen species.

References	Species	v_d (cm/s)	MMD (μm)	Number of samples averaged	Type of measurement	Time of exposure	Date
Bytnerowicz et al. (1987)	NH_4^+	0.44			Foliar extraction of *Ceanothus crassifolius* leaves	2 weeks	May–Oct. 1985
	NO_3^-	0.41					
	NH_4^+	0.60			Nylon filter over *Ceanothus crassifolius* canopy		
	NO_3^-	0.68					
	NH_4^+	0.07			Polycarbonate petri dish over *Ceanothus crassifolius* canopy		
	NO_3^-	0.13					
Cadle et al. (1985)	NH_4^+	0.10 ± 0.11	0.46	21	Snow sampling	7 days	Winter 1982–84
	HNO_3	1.4 ± 1.03		21			
Dasch and Cadle (1985)	NO_3^-	0.69 ± 0.35 0.48 (summer) 0.70 (fall) 0.89 (winter) 0.76 (spring)			Polyethylene bucket	5 days	June 15, 1981–July 24, 1983
Dasch and Cadle (1986)	NH_4^+	0.083 ± 0.083	0.43	9	Snow sampling	2 weeks	Dec. 18, 1983–Apr. 6, 1984
		0.13 ± 0.17	0.43	6	Cutoff polyethylene bucket with snow/water	A few days	
		0.33 ± 0.18	0.43	5	Cutoff polyethylene bucket with water	A few days	
	HNO_3	2.0 ± 1.5		9	Snow sampling	2 weeks	
		2.6 ± 7.1		6	Cutoff polyethylene bucket with snow/water	A few days	

Table 5-12. (*Continued*)

References	Species	v_d (cm/s)	MMD (μm)	Number of samples averaged	Type of measurement	Time of exposure	Date
		6.2 ± 4.8		5	Cutoff polyethylene bucket with water	A few days	
Delany (1983)	NO_x	0.30 ± 0.14 (0.1–0.6)		10	Gradient over cut grass $z = 5$ cm	25 min	March 1982
Duyzer et al. (1983)	NO_x	−2.6–1.5 (afternoon)		31	Gradient over pasture $z = 0.4$–2.5 m $h = 10$ cm	30 min	Aug.–Dec. 1982
Granat and Johansson (1983)	NO_x	<0.09			Chamber with spruce or pine		
		<0.03			Chamber with snow		
Gravenhorst et al. (1983)	NH_4^+	0.6–1.3 (summer) 0.2–0.8 (winter)			Throughfall in *Fagus silvatica* canopy		
	NO_3^-	0.7–1.7 (summer) 0.6–1.6 (winter)					
	NH_4^+	0.7–2.1 (summer) 0.6–1.6 (winter)			Throughfall in *Picea abies* canopy		
	NO_3^-	1.1–3.7 (summer) 1.3–3.2 (winter)					
Grennfelt et al. (1983)	NO_2	0.4–0.5			Chamber with Scots pine		
Hofken and Gravenhorst (1982)	NH_4^+	1.0 (summer)			Throughfall in *Fagus silvatica* canopy		
	NO_3^-	1.3 (summer)					
Hofken et al. (1983)	NO_3^-	1.3 ± 0.5	2.4		Throughfall in beech canopy		

		2.5 ± 1.6	2.4		Throughfall in spruce canopy		1980
	NH_4^+	1.0 ± 0.3	0.70		Throughfall in beech canopy		
		1.4 ± 0.7	0.70		Throughfall in spruce canopy $h = 25$–30 m		
Huebert (1983) Huebert and Robert (1985)	HNO_3	2.9 ± 1.1 (1.1–4.9)		21	Modified Bowen ratio over grass $z = 0.75, 5.8$ m $h = 30$ cm	90 min	June 1982
Japar et al. (1985)	HNO_3	2.5 ± 1.5			Nylon filter	5 days	Aug. 8–25, 1983
Johannes and Altwicker (1983)	NO_3^-	18 (1.6–37)			Polyethylene bucket	7 days	
Johansson and Granat (1986)	HNO_3	0 (−0.03–0.03)	(T = −18°C)	10	Chamber with snow	24 hr	
		0.04 (0.02–0.08)	(T = −8°C)	6			
		0.04 (0.01–0.12)	(T = −5°C)	6			
		0.12 (0.09–0.15)	(T = −4°C)	2			
		0.10 (0.07–0.13)	(T = −3°C)	2			
		0.57 (0.47–0.67)	(T = −2°C)	2			
John et al. (1985)	NO_3^-	0.22 ± 0.03	49% >2.5		Foliar extraction of *Ligustrum japonicum* leaves	4 days	Aug. 9–12, 1983
		0.14 ± 0.02	49% >2.5		Foliar extraction of *Ligustrum ovalifolium* leaves		

Table 5-12. (*Continued*)

References	Species	v_d (cm/s)	MMD (μm)	Number of samples averaged	Type of measurement	Time of exposure	Date
		0.18 ± 0.03	49% >2.5		Artificial leaves in ambient atmosphere $z = 50$ cm		
		0.26 ± 0.04	60% >2.5		Foliar extraction of *Ligustrum japonicum* leaves	4 days	Aug. 21–25, 1983
		0.13 ± 0.02	60% >2.5		Foliar extraction of *Ligustrum ovalifolium* leaves		
		0.13 ± 0.02	60% >2.5		Artificial leaves in ambient atmosphere $z = 2$ m		
Lovett and Lindberg (1984)	NO_3^-	0.55 (growing season) 0.71 (dormant season)			Throughfall in *Quercus prinus* canopy		
Mulawa et al. (1986)	HNO_3	0.39 ± 0.31 (dew)			Teflon surface; $z = 1$ m	17 hr	June 1981–July 1983
	NH_3	1.9 ± 1.55 (dew)					
	NO_3^-	0.33 ± 0.22 (dry)					
	NH_4^+	0.06 ± 0.09 (dry)					
Stedman et al. (1987)	NO_x	0.04 ± 0.01			Eddy correlation over wheat	30 min	Jan.–May 1986
Stocker et al. (1987)	NO_2	0.012 ± 0.088			Eddy correlation		June–July 1986
Wesely et al. (1982)	NO_x	0.05 (night) 0.6 (maximum in day)			Eddy correlation over soybean $z = 5$–6 m		Aug. 1979

Table 5-13. Dry deposition data for chloride.

References	v_d (cm/s)	MMD (μm)	Number of samples averaged	Type of measurement	Time of exposure	Date
Dasch and Cadle (1985)	3.1 ± 1.8 2.3 (summer) 2.7 (fall) 5.0 (winter) 2.7 (spring)			Polyethylene bucket	5 days	June 15, 1981–July 24, 1983
Dasch and Cadle (1986)	4.3 ± 6.1	3.8	9	Snow sampling	2 weeks	Dec. 18, 1983–Apr. 6, 1984
	5.1 ± 4.0	3.8	6	Cutoff polyethylene bucket with snow/water	A few days	
	4.9 ± 1.5	3.8	5	Cutoff polyethylene bucket with water	A few days	
Hofken et al. (1983)	1.0 ± 0.4	1.5		Throughfall in beech canopy		1980
	1.9 ± 0.7	1.5		Throughfall in spruce canopy $h = 25$–30 m		
Mulawa et al. (1986)	2.4 ± 1.8 (dry) 0.73 ± 0.11[a] (dew)			Teflon plate $z = 1$ m	17 hr	June 1981–July 1983

[a] HCl.

Table 5-14. Dry deposition data for ozone.

References	v_d (cm/s)	Number of samples averaged	Type of measurement	Time of exposure	Date
Colbeck and Harrison (1985)	0.53 ± 0.22 (0.08–0.91)	76	Gradient over grass $z = 0.25, 0.5, 1, 2$ m $h = 5$–15 cm	24–60 min	Jan.-Oct. 1983
Delany (1983)	0.06–1.0	9	Gradient over cut grass	25 min	Mar. 1982
Delany et al. (1986)	0.6		Eddy correlation over wheat $h = 0.75$ m	20 min	June–July 1983
Doran and Droppo (1983); Droppo and Doran (1983); Droppo (1985)	0.49		Eddy correlation over grass $h = 25$–30 cm	0.5–1 hr	June 1982
	0.49 (heat)		Gradient over grass $h = 25$–30 cm	0.5–1 hr	June 1982
	0.29 (momentum)		Gradient over grass $h = 25$–30 cm	0.5–1 hr	June 1982
Duyzer et al. (1983)	0–1.5 (afternoon)	27	Gradient over pasture $z = 0.4$–2.5 m $h = 10$ cm	30 min	Aug.-Dec. 1982
Hicks and Wesely (1982)	0.5 (day) 0.1 (night)		Eddy correlation over loblolly pine $z = 23$ m		July 1977
Neumann and den Hartog (1985)	0.47 (unstable, day) 0.34 (neutral, day) 0.35 (stable, day)		Eddy correlation over grass		June 1982
Stedman et al. (1987)	0.19		Eddy correlation	30 min	Jan.-May 1986
Stocker et al. (1987)	0.11 ± 0.01 (day) 0.03 ± 0.01 (night)		Eddy correlation		June–July 1986
Wesely et al. (1982)	0.3 (night) 0.8 (maximum in day)		Eddy correlation over soybean $z = 5$–6 m		Aug. 1979

5. Trace Elements

Dry deposition sampling for trace elements has involved primarily surrogate surface techniques. These have included dustfall buckets, petri dishes, filter paper, and flat plates made of Teflon, plexiglass, and polyethylene. In some instances, investigators have employed foliar extraction, throughfall, snow sampling, mass balances, and micrometeorological methods.

Results of a variety of studies are shown in Table 5–15. The elements represented in the table can be conveniently categorized on the basis of the crustal enrichment factor, defined by:

$$EF_{\text{crust}} = \frac{X_{\text{air}}/Al_{\text{air}}}{X_{\text{crust}}/Al_{\text{crust}}} \tag{61}$$

where X_{air} and Al_{air} represent the airborne concentrations of any element X and of aluminum, respectively, and X_{crust} and Al_{crust} are the respective concentrations in the earth's crust. Several investigators have shown that values of EF_{crust} near unity imply the earth's crust as a source. Elements with airborne concentrations that are enriched relative to crustal composition ($EF_{\text{crust}} >> 1$) are likely to have predominantly noncrustal sources, such as anthropogenic activities, natural combustion (volcanism, forest fires), biological emissions, and seaspray (e.g., Duce et al., 1975; Rahn, 1976). Those elements in Table 5–14 having large values of EF_{crust} commonly reported in the literature include Ag, As, Cd, Cu, In, Pb, Sb, Se, and Zn; Ni and V are marginally enriched. The other elements in the table are mainly soil derived.

Several investigators have shown that the enriched elements are associated with smaller particle sizes and generally have smaller dry deposition velocities than crustal elements (e.g., Milford and Davidson, 1985). This is apparent in Table 5–15, despite the variety of measurement techniques represented. Values of v_d for the enriched elements are generally smaller than 1 cm/s, whereas deposition velocities for the crustal elements often exceed this value. As a further illustration of the influence of trace element particle size on v_d, the individual data points from the studies in Table 5–15 have been combined with data from several earlier studies and plotted in Figure 5–25. The graph shows v_d versus MMD for roughly 800 measurements, representing several types of natural and surrogate surfaces. The value of MMD for each data point has been taken from the original reference if given. Otherwise, the overall average MMD for that element has been taken from Milford and Davidson (1985). The graph shows generally increasing values of deposition velocity with increasing MMD. A least-squares regression line through the points is:

$$v_d = 0.388\ \text{MMD}^{0.76} \tag{62}$$

where v_d is in centimeters per second and MMD is in micrometers. As noted earlier, the deposition velocity is most affected by the upper end of the size spectrum; the above relationship reflects the fact that trace elements with large MMD are likely to have a greater fraction of mass associated with coarse particles.

Table 5-15. Dry deposition data for trace elements.

References	Species	v_d (cm/s)	MMD (μm)	Number of samples averaged	Type of measurement	Time of exposure	Date
Cadle et al. (1985)	Ca	2.1 ± 1.8	4.4	24	Snow sampling	7 days	Winter 1982–84
	Mg	1.5 ± 1.3	2.7	23			
	Na	0.44 ± 0.48	1.8	23			
	K	0.51 ± 0.60	0.9	21			
Dasch and Cadle (1985)	Ca	2.3 ± 0.88 (full year)			Polyethylene bucket	5 days	June 15, 1981–July 24, 1983
		2.1 (summer)					
		2.1 (fall)					
		3.2 (winter)					
		1.7 (spring)					
	Mg	1.9 ± 1.0 (full year)					
		1.1 (summer)					
		2.1 (fall)					
		2.7 (winter)					
		2.0 (spring)					
	Na	2.3 ± 1.6 (full year)					
		1.7 (fall)					
		2.9 (winter)					
		2.0 (spring)					
	K	1.9 ± 1.8 (full year)					
		0.96 (summer)					
		0.51 (fall)					
		1.9 (winter)					
		2.4 (spring)					
Dasch and Cadle (1986)	Ca	2.0 ± 1.8	5.4	9	Snow sampling	2 weeks	Dec. 18, 1983–Apr. 6, 1984
		2.7 ± 2.6		6	Cutoff polyethylene bucket with snow/water	A few days	

		4.2 ± 0.69		5	Cutoff polyethylene bucket with water	A few days	
Davidson and Elias (1980, 1982); Elias and Davidson (1980)	K	1.2 ± 0.93		6	Teflon plates in meadow $z = 0.25, 1.5, 10$ m	3 days	June and Aug. 1976, Aug. 1977
	Rb	1.6 ± 1.3	34% >4 (ave.)	6			
	Cs	1.5 ± 1.8		6			
	Ca	1.9 ± 1.3		6			
	Sr	2.3 ± 1.7		6			
	Ba	1.7 ± 1.7		6			
	Pb	0.14 ± 0.13	10% >4	6			
Davidson et al. (1985a)	Al	7.9 ± 2.3	3.9	4	Teflon plates in meadows and forests $z = 1$ m	7 days	July–Aug. 1980
	Ba	2.3 ± 0.79	—	4			
	Ca	7.0 ± 2.4	3.7	4			
	Fe	7.5 ± 1.3	—	4			
	Mg	11.0 ± 6.8	—	4			
	Na	7.4 ± 2.6	6.9	4			
	Ti	5.6 ± 2.7	—	4			
	Ag	0.30 ± 0.23	—	4			
	As	0.18 ± 0.14	—	4			
	Cd	0.20 ± 0.11	0.61	4			
	Cu	0.29 ± 0.26	0.38	4			
	Pb	0.15 ± 0.07	0.87	4			
	Zn	0.40 ± 0.29	0.28	4			
Davidson et al. (1985c)	Cu	0.08 ± 0.04		1	Snow sampling in Arctic	3 days	May 10–13, 1983
	Al	0.2 ± 0.06		1			
	Fe	0.6 ± 0.09		1			
	K	0.05 ± 0.02		1			
	Mg	0.2 ± 0.06		1			
	Mn	0.3 ± 0.1		1			
	Na	0.2 ± 0.02		1			

Table 5-15. *(Continued)*

References	Species	v_d (cm/s)	MMD (μm)	Number of samples averaged	Type of measurement	Time of exposure	Date
Dedeurwaerder et al. (1983)					Vaseline-coated plexiglas over:	12 days	1980 and 1981
	Cu	0.19	0.38	7	Water		
		1.0	1.2	2	Water		
		1.6	0.50	2	Land		
	Zn	0.20	0.20	7	Water		
		0.23	0.66	2	Water		
		0.05	0.22	2	Land		
	Pb	0.41	0.68	7	Water		
		0.43	0.75	2	Water		
		0.19	0.70	2	Land		
	Cd	0.04	0.52	7	Water		
		0.10	1.1	2	Water		
		0.05	0.42	2	Land		
	Fe	1.3	1.2	7	Water		
		1.4	1.3	2	Water		
		0.51	1.7	2	Land		
	Mn	0.49	0.84	7	Water		
		0.77	0.98	2	Water		
		0.27	0.91	2	Land		
	Na	1.7	5.0	7	Water		
		3.9	6.2	2	Water		
		0.30	4.1	2	Land		

Dolske and Sievering (1979)	Al, Ca, Cu, Fe, Mg, Mn, Pb, Ti, Zn	≃0.5	0.1–2.0		Diabatic drag coefficient over water $z = 5$ m		May–Aug. 1977
El-Shobokshy (1985)	Pb			2	Leaf sprayed with oil:		
		0.33 ± 0.03 (stable)			Alfalfa		Feb. 2 and 9, 1984
		0.31 ± 0.02 (unstable)		2	Alfalfa		Mar. 15 and 20, 1984
		0.37 ± 0.04 (stable)		2	Grass		Feb. 12 and 16, 1984
		0.31 ± 0.02 (unstable)		2	Grass		Mar. 29, and Apr. 12, 1984
		0.28 ± 0.05 (stable)		1	Soil		Feb. 19, 1984
		0.34 ± 0.05 (unstable)		1	Soil		Apr. 20, 1984
					Throughfall for:		
Hofken et al. (1983)	Cd	1.8 ± 0.7	0.25		Beech canopy		1980
		2.2 ± 0.7	0.25		Spruce ″		
	Pb	0.9 ± 0.3	0.79		Beech ″		
		1.3 ± 0.5	0.79		Spruce ″		
	Mn	0.7 ± 0.3	1.1		Beech ″		
		0.6 ± 0.3	1.1		Spruce ″		
	Fe	1.0 ± 0.4	2.2		Beech ″		
		2.4 ± 0.9	2.2		Spruce ″		
Lindberg and Harriss (1981)	Cd	0.33	1.5	1	Polyethylene petri dish	7 days	May–July 1977

Table 5-15. (*Continued*)

References	Species	v_d (cm/s)	MMD (μm)	Number of samples averaged	Type of measurement	Time of exposure	Date
		0.23	1.5	1	Foliar extraction: oak	7 days	
		0.37 ± 0.18	1.5	4	Polyethylene petri dish	4–7 days	
	Mn	1.7	3.4	1	Polyethylene petri dish	7 days	
		0.8	3.4	1	Foliar extraction: oak	7 days	
		6.4 ± 3.6	3.4	4	Polyethylene petri dish	4–7 days	
	Pb	0.05	0.5	1	Polyethylene petri dish	7 days	
		0.005	0.5	1	Foliar extraction: oak	7 days	
		0.06 ± 0.01	0.5	4	Polyethylene petri dish	4–7 days	
	Zn	0.66	0.9	1	Polyethylene petri dish	7 days	
		0.46	0.9	1	Foliar extraction: oak	7 days	
		0.38 ± 0.10	0.9	4	Polyethylene petri dish	4–7 days	
Lindberg and Lovett (1983)	K	0.75 ± 0.24		15	Polycarbonate petri dish in oak canopy $h \simeq 20$ m $z = 15.2, 18.0, 18.3, 19.8$ m	80–340 hrs	July 1981–Oct. 1982
	Ca	1.1 ± 0.1		15			
McDonald et al. (1982)	Na	0.8–8.2	0.94–64		Polyethylene plate $z = 10, 20$ m		May 1978, Apr.–July 1979

Mulawa et al. (1986)	Ca	0.23 ± 0.18 (dry)		Teflon plates $z = 1$ m	≃17 hrs (ave.)	June 1981–July 1983
		0.46 ± 0.36 (dew)				
	Mg	0.15 ± 0.12 (dry)				
		0.41 ± 0.20 (dew)				
	Na	0.33 ± 0.34 (dry)				
		0.42 ± 0.67 (dew)				
	K	0.68 ± 0.72 (dry)				
		0.88 ± 0.89 (dew)				
Pattenden et al. (1982)	Pb	0.46		Filter paper	≃30 days	Aug. 1980–Jan. 1981
	Sb	0.47				
	As	0.19				
	Cr	1.5				
	Co	1.4				
	Cu	1.4				
	Fe	1.6				
	Mn	0.43				
	Ni	1.0				
	Se	0.10				
	Ag	0.23				
	V	0.46				
	Zn	0.20				
	Al	2.0				
	In	0.79				
	Sc	1.1				
	Na	0.16				
Rohbock (1982)	Pb	0.06	0.3	Bucket $z = 1$ m	14 days	Summer 1979–1981
	Mn	2.2	2.0			
	Fe	1.6	3.5			
	Cu	1.4	1.8			
	Na	1.5	3.4			

Table 5-15. (*Continued*)

References	Species	v_d (cm/s)	MMD (μm)	Number of samples averaged	Type of measurement	Time of exposure	Date
	Al	1.2	6.5				
	Ca	2.0	6.5				
	Cd	0.3	3.0				
	K	1.2	3.6				
	Co	1.0	5.2				
	Cr	1.0	3.6				
Sievering et al. (1979)	Pb	0.13	82% <1		Aerometric mass balance over water $z = 5$ m	5 hr.	May 18–20, 1977
	Fe	0.65	47% <1				
	Mn	0.55	49% <1				
Sievering (1983)	Al	3.1 ± 2.8 (stable, night) 3.5 ± 2.2 (unstable, day)	>3		Gradient over wheat $h = 0.3$–0.7 m $z = 5.5$ and 15.8 m		June 1982
	Ca	4.6 ± 3.0 (stable, night) 5.3 ± 2.1 (unstable, day)	>3				
	Fe	3.3 ± 1.9 (stable, night) 4.7 ± 2.4 (unstable, day)	>3				

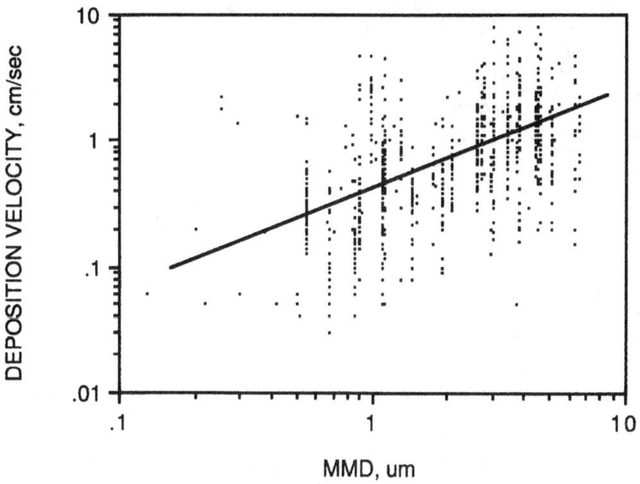

Figure 5–25. Measured dry deposition velocity versus MMD for trace elements, based on data reported in the literature. References from several sources are included in this dataset: all of the studies in Table 5–15, the reference list associated with Figure 5–26, and two additional references (Crecelius et al., 1978; White and Turner, 1970).

In addition to the Table 5–15 data based on measurement methods discussed at the NAPAP Workshop, deposition velocities using other techniques have appeared in the literature. For example, Friedlander and others (1986) have estimated v_d for Pb in the Los Angeles Basin using a modification of the artificial tracer method in which the atmosphere is assumed to be continuously stirred. The authors further assume that ambient CO and Pb result entirely from automobile emissions. The Pb/CO airborne concentration ratio in tunnels, characteristic of the source emissions, is greater than the average value of this ratio in ambient air because Pb is deposited while CO is conserved. Using this information with the average residence time and assuming uniform deposition of Pb throughout the area of the Basin yields the deposition velocity. The method provides an estimated v_d of 0.26 cm/s, in reasonable agreement with other values for Pb in Table 5–15.

As a further illustration of the deposition velocities for Pb, Figure 5–26 presents a summary of values reported in the literature. The figure includes all of the entries for the v_d of Pb included in Figure 5–25. The data are mostly in the range 0.1 to 1 cm/s, with a few values outside of this interval.

6. Particles

Table 5–16 lists the results of five field studies involving the deposition of submicron particles. All of the studies used micrometeorological methods to estimate v_d. The deposition velocities are generally less than 1 cm/s, in agreement with values of v_d in previous tables for submicron chemical species such as SO_4^{2-} and enriched trace elements.

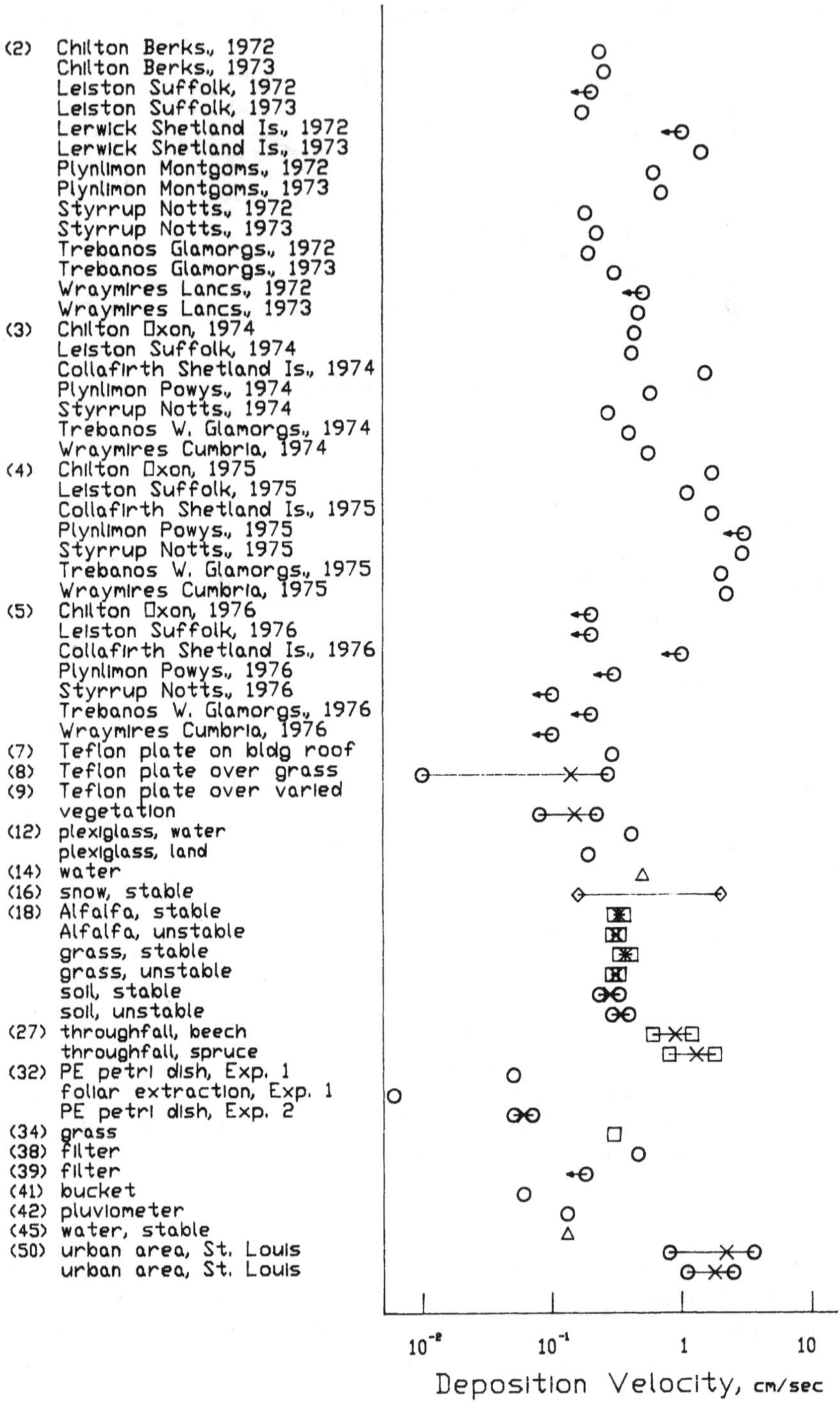
(2) Chilton Berks., 1972
Chilton Berks., 1973
Leiston Suffolk, 1972
Leiston Suffolk, 1973
Lerwick Shetland Is., 1972
Lerwick Shetland Is., 1973
Plynlimon Montgoms., 1972
Plynlimon Montgoms., 1973
Styrrup Notts., 1972
Styrrup Notts., 1973
Trebanos Glamorgs., 1972
Trebanos Glamorgs., 1973
Wraymires Lancs., 1972
Wraymires Lancs., 1973
(3) Chilton Oxon, 1974
Leiston Suffolk, 1974
Collafirth Shetland Is., 1974
Plynlimon Powys., 1974
Styrrup Notts., 1974
Trebanos W. Glamorgs., 1974
Wraymires Cumbria, 1974
(4) Chilton Oxon, 1975
Leiston Suffolk, 1975
Collafirth Shetland Is., 1975
Plynlimon Powys., 1975
Styrrup Notts., 1975
Trebanos W. Glamorgs., 1975
Wraymires Cumbria, 1975
(5) Chilton Oxon, 1976
Leiston Suffolk, 1976
Collafirth Shetland Is., 1976
Plynlimon Powys., 1976
Styrrup Notts., 1976
Trebanos W. Glamorgs., 1976
Wraymires Cumbria, 1976
(7) Teflon plate on bldg roof
(8) Teflon plate over grass
(9) Teflon plate over varied vegetation
(12) plexiglass, water
plexiglass, land
(14) water
(16) snow, stable
(18) Alfalfa, stable
Alfalfa, unstable
grass, stable
grass, unstable
soil, stable
soil, unstable
(27) throughfall, beech
throughfall, spruce
(32) PE petri dish, Exp. 1
foliar extraction, Exp. 1
PE petri dish, Exp. 2
(34) grass
(38) filter
(39) filter
(41) bucket
(42) pluviometer
(45) water, stable
(50) urban area, St. Louis
urban area, St. Louis
10^{-2}
10^{-1}
1
10
Deposition Velocity, cm/sec

B. Comparison between Measured and Predicted Deposition Velocities

Reasonably large databases exist for the dry deposition of SO_4^{2-} and trace elements, and hence we will focus on comparing measured data with model predictions for these species. We will rely on published studies that have used airborne size distribution data for SO_4^{2-} and trace elements with some of the models in section III, B to predict overall dry deposition velocities.

For SO_4^{2-}, a recent literature review has identified 58 size distributions measured at various locations around the world (Milford and Davidson, 1987). This collection of data includes 49 complete distributions obtained with impactors and backup filters, out of which 42 are from continental locations and 7 are from marine sites. The size data have been used with the six models shown in Figure 5–17 to estimate the expected dry deposition velocities for SO_4^{2-} for the conditions defined by each model. The following equation is used to calculate the deposition velocity for each individual distribution:

$$V_d = \sum_i v_{di}\Delta C_i / \sum_i \Delta C_i \qquad (63)$$

Figure 5–26. Measured dry deposition velocities for Pb reported in the literature. The symbols are defined in the same manner as in Figure 5–24. Each study includes a reference number from the list below.

Ref.	Author(s)	Ref.	Author(s)
1	Bytnerowicz et al. (1987)	26	Hofken and Gravenhorst (1982)
2	Cawse (1974)	27	Hofken et al. (1983)
3	Cawse (1975)	28	Ibrahim et al. (1983)
4	Cawse (1976)	29	Japar et al. (1985)
5	Cawse (1977)	30	Johannes and Altwicker (1983)
6	Dasch and Cadle (1985)	31	John et al. (1985)
7	Davidson (1977)	32	Lindberg and Harriss (1981)
8	Davidson and Elias (1982)	33	Lindberg and Lovett (1985)
9	Davidson et al. (1985a)	34	Little and Wiffen (1977)
10	Davidson et al. (1985b)	35	Lovett and Lindberg (1984)
11	Davidson et al. (1985c)	36	Mulawa et al. (1986)
12	Dedeurwaerder et al. (1983)	37	Neumann and den Hartog (1985)
13	Dolske and Gatz (1982)	38	Pattenden et al. (1982)
14	Dolske and Sievering (1979)	39	Peirson et al. (1973)
15	Doran and Droppo (1983)	40	Prahm et al. (1976)
16	Dovland and Eliassen (1976)	41	Rohbock (1982)
17	Droppo (1980)	42	Servant (1974)
18	El-Shobokshy (1985)	43	Shanley (1988)
19	Everett et al. (1979)	44	Sievering (1986)
20	Feely et al. (1985)	45	Sievering et al. (1979)
21	Garland (1978)	46	Smith and Friedman (1982)
22	Gravenhorst et al. (1983)	47	Vandenberg and Knoerr (1985)
23	Hicks and Wesely (1982)	48	Wesely et al. (1983b; 1985)
24	Hicks et al. (1983)	49	Wesely and Hicks (1979)
25	Hicks et al. (1986)	50	Young (1978)

Table 5-16. Dry deposition data for atmospheric particles.

References	v_d (cm/s)	MMD (μm)	Number of samples averaged	Type of measurement	Time of exposure	Date
Delumyea and Petel (1979)	0.57 ± 0.16 (0.01-1.8)	≃1	25	Mixing box model over water, phosphorus-containing particles		Apr.-Oct. 1975
Garland and Cox (1982)	<0.1	0.05-1.0		Gradient over grass $z = 0.2, 1$ m	30 min	
Neumann and den Hartog (1985)	<0.05	0.1-0.5		Eddy correlation over grass		June 1982
Sievering (1982)	0.38 ± 0.29 (0.1-1.9)	0.15-0.30	24	Gradient over rye and wheat $z = 4.5, 9, 18, 36$ m $h = 1$ m	20 min	Sept.-Oct. 1979
Sievering (1983)	1.19 ± 0.18 (unstable) 0.37 ± 0.04 (stable)	0.25		Eddy correlation	20 min	June 1982

where V_d is the overall deposition velocity for the distribution, v_{di} is the deposition velocity corresponding to particle size range i, taken as the deposition velocity at the mass median diameter of the size range, and ΔC_i is the mass concentration in size range i. The values of V_d for the separate distributions have been combined to determine the grand average and standard deviation for the full set of size data using each model. The continental distributions have been used with all six models, and the marine distributions have been used with three models: two for a water surface plus sedimentation (Equation 11).

Results are shown in Table 5–17. The larger values of V_d for the marine distributions reflect the large particle size near the ocean: The continental distributions have an average MMD of 0.52 ± 0.23 μm, and the marine distributions have an average of 2.3 ± 1.8 μm. The predicted deposition velocities for each model have large standard deviations, mostly because of differences in the shapes of the various distributions. These results may be compared with the measured v_d in Table 5–10. The average surrogate surface, foliar extraction, and micrometeorological values are 0.26, 0.23, and 0.55 cm/s, which are somewhat larger than the predicted values based on continental distributions in Table 5–17. Good agreement is not expected, however, because of differences in the atmospheric and surface conditions between the models and the field experiments.

A similar set of calculations has been performed for trace elements. Milford and Davidson (1985) have summarized the results of 55 separate studies involving size distributions for 38 elements. The same data set with minor additions has been used to provide size data for 41 elements. The data have been used with Equation 63 to predict overall dry deposition velocities for each element. Values of v_{di} for each of the six models in Figure 5–17 have been applied.

Results are shown in Figure 5–27. The graph indicates v_d versus MMD for the six models, where MMD represents the geometric mean of the mass median aerodynamic diameters from the original size distributions for each element. These MMD values are similar to the MMDs determined by Milford and Davidson (1985); the latter were calculated from a single normalized mass distribution representing the average of the original distributions for each element. Only one curve from Figure 5–12, namely curve C, has been used as representative of the model of Davidson and others (1982). The figure indicates that v_d increases nearly monotonically with MMD. Curve C from Figure 5–12 gives the largest deposition velocity for each element, whereas the model of Slinn and Slinn (1981) for a water surface gives the smallest. If it is assumed that these two models represent a range of deposition velocities encountered under field conditions to a variety of surfaces, the calculated results can be compared to the measurements in Table 5–15. As in the case of SO_4^{2-}, however, it must be cautioned that this comparison is highly tenuous because of the wide variety of atmospheric and surface conditions represented in the field data that are not accounted for in the models.

Results of the comparison are shown in Table 5–18. The predicted deposition velocities for the two models are shown for each element, along with the number of size distributions providing input data for Equation 63. The measured deposition velocities averaged from Table 5–15 are also shown. Of the 24 elements with

Table 5-17. Predicted dry deposition velocities for SO_4^{2-} using the models of Figure 5-17 with size distributions from the literature review of Milford and Davidson (1987).

Mathematical model	Surface type	Size distribution category	Number of distributions	Deposition velocity (cm/s)
Davidson et al. (1982)	*Dactylus glomerata* (orchard grass)	Continental	42	0.20 ± 0.18
Slinn (1982)	*Eucalyptus* forest	Continental	42	0.14 ± 0.12
Ibrahim et al. (1983)	Snow	Continental	42	0.11 ± 0.11
Williams (1982)	Water	Continental	42	0.21 ± 0.10
Slinn and Slinn (1981)	Water	Continental	42	0.22 ± 0.23
Sedimentation	—	Continental	42	0.02 ± 0.02
Williams (1982)	Water	Marine	7	0.63 ± 0.27
Slinn and Slinn (1981)	Water	Marine	7	0.54 ± 0.26
Sedimentation	—	Marine	7	0.14 ± 0.13

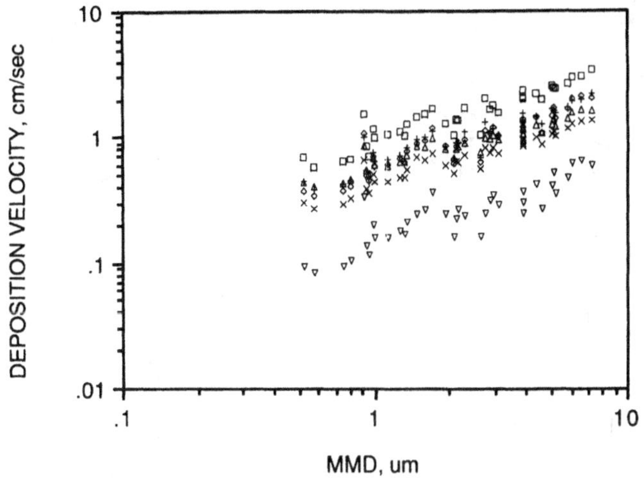

Figure 5–27. Predicted dry deposition velocity versus MMD for trace elements, using models for several surfaces with size distributions from the literature review of Milford and Davidson (1985). The symbols refer to the following models taken from Figure 5–17: squares, orchard grass from Davidson et al. (1982); crosses, *Eucalyptus* forest from Slinn (1982); diamonds, snow surface from Ibrahim et al. (1983); triangles pointing upward, water surface from Williams (1982); *X*s, water surface from Slinn and Slinn (1981); triangles pointing downward, sedimentation according to Equation 11.

both modeled and measured values, 11 elements have more than one or two measured data points. For 7 of these elements, the averaged measured v_d falls within the range of the predicted values. In three cases (Al, Ca, and Fe), the average measured v_d is slightly greater than the predicted range, whereas in one case (Zn) the measured values is less than the predicted range. Much poorer agreement is observed for the 13 elements with only one or two measured values.

The differences between the measured and predicted dry deposition velocities reported above are common in the literature: Uncertainties in both field measurements and modeling that are responsible for these differences have been noted by several authors. For example, Slinn (1983) points out that certain micrometeorological methods rely on extrapolation of data obtained well above a canopy to infer a deposition velocity at the surface, rather than providing a direct measurement of the desired v_d. He further questions the resolution of the methods at small deposition velocities often encountered in the field. Hicks and others (1980) note that substantial differences exist in airflow patterns surrounding surrogate surfaces compared with vegetation and conclude that it is difficult to use surrogate surface data to infer fluxes onto surfaces of interest. Milford and Davidson (1985) point out that size distribution data are often inaccurate for large particle sizes with the highest deposition velocities, leading to uncertain predictions when used with available models. Many additional discussions of uncertainties in measured and predicted dry deposition velocities can be found in the literature.

Table 5-18. Comparison of predicted and measured dry deposition velocities of trace elements.

		MMD (μm)		Predicted v_d^a (cm/s)		Predicted v_d^b (cm/s)			Measured v_d in Field (cm/s)		
Species	N	Geom. mean	GSD	Mean	SD	Mean	SD	N	Mean	SD	Range
Al	21	4.80	1.03	2.41	1.41	0.98	0.45	7	2.63	2.63	0.2–7.9
Ag	—	—	—	—	—	—	—	2	0.27	0.05	0.23–0.3
As	10	1.31	1.05	1.21	1.12	0.52	0.38	2	0.19	0.01	0.18–0.19
Ba	6	2.85	1.32	1.67	0.92	0.77	0.35	2	2.00	0.42	1.7–2.3
Br	30	0.51	1.21	0.67	0.85	0.29	0.30				
Ca	20	4.96	1.05	2.26	0.95	0.94	0.33	14	2.60	2.00	0.23–7
Cd	17	1.10	1.04	1.00	0.86	0.42	0.28	10	0.56	0.77	0.04–2.2
Ce	5	4.90	1.26	2.31	2.23	1.02	0.81				
Cl	32	2.21	1.14	1.61	0.97	0.67	0.32				
Co	9	2.78	1.33	1.56	1.15	0.70	0.42	2	1.20	0.28	1.0–1.4
Cr	10	0.96	1.55	1.11	0.78	0.48	0.33	2	1.25	0.35	1.0–1.5
Cs	5	2.01	1.47	0.97	0.49	0.48	0.23	1	1.5		
Cu	23	1.28	1.02	0.98	0.77	0.45	0.31	8	0.81	0.61	0.08–1.6
Eu	2	2.55	1.10	0.98	0.70	0.52	0.18				
Fe	41	3.73	1.04	1.89	1.02	0.82	0.36	13	2.08	2.04	0.5–7.5
Ga	2	6.30	1.18	2.84	0.77	1.24	0.26				
Hf	5	5.78	1.58	2.80	2.58	1.19	0.93				
Hg	5	0.78	1.22	0.64	0.56	0.31	0.25				
I	8	0.97	1.44	0.95	0.53	0.42	0.22				
In	4	1.86	1.16	1.21	0.68	0.56	0.28	1	0.79		
K	24	3.73	1.26	2.22	1.68	0.89	0.59	8	0.90	0.55	0.05–1.9

Mg	8	5.56	1.03	2.49	1.49	1.09	0.53	7	2.24	3.92	0.15–11
Mn	29	2.05	1.05	1.31	0.93	0.60	0.35	13	1.21	1.66	0.27–6.4
Na	18	3.73	1.13	1.95	1.26	0.86	0.47	12	1.93	2.27	0.16–7.4
Ni	15	0.73	1.34	0.62	0.47	0.28	0.20	1	1.0		
P	2	3.73	1.02	1.91	0.10	0.79	0.02				
Pb	43	0.56	1.20	0.56	0.44	0.26	0.17	22	0.32	0.30	0.005–1.3
Rb	2	4.39	1.18	1.86	1.07	0.81	0.32	1	1.6		
Sb	13	0.90	1.06	0.81	0.68	0.37	0.24	1	0.47		
Sc	9	4.20	1.24	2.08	1.69	0.92	0.63	1	1.1		
Se	12	0.92	1.37	0.67	0.56	0.34	0.24	1	0.1		
Si	4	2.67	1.65	1.91	0.82	0.77	0.30				
Sm	2	1.43	2.25	1.37	0.30	0.65	0.23				
Sr	2	11.90	1.29	4.38	1.31	1.78	0.44	1	2.3		
Ta	4	2.08	1.22	1.28	0.96	0.59	0.36				
Ti	7	6.92	1.00	3.22	1.10	1.26	0.39	2	3.05	3.61	0.5–5.6
Th	6	3.00	1.93	1.46	1.82	0.69	0.68				
U	2	1.66	1.66	1.60	1.03	0.70	0.42				
V	15	1.55	1.35	1.45	1.27	0.62	0.50				
W	2	0.88	2.63	1.47	1.61	0.63	0.64				
Zn	37	1.23	1.01	1.05	1.09	0.45	0.41	9	0.34	0.19	0.05–0.66

[a] Model of Davidson et al. (1982).

[b] Model of Slinn and Slinn (1981).

All values are arithmetic means and standard deviations, except for values of mass median aerodynamic diameter (MMD), which are geometric means and geometric standard deviations.

Overall, the comparisons between measurements and model results show that only rough agreement is obtained; we do not have the ability to measure routinely or predict deposition velocities with high accuracy. Further developments in both of these areas are underway.

VI. Summary

Our current knowledge of dry deposition of contaminant particles and gases from the atmosphere has been reviewed. The emphasis of the chapter is on vegetative canopies, although other surfaces are considered briefly.

The process of dry deposition is conveniently divided into three steps: aerodynamic transport, boundary layer transport, and interactions with the surface. The first step involves contaminant transport from the free atmosphere down to the quasi-laminar sublayer immediately adjacent to the surface. Movement across the sublayer comprises boundary layer transport, and physical and chemical reactions between the contaminant and the surface make up the final transport step.

Mathematical relations describing the dry deposition process make use of the same steps. The aerodynamic resistance r_a is used to define contaminant transport through the atmosphere, whereas the boundary layer resistance r_b defines contaminant transport across the sublayer. The canopy resistance r_c represents interactions with the surface.

Two methods for developing predictions of dry deposition are discussed. The first method involves wind tunnel deposition studies covering a range of wind speeds, turbulence intensities, surface characteristics, and contaminant properties. Data from these studies have been used to derive empirical models for application to the field. The second method involves more detailed mathematical models based on theoretical formulations, laboratory data, and field data to predict dry deposition for specific canopies and atmospheric conditions.

Dry deposition measurement methods include surface analysis techniques for assessing accumulation of a contaminant on a surface and techniques for inferring the flux based on measurement of airborne concentrations and atmospheric properties. A wide variety of methods exist in each category. Examples of surface analysis techniques include foliar extraction, where deposited contaminants are washed off leaves after a desired exposure period, and the throughfall and stemflow method, which involves analysis of rain that has removed dry-deposited contaminants. Other surface analysis methods include collection of cloud droplets, watershed mass balance, use of isotopic tracers, snow sampling, and exposure of surrogate surfaces. This section also includes a brief discussion of application of surface analysis methods to estimate deposition onto man-made structures. Examples of atmospheric flux techniques include eddy correlation, in which turbulent fluctuations in vertical wind velocity and in contaminant concentration are measured, and the gradient method, which involves measurement of contaminant concentration as a function of height. Other atmospheric flux methods include aerometric mass balance, use of multiple artifical tracers, eddy accumulation, and variance.

Dry deposition data from published field and chamber studies are presented in the last section of the chapter. Data are reported for SO_2 and other sulfur-containing gases, SO_4^{2-}, nitrogen species, chloride species, O_3, trace elements, and atmospheric particles. The data for SO_4^{2-} and trace elements are compared with deposition velocities predicted with the models discussed earlier, using published airborne size distributions as model inputs. Only rough agreement between measured and predicted values is obtained, suggesting limitations in our ability to predict and measure dry deposition accurately.

Acknowledgments

The authors gratefully acknowledge the suggestions of D. H. Bache, A. C. Chamberlain, and an anonymous reviewer. The manuscript was prepared by S. A. Knapp. This work was funded in part by NSF Grant DPP-8618223, NOAA Contract NA-85-WC-C-06193, and California Air Resources Board Contract A6-186-32. A scholarship from the Claude Worthington Benedum Foundation is also greatly appreciated.

References

Aylor, D. E., and F. J. Ferrandino. 1985. Atmos Environ 19:803–806.
Bache, D. H. 1977. J Applied Ecology 14:881–895.
Bache, D. H. 1979a. Atmos Environ 13:1257–1262.
Bache, D. H. 1979b. Atmos Environ 13:1681–1687.
Bache, D. H. 1984. Atmos Environ 18:2517–2519.
Bache, D. H. 1986a. Atmos Environ 20:1369–1378.
Bache, D. H. 1986b. Atmos Environ 20:1379–1388.
Baldocchi, D. D., B. B. Hicks, and P. Camara. 1987. Atmos Environ 21:91–101.
Barrie, L. A., and J. L. Walmsley. 1978. Atmos Environ 12:2321–2332.
Belot, Y., A. Baille, and J. L. Delmas. 1976. Atmos Environ 10:89–98.
Belot, Y., and D. Gauthier. 1975. *In* D. A. de Vries and N. H. Afgan, eds. *Heat and mass transfer in the biosphere,* Scripta, WA.
Brutsaert, W. P. 1975. J Atmos Sci 32:2028–2031.
Businger, J. A., J. C. Wyngaard, Y. Izumi, and E. F. Bradley. 1971. J Atmos Sci 28:181–189.
Bytnerowicz, A., P. R. Miller, and D. M. Olszyk. 1987. Atmos Environ 21:1740–1757.
Cadle, S. H., and J. M. Dasch. 1985. *Wintertime wet and dry deposition in Northern Michigan,* GMR 5000. General Motors Corp., Warren, MI.
Cadle, S. H., J. M. Dasch, and P. A. Mulawa. 1985. Atmos Environ 19:1819–1827.
Cawse, P. A. 1974. *A survey of atmospheric trace elements in the U.K. (1972–73),* AERE-R 7669. Environmental and Medical Sciences Division, AERE Harwell, Oxfordshire, England.
Cawse, P. A. 1975. *A survey of atmospheric trace elements in the U.K.: Results for 1974,* AERE-R 8038. Environmental and Medical Sciences Division, AERE Harwell, Oxfordshire, England.

Cawse, P. A. 1976. *A survey of atmospheric trace elements in the U.K.: Results for 1975*, AERE-R 8398. Environmental and Medical Sciences Division, AERE Harwell, Oxfordshire, England.

Cawse, P. A. 1977. *A survey of atmospheric trace elements in the U.K.: Results for 1976*, AERE-R 8869. Environmental and Medical Sciences Division, AERE Harwell, Oxfordshire, England.

Chalmers, J. A. 1949. *Atmospheric electricity*. Clarendon Press, Oxford.

Chamberlain, A. C. 1953. Phil Mag 44:1145–1153.

Chamberlain, A. C. 1960. *In* E. G. Richardson, ed. *Aerodynamic capture of particles*, 63–88. Pergamon, New York.

Chamberlain, A. C. 1966. Proc Roy Soc Ser A 290:236–265.

Chamberlain, A. C. 1967. Proc Roy Soc Ser A 296:45–70.

Chamberlain, A. C. 1968. Q J R Met Soc 94:318–332.

Chamberlain, A. C. 1974. Boundary-Layer Meteorol 6:477–486.

Chamberlain, A. C., and R. C. Chadwick. 1972. Ann Appl Biol 71:141–158.

Clough, W. S. 1975. Atmos Environ 9:1113–1119.

Coe, J. M., and S. E. Lindberg. 1987. J APCA 37:237–243.

Colbeck, I., and R. M. Harrison. 1985. Atmos Environ 19:1807–1818.

Cope, D. M., and D. J. Spedding. 1982. Atmos Environ 16:349–352.

Crecelius, E. A., D. E. Robertson, K. H. Abel, D. A. Cochran, and W. C. Weimer. 1978. *Atmospheric deposition of ^{7}Be and other elements on the Washington coast*, Pacific Northwest Laboratory Annual Report for 1977 to the DOE Assistant Secretary for Environment: Ecological Sciences, PNL-2500 PT2, 7.25–7.26. Battelle, Pacific Northwest Laboratory, Richland, WA.

Dasch, J. M. 1985a. *Measurement of dry deposition to a deciduous canopy*. GMR 5019. General Motors Corp., Warren, MI.

Dasch, J. M. 1985b. Environ Sci Technol 19:721–725.

Dasch, J. M., and S. H. Cadle. 1985. Atmos Environ 19:789–796.

Dasch, J. M., and S. H. Cadle. 1986. Water Air Soil Pollut 29:297–308.

Davidson, C. I. 1977. Powder Tech 18:117–126.

Davidson, C. I., and R. W. Elias. 1980. AlChE symposium series 76:154–157.

Davidson, C. I., and R. W. Elias, 1982. Geophys Res Letters 9:91–93.

Davidson, C. I., and S. K. Friedlander. 1978. J Geophys Res 83:2343–2352.

Davidson, C. I., W. D. Goold, T. P. Mathison, G. B. Wiersma, K. W. Brown, and M. T. Reilly. 1985a. Environ Sci Technol 19:27–35.

Davidson, C. I., W. D. Goold, and G. B. Wiersma. 1983. *In* H. R. Pruppacher, R. G. Semonin, and W. G. N. Slinn, eds. *Precipitation scavenging, dry deposition and resuspension*, 871–882. Elsevier, New York.

Davidson, C. I., S. E. Lindberg, J. A. Schmidt, L. G. Cartwright, and L. R. Landis. 1985b. J Geophys Res 90:2123–2130.

Davidson, C. I., J. M. Miller, and M. A. Pleskow. 1982. Water Air Soil Pollut 18:25–43.

Davidson, C. I., S. Santhanam, R. C. Fortmann, and M. P. Olson. 1985c. Atmos Environ 19:2065–2081.

Davies, T. D., and J. R. Mitchell. 1983. *In* H. R. Pruppacher, R. G. Semonin, and W. G. N. Slinn, eds. *Precipitation scavenging, dry deposition, and resuspension*, 795–805. Elsevier, New York.

Davis, C. S., and R. G. Wright. 1985. J Geophys Res 90:2091–2095.

Dedeurwaerder, H. L., F. A. Dehairs, G. G. Decadt, and W. F. Baeyens. 1983. *In* H. R. Pruppacher, R. G. Semonin, and W. G. N. Slinn, eds. *Precipitation scavenging, dry deposition, and resuspension*, 1219–1231. Elsevier, New York.

de la Mora, J. F., and S. K. Friedlander. 1982. Int J Heat Mass Transfer 25:1725–1735.
Delany, A. C. 1983. Atmos Environ 17:1391–1394.
Delany, A. C., D. R. Fitzjarrald, D. H. Lenschow, R. Pearson, Jr., G. J. Wendel, and B. Woodruff. 1986. J Atmos Chem 4:429–444.
Delumyea, R., and R. L. Petel. 1979. Atmos Environ 13:287–294.
Dolske, D. A., and D. F. Gatz. 1982. A field intercomparison of sulfate dry deposition monitoring and measurement methods: Preliminary results. ACS Acid Rain Symposium, Las Vegas, NV, March 30.
Dolske, D. A., and D. F. Gatz. 1985. J Geophys Res 90:2076–2084.
Dolske, D. A., and H. Sievering. 1979. Water Air Soil Pollut 12:485–502.
Doran, J. C., and J. G. Droppo. 1983. *In* H. R. Pruppacher, R. G. Semonin, and W. G. N. Slinn, eds. *Precipitation scavenging, dry deposition, and resuspension,* 1003–1012. Elsevier, New York.
Dovland, H., and A. Eliassen. 1976. Atmos Environ 10:783–785.
Droppo, J. G. 1980. *In* D. S. Shriner, C. R. Richmond, and S. E. Lindberg, eds. *Atmospheric sulfur deposition,* 209–220. Ann Arbor Science, Ann Arbor, MI.
Droppo, J. G., and J. C. Doran. 1983. *In* H. R. Pruppacher, R. G. Semonin, and W. G. N. Slinn, eds. *Precipitation scavenging, dry deposition, and resuspension,* 807–815. Elsevier, New York.
Droppo, J. G., Jr. 1985. J Geophys Res 90:2111–2118.
Duce, R. A., G. L. Hoffman, and W. H. Zoller. 1975. Science 187:59–61.
Duyzer, J. H., G. M. Meyer, and R. M. van Aalst. 1983. Atmos Environ 17:2117–2120.
Dyer, A. J., and B. B. Hicks. 1970. Quart J Roy Meteorol Soc 96:715–721.
Edwards, G. C., and G. L. Ogram. 1983. *In* H. R. Pruppacher, R. G. Semonin, and W. G. N. Slinn, eds. *Precipitation scavenging, dry deposition, and resuspension,* 817–824. Elsevier, New York.
Elias, R. W., and C. I. Davidson. 1980. Atmos Environ 14:1427–1432.
El-Shobokshy, M. S. 1985. Atmos Environ 19:1191–1197.
Esmen, N. A., P. Ziegler, and R. Whitfield. 1978. Atmos Environ 9:547–556.
Everett, R. G., B. B. Hicks, W. W. Berg, and J. W. Winchester. 1979. Atmos Environ 13:931–934.
Fairall, C. W., and S. E. Larsen. 1984. Atmos Environ 18:69–77.
Feely, H. W., D. C. Bogen, S. J. Nagourney, and C. C. Torquato. 1985. J Geophys Res 90:2161–2165.
Fisher, M. S., D. A. Charles-Edwards, and M. M. Ludlow. 1981. Aust J Plant Physiol 8:347–357.
Fowler, D. 1978. Atmos Environ 12:369–373.
Fowler, D., and J. N. Cape. 1983. *In* H. R. Pruppacher, R. G. Semonin, and W. G. N. Slinn, eds. *Precipitation scavenging, dry deposition, and resuspension,* 763–773. Elsevier, New York.
Friedlander, S. K. 1977. *Smoke, dust, and haze: Fundamentals of aerosol behavior*. John Wiley and Sons, New York.
Friedlander, S. K., and H. F. Johnstone. 1957. Ind Eng Chem 49:1151–1156.
Friedlander, S. K., J. R. Turner, and S. V. Hering. 1986. J Aerosol Sci 17:240–244.
Fuchs, N. A. 1964. *Mechanics of aerosols*. Pergamon, New York.
Galbally, I. E. 1971. Quart J Roy Meteorol Soc 97:18–29.
Galloway, J. N., D. M. Whelpdale, and G. T. Wolff. 1984. Atmos Environ 18:2595–2607.
Gamble, J. S., and C. I. Davidson. 1985. *In* R. Baboian, ed. *Materials degradation caused by acid rain,* 42–63. Proceedings, 20th State of the Art Symposium, American Chemical Society, Washington, DC, June 17–19.

Garland, J. A. 1978. Atmos Environ 12:349–362.
Garland, J. A. 1982. *In* H. W. Georgii and J. Pankrath, eds. *Deposition of atmospheric pollutants,* 9–16. Reidel, Dordrecht, Netherlands.
Garland, J. A. 1983. *In* H. R. Pruppacher, R. G. Semonin, and W. G. N. Slinn, eds. *Precipitation scavenging, dry deposition, and resuspension,* 849–858. Elsevier, New York.
Garland, J. A., and L. C. Cox. 1982. Atmos Environ 16:2699–2702.
Granat, L., and C. Johansson. 1983. Atmos Environ 17:191–192.
Gravenhorst, G., K. D. Hofken, and H. W. Georgii. 1983. *In* S. Beilke and A. J. Elshout, eds. *Acid deposition,* 155–171. Reidel, Dordrecht, Netherlands.
Grennfelt, P., C. Bengtson, and L. Skarby. 1983. *In* H. R. Pruppacher, R. G. Semonin, and W. G. N. Slinn, eds. *Precipitation scavenging, dry deposition, and resuspension,* 753–762. Elsevier, New York.
Hales, J. M., B. B. Hicks, and J. M. Miller. 1987. Bull Am Meteorol Soc 68:216–225.
Hesketh, H. E. 1981. *Air pollution control,* 117. Ann Arbor Science, Ann Arbor, MI.
Hicks, B. B., D. D. Baldocchi, R. P. Hosker, Jr., B. A. Hutchison, D. R. Matt, R. T. McMillen, and L. C. Satterfield. 1985. *On the use of monitored air concentrations to infer dry deposition,* NOAA Technical Memorandum ERL ARL-141, Silver Spring, MD.
Hicks, B. B., and P. S. Liss. 1976. Tellus 28:348–354.
Hicks, B. B., and M. L. Wesely. 1982. Atmos Environ 16:2899–2903.
Hicks, B. B., M. L. Wesely, R. L. Coulter, R. L. Hart, J. L. Durham, R. E. Speer, and D. H. Stedman. 1983. *In* H. R. Pruppacher, R. G. Semonin, and W. G. N. Slinn, eds. *Precipitation, scavenging, dry deposition, and resuspension,* 933–942. Elsevier, New York.
Hicks, B. B., M. L. Wesely, R. L. Coulter, R. L. Hart, J. L. Durham, R. E. Speer, and D. H. Stedman. 1986b. Boundary-Layer Meteorol 34:103–121.
Hicks, B. B., M. L. Wesely, and J. L. Durham. 1980. *Critique of methods to measure dry deposition: Workshop summary,* EPA-600/9-80-050.
Hicks, B. B., M. L. Wesely, S. E. Lindberg, and S. M. Bromberg. 1986a. Proceedings of the NAPAP Workshop on Dry Deposition, Harpers Ferry, WV, March 25–27.
Hofken, K. D., and G. Gravenhorst. 1982. *In* H. W. Georgii and J. Pankrath, eds. *Deposition of atmospheric pollutants,* 187–190. Reidel, Dordrecht, Netherlands.
Hofken, K. D., F. X. Meixner, and D. H. Ehhalt. 1983. *In* H. R. Pruppacher, R. G. Semonin, and W. G. N. Slinn, eds. *Precipitation scavenging, dry deposition, and resuspension,* 825–835. Elsevier, New York.
Holman, J. P. 1972. *Heat transfer*. McGraw-Hill, New York.
Hosker, R. P., and S. E. Lindberg. 1982. Atmos Environ 16:889–910.
Huebert, B. J. 1983. *In* H. R. Pruppacher, R. G. Semonin, and W. G. N. Slinn, eds. *Precipitation scavenging, dry deposition, and resuspension,* 785–794. Elsevier, New York.
Huebert, B. J., and C. H. Robert. 1985. J Geophys Res 90:2085–2090.
Ibrahim, M., L. A. Barrie, and F. Fanaki. 1983. Atmos Environ 17:781–788.
Japar, S. M., W. W. Brachaczek, R. A. Gorse, Jr., J. M. Norbeck, and W. R. Peirson. 1985. Dry deposition of nitric acid and sulfur dioxide to surrogate nylon surfaces, presented at Muskoka 85 International Symposium on Acidic Precipitation, Minett, Ontario, September 1985.
Jarvis, P. G., G. B. James, and J. J. Landsberg. 1976. *In* Vegetation and the atmosphere, Vol 2:171–240. Academic Press, London.

Johannes, A. H., and E. R. Altwicker. 1983. *In* H. R. Pruppacher, R. G. Semonin, and W. G. N. Slinn, eds. *Precipitation scavenging, dry deposition, and resuspension,* 903–912. Elsevier, New York.

Johansson, C., and L. Granat. 1986. Atmos Environ 20:1165–1170.

Johansson, C., A. Richter, and L. Granat. 1983. *In* H. R. Pruppacher, R. G. Semonin, and W. G. N. Slinn, eds. *Precipitation scavenging, dry deposition, and resuspension,* 775–784. Elsevier, New York.

John, W., S. M. Wall, and J. L. Ondo. 1985. *Dry acid deposition on materials and vegetation: Concentrations in ambient air,* Final Report, California Air Resources Board, A1-160-32.

Katen, P. C., and J. M. Hubbe. 1983. *In* H. R. Pruppacher, R. G. Semonin, and W. G. N. Slinn, eds. *Precipitation scavenging, dry deposition, and resuspension,* 953–962. Elsevier, New York.

Katen, P. C., and J. M. Hubbe. 1985. J Geophys Res 90:2145–2160.

Kluczcwski, S. M., K. A. Brown, and J. N. B. Bell. 1985. Atmos Environ 19:1295–1299.

Lassey, K. R. 1982. Atmos Environ 16:13–24.

Lassey, K. R. 1983. Atmos Environ 17:2303–2310.

Legg, B. J., and R. I. Price. 1980. Atmos Environ 14:305–309.

Leuning, R., and P. N. Attiwill. 1978. Agric Meteorol 19:215–241.

Lindberg, S. E., and R. C. Harriss. 1981. Water Air Soil Pollut 16:13–31.

Lindberg, S. E., and G. M. Lovett. 1983. *In* H. R. Pruppacher, R. G. Semonin and W. G. N. Slinn, eds. *Precipitation scavenging, dry deposition, and resuspension,* 837–848. Elsevier, New York.

Lindberg, S. E., and G. M. Lovett. 1985. Environ Sci Technol 19:238–244.

Lindberg, S. E., G. M. Lovett, D. D. Richter, and D. W. Johnson. 1986. Science 231:141–145.

Little, P. J., and R. D. Wiffen. 1977. Atmos Environ 11:437–447.

Lorenz, R., and C. E. Murphy, Jr. 1985. Atmos Environ 19:797–802.

Lovett, G. M., and S. E. Lindberg. 1984. J Appl Ecol 21:1013–1027.

Lumley, J. L., and H. A. Panofsky. 1964. *In Interscience monographs and texts in physics and astronomy,* vol. XII. Wiley, New York.

May, K. R., and R. Clifford. 1967. Ann Occup Hyg 10:83–95.

McDonald, R. L., C. K. Unni, and R. A. Duce. 1982. J Geophys Res 87:1246–1250.

McMahon, T. A., and P. J. Denison. 1979. Atmos Environ 13:571–585.

McRae, G. J., and A. G. Russell. 1984. *In* B. B. Hicks, ed. *Deposition both wet and dry,* 153–192. Ann Arbor Science, Ann Arbor, MI.

Meyers, T. P., and T. S. Yuen. 1987. J Geophys Res 92:6705–6712.

Milford, J. B., and C. I. Davidson. 1985. J APCA 35:1249–1260.

Milford, J. B., and C. I. Davidson. 1987. J APCA 37:125–134.

Milne, J. W., D. B. Roberts, and D. J. Williams. 1979. Atmos Environ 13:373–379.

Moller, U., and G. Schumann. 1970. J Geophys Res 75:3013–3019.

Mulawa, P. A., S. H. Cadle, F. Lipari, C. C. Ang, and R. T. Vandervennet. 1986. *In* R. Baboian, ed. *Materials degradation caused by acid rain,* Proceedings, 20th State of the Art Symposium, 92–101. American Chemical Society, Washington, DC, June 17–19.

Munn, R. E. 1966. *Descriptive Micrometeorology*. Academic Press, New York.

Murphy, C. E., Jr., and J. T. Sigmon. 1989. *In* S. E. Lindberg, A. L. Page, and S. A. Norton, eds. *Acidic precipitation: Sources, deposition, and canopy interactions, mitigation*. Springer-Verlag, New York.

Neumann, H. H., and G. den Hartog. 1985. J Geophys Res 90:2097–2110.

O'Dell, R. A., M. Taheri, and R. L. Kabel. 1977. J APCA 27:1104–1109.

Owen, P. R., and W. R. Thompson. 1963. J Fluid Mech 15:321–334.

Pattenden, N. J., J. R. Branson, and E. M. R. Fisher. 1982. *In* H. W. Georgii and J. Pankrath, eds. *Deposition of atmospheric pollutants*, 173–184. Reidel, Dordrecht, Netherlands.

Peirson, D. H., P. A. Cawse, L. Salmon, and R. S. Cambray. 1973. Nature 241:252–256.

Plate, E. J. 1971. *Aerodynamic characteristics of atmospheric boundary layers*, U.S.A.E.C. Div. Tech. Inf., Oak Ridge, TN.

Platt, U. 1978. Atmos Environ 12:363–367.

Prahm, L. P., U. Torp, and R. M. Stern. 1976. Tellus 27:355–372.

Rahn, K. A. 1976. *The chemical composition of the atmospheric aerosol,* Technical Report, University of Rhode Island, July 1.

Rauner, J. L. 1976. *In Vegetation and the atmosphere,* Vol. 2:241–264. Academic Press, London.

Raynor, G. S. 1976. *Symposium on atmosphere-surface exchange of particulate and gaseous pollutants,* 264–279. September 4–6, 1974, National Technical Information Service, U.S. Department of Commerce, Springfield, VA.

Rohbock, E. 1982. *In* H. W. Georgii and J. Pankrath, eds. *Deposition of atmospheric pollutants,* 159–171. Reidel, Dordrecht, Netherlands.

Saugier, B. 1976. *In Vegetation and the atmosphere*, Vol. 2:87–118. Academic Press, London.

Schack, A. J., Jr., S. E. Pratsinis, and S. K. Friedlander. 1985. Atmos Environ 19:953–960.

Sehmel, G. A. 1972. *In* T. T. Mercer, P. E. Morrow, and W. Stober, eds. *Assessment of airborne particles,* 18–42. Charles C. Thomas Publishing Co., Springfield, IL.

Sehmel, G. A. 1973. Aerosol Sci 4:125–138.

Sehmel, G. A. 1980. Atmos Environ 14:983–1009.

Sehmel, G. A., and W. H. Hodgson. 1977. *Improved particle deposition model,* 61–65. Pacific Northwest Laboratory Annual Report for 1976, Atmospheric Sciences, BNWL-2100, Pt 3, Battelle, Pacific Northwest Laboratory, Richland, WA.

Sehmel, G. A., and W. H. Hodgson. 1978. *A model for predicting dry deposition of particles and gases to environmental surfaces,* PNL-SA-6721 Battelle, Pacific Northwest Laboratory, Richland, WA.

Sehmel, G. A., W. H. Hodgson, and S. L. Sutter. 1974. *Dry deposition of particles,* 157–162. Pacific Northwest Laboratory Annual Report for 1973, Atmospheric Sciences, BNWL-1850, Pt 3, Battelle, Pacific Northwest Laboratory, Richland, WA.

Sehmel, G. A., and S. L. Sutter. 1974. J Rech Atmos 8:911–920.

Sehmel, G. A., S. L. Sutter, and M. T. Dana. 1973. *Dry deposition processes,* 43–49. Pacific Northwest Laboratory Annual Report for 1972, Atmospheric Sciences, BNWL-1751, Pt 1, Battelle, Pacific Northwest Laboratory, Richland, WA.

Seinfeld, J. H. 1975. *Air pollution: Physical and chemical fundamentals,* 254. McGraw-Hill, New York.

Semonin, R. G., G. L. Stensland, V. C. Bowersox, M. E. Peden, J. M. Lockard, K. G. Doty, D. F. Gatz, L. Chu, S. R. Backman, and R. K. Stahlhut. 1984. *Study of atmospheric pollution scavenging,* Illinois State Water Survey Contract Report 347, 20th Progress Report. Champaign, IL.

Servant, J. 1976. Deposition of atmospheric lead particles to natural surfaces in field experiments, *Symposium on atmosphere-surface exchange of particulate and gaseous pollutants,* 87–95. September 4–6, 1974, National Technical Information Services, U.S. Department of Commerce, Springfield, VA.

Shanley, J. B. 1989. Atmos Environ 23:403–414.

Shannon, J. D. 1981. Atmos Environ 15:689–701.

Sheih, C. M., M. L. Wesely, and B. B. Hicks. 1979. Atmos Environ 13:1361–1368.

Shreffler, J. H. 1978. Atmos Environ 12:1497–1503.

Sievering, H. 1982. Atmos Environ 16:301–306.

Sievering, H. 1983. *In* H. R. Pruppacher, R. G. Semonin, and W. G. N. Slinn, eds. *Precipitation scavenging, dry deposition, and resuspension,*963–978. Elsevier, New York.

Sievering, H. 1986. Atmos Environ 20:341–345.

Sievering, H., M. Dave, P. McCoy, and N. Sutton. 1979. Atmos Environ 13:1717–1719.

Slinn, S. A., and W. G. N. Slinn, 1980. Atmos Environ 14:1013–1016.

Slinn, S. A., and W. G. N. Slinn, 1981. *In* S. J. Eisenreich, ed. *Atmospheric pollutants in natural waters,* 23–53. Ann Arbor Science, Ann Arbor, MI.

Slinn, W. G. N. 1982. Atmos Environ 16:1785–1794.

Slinn, W. G. N. 1983. *In* H. R. Pruppacher, R. G. Semonin, and W. G. N. Slinn, eds. *Precipitation scavenging, dry deposition, and resuspension,* 1361–1416. Elsevier, New York.

Smith, B. E., and E. J. Friedman. 1982. *The chemistry of dew as influenced by dry deposition: Results of Sterling, VA and Champaign, IL experiments,* WP-82W00141, MITRE Corp., McLean, VA.

Speer, R. E., K. A. Peterson, T. G. Ellestad, and J. L. Durham. 1985. J Geophys Res 90:2119–2122.

Sprugel, D. G., and J. E. Miller. 1979. Water Air Soil Pollut 12:233–236.

Stedman, D. H., B. Evilsizor, and D. W. Stocker. 1987. *Dry deposition of oxides of nitrogen,* 1987 EPA/APCA symposium on measurement of toxic and related air pollutants, May 3–6, 1987, at Research Triangle Park, NC.

Stocker, D. W., M. P. Burkhardt, M. Plooster, and D. H. Stedman. 1987. *Continuous eddy correlation dry deposition flux measurements,* Final Report to the Western Energy Supply and Transmission Associates.

Taylor, G. E., Jr., S. B. McLaughlin, Jr., D. S. Shriner, and W. J. Selvidge. 1983. Atmos Environ 17:789–796.

Taylor, G. E., Jr., and D. T. Tingey. 1983. Plant Physiol 72:237–244.

Taylor, G. E., Jr., D. T. Tingey, and H. C. Ratsch. 1982. Oecologia 53:179–186.

Thom, A. S. 1968. Q J Royal Meteorol Soc 94:44–55.

Thom, A. S. 1975. *In* J. L. Monteith, ed. *Vegetation and the atmosphere,* 57–109. Academic Press, London.

Turner, D. B. 1970. *Workbook of atmospheric dispersion estimates,* U.S. Environmental Protection Agency Office of Air Programs Publication No. AP-26, EPA, Research Triangle Park, NC.

Vandenberg, J. J., and K. R. Knoerr. 1985. Atmos Environ 19:627–635.

Voldner, E. C., L. A. Barrie, and A. Sirois. 1986. Atmos Environ 20:2101–2123.

Webb, E. K. 1970. Quart J Roy Meteorol Soc 96:67–90.

Wells, A. C., and A. C. Chamberlain. 1967. Brit J Applied Physics 18:1793–1799.

Welty, J. R., C. E. Wicks, and R. E. Wilson. 1969. *Fundamentals of momentum, heat, and mass transfer*. John Wiley and Sons, New York. 160 p.

Wesely, M. L., D. R. Cook, and R. L. Hart. 1983a. Boundary-Layer Meteorol 27: 237–255.

Wesely, M. L., D. R. Cook, and R. L. Hart. 1985. J Geophys Res 90:2131–2143.

Wesely, M. L., D. R. Cook, R. L. Hart, B. B. Hicks, J. L. Durham, R. E. Speer, D. H. Stedman, and R. J. Tropp. 1983b. *In* H. R. Pruppacher, R. G. Semonin, and W. G. N.

Slinn, eds. *Precipitation scavenging, dry deposition, and resuspension,* 943–952. Elsevier, New York.
Wesely, M. L., J. A. Eastman, D. H. Stedman, and E. D. Yalvac. 1982. Atmos Environ 16:815–820.
Wesely, M. L., and B. B. Hicks. 1977. J APCA 27:1110–1116.
Wesely, M. L., and B. B. Hicks. 1979. *Proc fourth symp. on turbulence, diffusion and air quality,* Reno, NV, January 15–18, 510–513. Am. Met. Soc., Boston, MA.
Wesely, M. L., B. B. Hicks, W. D. Dannevik, S. Frisella, and R. B. Husar. 1977. Atmos Environ 11:561–563.
White, E., and F. Turner. 1970. J Appl Ecol 7:441–461.
Williams, R. M. 1982. Atmos Environ 16:1933–1938.
Young, J. A. 1978. *The rates of change of pollutant concentrations downwind of St. Louis,* Pacific Northwest Laboratory Annual Report for 1977 to the DOE Assistant Secretary for Environment, Atmospheric Sciences, PNL-2500PT3 Battelle, Pacific Northwest Laboratory, Richland, WA.
Young, J. R., C. Ellis, and G. M. Hidy. 1987. Environ Qual 17:1–26.

Dry Deposition of Sulfur and Nitrogen Oxide Gases to Forest Vegetation

C.E. Murphy, Jr.,* and J.T. Sigmon†

Abstract

The study of low-level, chronic pollution of forest stands and entire watersheds requires the estimation of the deposition of the pollutants in question. A great deal of effort has been expended in determining dry deposition rates of the acid gases of nitrogen and sulfur. The "big leaf" model is used in this paper to partition the influences of the atmosphere, the forest stand, and the surfaces in the forest stand in order to gain a better understanding of what processes limit gaseous pollutant uptake. The estimates of deposition based on the influence of particular parts of the plant-atmosphere system are integrated to provide estimates of deposition for entire stands. These estimates are compared to deposition rates measured for entire stands. The agreement is good for sulfur dioxide where there is enough data to make a valid comparison. The general magnitude of deposition of the other gases appears to be of the correct size. However, the data and the model point out the inherent variability in the deposition process caused by atmospheric and vegetation differences and suggest that either intensive measurements are necessary or the investigator will have to accept moderate to large uncertainties, depending on the circumstance under which the estimates are made.

I. Introduction

The acute toxicity effects of air pollutants on vegetation have been studied for over a century. Recently, concern over forest dieback has focused attention on the effects of both acidic deposition and chronic low-level pollution concentrations. The study of chronic low-level pollution concentration effects raises questions of tree physiology, soil chemistry, and ecosystem processes not necessarily impor-

*Westinghouse Savannah River Co., Savannah River Laboratory, Aiken, SC 29802, USA.

†Department of Environmental Science, University of Virginia, Charlottesville, VA 22903, USA.

Definitions of Symbols

a = an empirical proportionality constant in the stomatal conductance equation
c_p = specific heat of air at constant pressure
d = zero-plane displacement height for gradient measurements
e_a = the vapor pressure of the air
e_s = the saturation vapor pressure of the air at air temperature
h = the stomatal conductance
h_{mn} = the stomatal conductance when the stomata are fully closed
k = von Karman's constant
l = aerodynamic dimension of a canopy element, such as a leaf or a stem
s = Henry's law solubility (mass liquid/volume gas)
r_{bl} = the boundary layer resistance to diffusion near an element surface
r_{blb} = the boundary layer resistance over the surface of stem or branches
r_{bll} = the boundary layer over the surface of the leaves
r_s = the leaf stomatal resistance
r_L = the total average resistance for an single leaf
r_m = the leaf mesophyll (internal) resistance to diffusion in the cell solution
r_{sf} = the resistance to surface uptake
r_{sfb} = the resistance to uptake by the stem or branch surface
r_{sfl} = the resistance to uptake by the leaf surface
t = time
u = wind speed
u_* = friction velocity, the surface shear stress divided by the density of air
z_h = roughness length for the logarithmic temperature profile
z_o = roughness length for the logarithmic wind profile
A_L = the leaf area of the vegetation per unit ground area
A_b = the stem and branch area of the vegetation per unit ground area
C = constant in the canopy air space equation
C_a = air concentration of a pollutant gas
C_i = liquid concentration of a pollutant gas
C_{sf} = the quasi-steady-state concentration of a pollutant on a solid surface
D_g = molecular diffusivity of the gas of interest in air
D_w = molecular diffusivity of water vapor in air
E_s = the greatest evaporation rate at which the soil can supply the vegetation
H = height of the top of the forest canopy above the ground
H_l = the radiant heat load of the vegetation
L = the latent heat of vaporization of water
R_a = the forest stand boundary layer resistance
R_{ac} = the resistance to diffusion in the forest air space
R_b = the bulk aerodynamic canopy resistance
R_L = the bulk diffusion resistance to the leaves
R_{sf} = the bulk diffusion resistance to the stem and branches
Re = Reynolds number = $\rho\ 1/\mu$
Sh = Sherwood number = $l/(r_{bl} D_g)$
Sc = Schmidt number = $\mu/(\rho D_g)$
S = the light intensity above the leaf

SL = the sulfur loading of the vegetation surface
SS = the saturation sulfur loading of the vegetation surfaces
T_s = the time to reach 63% saturation of the vegetation surfaces from an initially clean surface
U = uptake flux density of a pollutant gas
Δ = the first derivative of saturation vapor pressure with temperature
γ = the psychometric constant
μ = dynamic viscosity of air
θ = the soil-water content; the subscripts indicate maximum (mx) and minimum (mn) found in field for the soil type
ρ = density of air

tant to the study of acute pollutant effects. Acidic oxides of nitrogen and sulfur have been identified as potential contributors to forest dieback, the most important of which are nitric oxide (NO), nitrogen dioxide (NO_2), nitric acid vapor (HNO_3), and sulfur dioxide (SO_2).

In this chapter we discuss the transport mechanisms that move the sulfur and nitrogen oxide gases from the atmosphere to the vegetation and the soil surface. At these sites the oxidizing gases chemically act to produce damage and/or become part of the nitrogen or sulfur cycles leading to acidification of forest soils. It is useful to discuss the transport of the acidic gases in terms of a model of the flow from the atmosphere to the vegetation and the soil. Davidson and Wu (see the previous chapter) have described a number of models that use the analogy between electrical current and resistance to divide the flow of gases into compartments where the flow processes can be identified and individually modeled.

The organization of this chapter is based on the flows and resistances illustrated in Figure 6–1. Two resistance networks are shown. The more complex network, at the left, takes into account the vertical distribution of canopy and air transport structure. This is certainly closer to the actual transport processes in a real canopy and has been used to predict pollutant uptake in the models of Murphy and others (1977) and Shreffler (1978). The simpler network, on the right, is based on the "big leaf" concept first defined by Monteith (1965). It idealizes the canopy as an average leaf with average transport resistances from the atmosphere to the leaf surface. Use of this model, and also most of the multi-level models, assumes that near the surface of the earth vertical transport from the atmosphere to the surface dominates deposition.

Two slightly different approaches to the big leaf model have been developed. The first, which we shall call the *top down* approach has been developed as a means of interpreting and extrapolating gas and particle transport data from micrometeorological experiments. The models developed by Wesely and Hicks (1977), Unsworth (1981), and Hicks and others (1987) are examples of this approach. The second type of big leaf model, which could be called the *bottom up* approach, was developed to integrate gas and particle exchange data collected from experiments with individual leaves and other plant surfaces for the purpose of

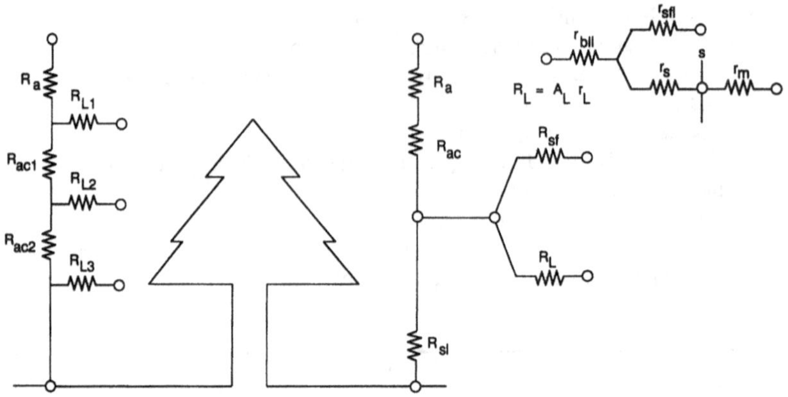

Figure 6–1. A simple resistance network describing the path of a gas from the atmosphere to uptake by the vegetation. The processes related to each resistance are explained in the text.

calculating whole plant or whole forest transport. The models developed by Stewart and Thom (1973) and Sinclair and others (1976) are examples of the bottom up approach.

The two types of models are similar, and the deposition rates calculated from them should be very similar. The difference results from a different assignment of the resistances, largely due to different means of measuring the processes involved. Figure 6–1 illustrates the resistance network for the bottom up approach. The differences in the two approaches and how the two different systems of resistance can be reconciled will be addressed when discussing the processes involved in the transport of the oxides of nitrogen and sulfur.

The methods used to predict the effect of atmospheric processes have been thoroughly developed by Davidson and Wu (see previous chapter) and are discussed here only briefly. In this chapter we emphasize the processes that determine the value of the resistances to transport from the canopy air space to the vegetation and the soil surface, and the transport from the surfaces into the vegetation and the soil.

After discussing the individual transport processes with the objective of defining procedures to estimate the transport resistances, we review the literature to determine representative estimates for each of the resistances for the acidic gases of interest as the basis for crude estimates of the deposition of the acidic gases to forests.

II. Mechanisms of Dry Deposition of Gases

Transport of pollutants from the source to the forest begins with turbulent diffusion into the atmosphere at the source. This is followed by transport of the pollutant by the general circulation of the atmosphere. During transport, substantial transformation of chemical form can take place (Urone and Schroeder, 1978). For the oxide gases of sulfur and nitrogen, most of the transformation is further oxidation of the gases: NO oxidizes fairly rapidly to NO_2, NO_2 interacts further with atmospheric oxidants and water vapor to produce HNO_3 and particulate nitrates, and SO_2 will oxidize to SO_3, which is rapidly further oxidized to sulfate. The result of long-distance transport is a decrease in the concentration of most of the gases of interest to us in this volume, with the exception of HNO_3 vapor.

A. Transport through the Atmosphere Near the Forest

As discussed by Davidson and Wu, the three primary pathways of dry deposition are through the atmosphere above the surface (termed *aerodynamic* transport) and through the canopy air space and the boundary layer at the surface (termed *quasi-laminar sublayer* transport), and through interactions with the surface. For dry deposition to forests, the variable structure of the "surface" complicates the dry deposition process even further.

Transport in the atmospheric boundary layer and the stand space is primarily through turbulent diffusion. It is accomplished by the movement of air parcels (eddies) of higher concentration downward into the forest and eddies of lower concentration moving upward from the forest. The rate of transport through the boundary layer and stand space is controlled by the size and energy of the eddies and the concentration differences. Eddy size and energy, in turn, are controlled by the state of the larger-scale atmosphere and atmosphere-forest interactions. Turbulent transport in the atmospheric boundary layer is fairly well understood, and it is possible to estimate transfer functions under many circumstances.

Transport through the stand boundary layer is generally modeled as turbulent diffusion along a vertical concentration gradient. The turbulent diffusivity controlling the flow along the gradient is predicted from turbulent boundary layer theory as a function of wind speed and thermal stratification of the atmosphere (Busch, 1973; Businger, 1973). Transport in the stand boundary layer has also been modeled by second-order closure models (Shaw, 1977; Finnigan, 1985; Meyers, 1987), although it is generally agreed that the concentration gradient models are sufficient for most predictions (Denmead and Bradley, 1985). These considerations lead to the big leaf estimate of the aerodynamic resistance (R_a) as:

$$R_a = \frac{u}{u_*^2} \tag{1}$$

The friction velocity u_* should be suitably corrected for atmospheric stability.

Transport through the forest air space is not nearly so well understood, and an adequate theoretical basis does not exist at present. In some cases, transport appears to be intermittent, caused primarily by larger eddies that occasionally penetrate the forest air space from the atmosphere above the canopy (Denmead and Bradley, 1985).

Furthermore, there does not appear to be a uniform cascade of eddy size through the forest canopy. Baldocchi and Hutchison (1988), using spectral velocity analysis, showed that the crown space of an almond orchard contained more small eddies than the trunk space of the orchard. It would be reasonable to assume that the same situation would apply to mature forests that did not have a large number of understory trees. This intermittency in the transport process and discontinuity of transport eddy size can result in measurement-averaging anomalies such as apparent countergradient transport.

In spite of these problems, big leaf models estimate the in-canopy diffusion on the basis of a turbulent velocity scale and a length scale. The most common velocity scale is the friction velocity u_* above the canopy. The length scale is set on the assumption that the center of mass action for momentum transport in the stand is at the canopy level of the sum of the zero-plane displacement height d and the roughness length z_o (Thom, 1975). The length scale is then the difference between the height of the canopy H and the sum of d and z_o. This leads to an estimate of the canopy air space resistance of the form

$$R_{ac} = C \frac{H - (z_o + d)}{u_*\{H - (z_o + d)\}} = \frac{C}{u_*} \tag{2}$$

The constant C in Equation 2 has been estimated to have a value between 1 and 4, depending on the material being deposited. One source of the variation in the constant is the difference in the canopy level of the center of mass action of the different materials. Rapidly deposited materials may be deposited preferentially near the top of the canopy whereas less rapidly deposited materials may be deposited more evenly throughout the canopy.

Another approach to estimating the resistance to diffusion in the canopy is to use the transports of heat and momentum to estimate an additional resistance to heat transport that should include the canopy air space resistance. The relationship between the resistance for mass and heat transport is often defined in terms of a resistance ratio

$$B = k \ln \left(\frac{z_o}{z_h}\right) \tag{3}$$

The definition of a roughness length for heat, z_h, is an attempt to take into account differences in the transport processes for momentum and heat, such as differences in the center of mass action and differences in the diffusion processes at the surface of the forest elements (Owen and Thompson, 1963; Garratt and Hicks, 1973; Hicks et al., 1987). This top down approach to the big leaf model adds the aerodynamic resistance above (R_a) and within (R_b) the canopy.

Investigation of the relationship of R_b to friction velocity suggests that its value is influences by both the resistance to diffusion in the canopy space and diffusion through the forest surface element's boundary layer. The values R_{ac} and R_b are approximately related as:

$$R_b = R_{ac} + r_{bl}/A_L \tag{4}$$

Reasonable values of the input parameters used to estimate R_{ac} and r_{bl} and reasonable values of projected leaf area index A_L will produce values of R_b in the range of the data assembled by Owen and Thompson (1963) and Garratt and Hicks (1973).

The bottom up approach to estimating deposition sums R_a and R_{ac} and includes r_{bl} in the network used to calculate the leaf resistance R_L, as shown in Figure 6–1. There is no evidence that one approach is better than the other, given the problems of estimating the parameters used in the calculations. Which one is used usually depends on the data on which the estimates of parameters are based.

B. Transport at the Vegetation Surface

Leaves are the organs through which most gases reach the inside of the vegetation. Leaves have evolved to optimize light gathering and the exchange of CO_2 and water vapor. The structure of a leaf allows CO_2 to diffuse from the atmosphere to the site of carbon fixation in the leaf cells while allowing some control over the amount of water evaporated. Resistance models of leaf gas exchange have been developed by Bennett and others (1973) and O'Dell and others (1977). These models describe gas exchange along a concentration gradient from the atmosphere to sinks at the leaf surface and inside the leaf mesophyll cells. For most forest tree species, gas exchange is modeled by the equation:

$$U = \frac{C_a - C_i/s}{r_{bll} + r_s + r_m/s} + \frac{C_a - C_{sfl}}{r_{bll} + r_{sf}} \tag{5}$$

In Equation 5 the uptake flux density to the leaf (U) is proportional to the gradient of pollutant concentration between the air (C_a) and the concentration at leaf surface (C_{sfl}) or at the uptake sites inside the leaf (C_i). Uptake to the leaf surface is limited by the boundary layer resistance (r_{bll}) of the leaf and the additional resistance to uptake due to the leaf surface (r_{sfl}). Uptake to the internal sites is limited by the boundary layer resistance, the resistance through the stomata on the leaf surface (r_s), and the internal, mesophyll resistance (r_m). The internal concentration and internal resistance are divided by the Henry's law solubility (s) for the pollutant of interest to account for the change from gas to liquid diffusion.

In most models, the changes in the sink concentrations are included in the mesophyll and leaf surface resistances so that the internal and external surface pollutant concentrations can be set to zero. This approximation allows the calculation of a leaf resistance, r_L, from the network of surface and internal leaf resistances.

$$r_L = A_L R_L = \frac{\left(r_{bl} + r_s + \frac{r_m}{S}\right)(r_{bl} + r_{sf})}{2r_{bl} + r_s + \frac{r_m}{S} + r_{sf}} \tag{6}$$

This simplification can be justified when the resistances are the actual limiting factors to deposition.

Transport through the vegetative element boundary layer is a combination of turbulent and molecular diffusion. The boundary layer is established by the interaction of the surface (leaf, twig, branch, etc.) and the adjacent air to form a region where the size of the eddies are small compared to those in the forest air space. Surface drag tends to decrease the air speed at the surface to zero. Transport near the surface then becomes limited by molecular diffusion. The thickness of the boundary layer is a function of the element's size, shape, and orientation and of the wind speed. If the element's surface temperature departs significantly from air temperature, thermal "free" diffusion can increase the transport rate. Although a great deal is known about boundary layer transport over objects similar to leaves and other vegetative surfaces, the complexity of forest vegetation surfaces limits the accuracy with which these transports can be estimated in the field.

Wigley and Clark (1974) and Murphy and Knoerr (1977) summarized the available data on mass and heat transfer coefficients for broad-leaved species. Their results suggest that reasonable estimates of the quasi-laminar boundary layer resistance can be made from the equations:

$$Sh = 1.0\ Re^{1/2}\ Sc^{1/3} \qquad Re<7000 \qquad (7)$$

$$Sh = (8.96 + 0.044\ Re^{0.84})\ Sc^{1/3} \qquad Re>7000 \qquad (8)$$

In Equations 7 and 8, the Sherwood number (*Sh*) is the dimensionless ratio of the average length parallel to airflow (*l*) divided by the product of boundary layer resistance r_{bl} and molecular diffusivity of the gas of interest in air (*D*). The Reynolds number (*Re*) is the product of the air density (ρ), the average length to airflow, and wind speed (*u*) divided by the dynamic viscosity of air (μ). The Schmidt number (*Sc*) is the dynamic viscosity divided by the product of the air density and the molecular diffusivity for the gas in air. The equations predict higher transfer rates than those proposed for similar surfaces (metal plates). This is not surprising because leaves are not flat, have rough edges, and flutter in the wind.

Transport for needle-leaf species has been modeled by engineering transport equations for cylinder (wires). These have the form (Hilpert, 1933):

$$Sh = 0.615\ Re^{0.466}\ Sc^{0.333} \qquad (9)$$

As in the case of broadleaf vegetation, the effects of fluttering and changes in wind speed and attack angle probably increase the transport above that found in simpler engineering situations.

The factors that govern the control of leaf gas exchange by the stomata on the leaf surface have been extensively studied and are the subject of numerous reviews (see Jarvis and Manfield, 1981). The following discussion is a brief overview to establish the stomatal mechanisms that limit pollutant uptake. The size of the stomata can be altered by changes in the shape of surrounding guard cells. The shape of the guard cell and its neighbors is determined by their turgor, the pressure of water pushing against the walls. When turgor pressure in the guard cells is high, they assume a shape that opens the stomata. When guard cell turgor is low, the change in shape in conjunction with the pressure of the surrounding cells closes the stomata. Much of the research into guard cell physiology has been pointed at determining why guard cells become more turgid in light, leading to open stomata.

The guard cells, unlike most leaf surface cells, contain chloroplasts. Therefore, it is assumed that they photosynthesize in light. Early models attributed light-induced opening to photosynthetic use of CO_2, reducing cell pH, which, in turn, caused stored starches to be changed to soluble materials that would raise guard cell turgor by increased osmosis. Later research suggests that osmotic pressure is increased in the guard cells by means of a metabolic potassium ion pump either fed directly or triggered by photosynthesis in the cells (Allaway, 1981; MacRobbie, 1981).

Guard cells also open when exposed to decreased CO_2 levels in the ambient air. This could be triggered by the same mechanism that causes light opening because CO_2 is decreased in the cell by photosynthesis. Guard cells are also sensitive to leaf water stress, which is caused by an imbalance in the water leaving the leaf by evaporation and the water entering from the soil by way of the roots and stem. Therefore, if conditions favor high evaporation (i.e., high solar heat load, dry air, high wind, or low soil water availability), the stomata may close.

One factor affecting stomata size is the concentration of air pollutants such as SO_2 (Majernik and Mansfield, 1970; Unsworth et al., 1972; and Temple et al., 1985). At relative humidities above 40%, stomatal opening occurred at SO_2 levels as low as 25 ppb. At lower relative humidities, the stomata closed at concentrations in the same range of exposure level. When exposures were for a few days, permanent opening can be observed, which can result in guard cell death in some cases. In other cases, extended exposure leads to stomatal closure (Unsworth and Black, 1981). The result of leaf exposure to SO_2 can be a change in the plant water balance caused by the inability to control the stomata, which can finally lead to damage from desiccation.

The prediction of stomatal resistance is complex and difficult. As can be expected in such a complex physiological process, there are differences among species and even between individuals of the same species that have experienced different preconditioning. However, attempts to construct water balance models in agriculture, hydrology, and forestry have provided some models that can be used to estimate the stomatal resistance. Light response of stomata can be estimated using the equation:

$$h = \frac{1}{r_s} = h_{mn} + a\ S \tag{10}$$

Equation 10 has been found to predict the linear response of the stomatal conductance (h) to light intensity, from a minimum conductance (h_{mn}) at low light intensities to high conductances at a constantly increasing rate (a) for higher light intensities. Equation 10 has been modified to include the effects of temperature and atmospheric water stress for loblolly pine by adding linear terms for decreasing conductance with increasing air temperature and air water vapor deficit (Murphy, 1985).

Baldocchi and others (1987) have developed a somewhat more complex stomatal resistance model that is not limited to one species and that can be linked to the big leaf deposition model. The light response of the stomata is computed as a function of the photosynthetically active radiation using a radiative transfer submodel. In addition, the canopy stomatal resistance is related to temperature, vapor pressure deficit, and leaf water potential.

Modeling of soil water stress requires modeling the water balance of the forest. A simple but useful model (Cowan, 1965; Spittlehouse and Black, 1981) assumes that the forest is able to supply water to the leaves at a known maximum rate E_s. When meteorological conditions can cause evaporation greater than this rate, the stomata will close enough to maintain the maximum evaporation. The maximum evaporation is determined by soil-water conditions that can be estimated from water balance calculations. Using Monteith's equation (1965) to calculate the stomatal resistance for the soil conditions when water is limiting ($E = E_s$), the stomatal resistance is:

$$r_s = \left\{ \frac{\Delta H_l + \rho c_p \dfrac{(e_s - e_a)}{(R_a + R_{ac})} - \Delta L E_s}{L E_s \gamma - 1} \right\} \left\{ (R_a + R_{ac}) A_L \frac{D_g}{D_w} \right\} \tag{11}$$

In its original form, the Monteith equation solved the stand energy balance, under the big leaf assumption, to determine the stand evaporation. In Equation 11 the forest energy balance is solved for stomatal resistance. Equation 11 requires measured (or estimated) inputs of the radiant heat load on the forest (H_l), the water vapor pressure of the air (e_a), the saturation water vapor pressure at air temperature (e_{as}), and the sum of the aerodynamic resistances. For most applications the density of air, the specific heat of air at constant pressure (c_p), the latent heat of vaporization of water (L), and the ratio of the diffusivities of the gas of interest and water vapor in air (D_g/D_w) can be considered constant. The first derivative of saturation vapor pressure with temperature (Δ) and the psychrometric constant (γ) can be assumed to be constant.

The maximum evaporation rate for a given soil-water status can be estimated from:

$$E_s = b \frac{(\Theta - \Theta_{mn})}{(\Theta_{mx} - \Theta_{mn})} \tag{12}$$

In Equation 12 the slope (b) of the relationship between the normalized soil-water content (θ) may vary with soil type and climate. The maximum (θ_{mx}) and minimum (θ_{mm}) field water contents of the soil are close to the field capacity and permanent wilting points that have been developed for many field soils for agronomic growth estimates. When the supply of water from the soil is not limiting ($E < E_s$), then r_s can be estimated from Equation 10 or a similar function based on other environmental parameters.

Figure 6–2 illustrates the results of applying this model to two hypothetical conditions. The relative stomatal resistance (the calculated less the minimum resistance divided by the maximum less the minimum resistance) increases with

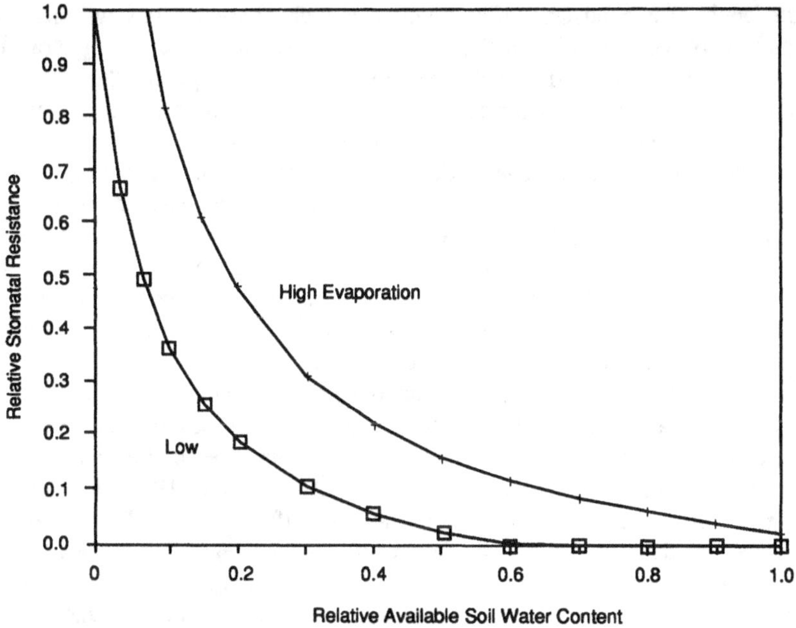

Figure 6–2. The response of the leaf stoma to soil and atmospheric moisture stress according to a simple model of water conduction and stomatal response to leaf moisture stress.

decreasing relative soil moisture content. The resistance increases slowly at higher moisture contents but increases more rapidly at lower moisture contents as the maximum resistance is approached. The effect of high evaporative demand is to close the stomata at higher water contents than at low water demand. This suggests that air pollution damage should be less under high evaporative demands and low soil moisture. This has been observed in crop plants (Heck et al., 1965).

Once in solution the gas, or its decomposition products, must reach the sinks in the cells by molecular diffusion in the cell sap. O'Dell and others (1977) have modeled the diffusion of gaseous pollutants into the mesophyll cells in the plant leaves. Their calculated values of mesophyll resistance were very close to the residual resistance to uptake measured under low pollutant concentrations. This suggests that the internal sinks are not limiting pollutant uptake under these conditions. The fact that pollutant uptake is closely related to solubility (Hill, 1971; Bennett and Hill, 1973) also suggests that uptake of pollutants by plants can often be estimated by assuming that the internal sink concentration is zero.

However, it has been observed that a great deal of the sulfur dioxide absorbed in plants can still be in the form of dissolved SO_2 during exposures under high concentrations (Thomas et al., 1944). The buildup of dissolved SO_2 suggests that vegetation uptake is falling behind diffusion into the plant. Furthermore, there is little doubt that exposures of vegetation at high concentrations will lead to saturation of the metabolic uptake processes and eventual decrease in the rate of deposition. However, it does not appear to be necessary to predict the effects of pollutants on the vegetation metabolism for the purpose of modeling the uptake of gaseous pollutants in forests distant from pollutant sources.

Sorption of sulfur or nitrogen oxide can take place on the vegetation surface. Surface sorption can be through physical adsorption or chemical reaction with the surface, although the actual processes are unknown. In either case, the processes will tend to saturation as sites for sorption are filled. However, it is possible for surfaces to be renewed by erosion or washing, or new sites can be formed through plant growth. The importance of surface sorption, relative to internal uptake, depends on the relationship of sorption to the capacity of the plant surface, and the relationship of the saturation time period to the frequency of surface renewal.

No data are available to determine the fate of absorbed gases. Experiments with labeled sulfur compounds show that some of the sulfur sorbed on the plant surface from SO_2 can be easily washed off (Fowler, 1978; Dollard, 1980). Fowler has estimated this to be as much as 50% of the total sulfur uptake by the vegetation. However, no data are available on the rate of loss from vegetation or the renewal of the surface by precipitation. We can get some insight into the problem by looking at a simple model of this process. If we assume that the uptake can be modeled by simple first-order kinetics, then the sulfur loading on the surface (*SL*) at any time after the surface has been washed (t) can be estimated as a function of the saturation sulfur loading and the time constant of loading from a clean surface (T_s). This relationship is:

$$SL = SS\,[1.0 - \exp(-t/T_s)] \qquad (13)$$

The relationship between the relative sulfur dioxide uptake rate, $d(SL/SS)/dt$, the time to reach 63% saturation, and the average duration between precipitation events can easily be calculated and is shown in Figure 6–3. It is likely that the time to saturation is a function of the air concentration of sulfur dioxide. If the rate of uptake is a linear function of concentration, the deposition velocity assumption, then T_s is linearly proportional to air concentration.

Once the sulfur or nitrogen oxide is sorbed onto the leaf surface, it can remain there either until it is washed off or until the leaf dies and decomposes. In either case, some of the sulfur or nitrogen will be delivered to the soil. An alternate fate is internal uptake by the leaf. The fact that foliar fertilization is successful suggests that internal uptake is possible. Although the rate of uptake has not been documented, transport of sulfur and nitrogen into plants under conditions of closed stomata (dark) is low and suggests that internal uptake by the leaf is small compared to uptake through the stomata (Johansson et al., 1983).

The sorption processes on the vegetation surfaces are changed considerably when the surfaces are wet. The deposition of the gases is related to their solubilities. The solubility of $HNO_3 >> SO_2 > NO_2 > NO$ (Finlayson-Pitts and Pitts, 1986).

The high solubility of HNO_3 results in the surface resistance being negligible,

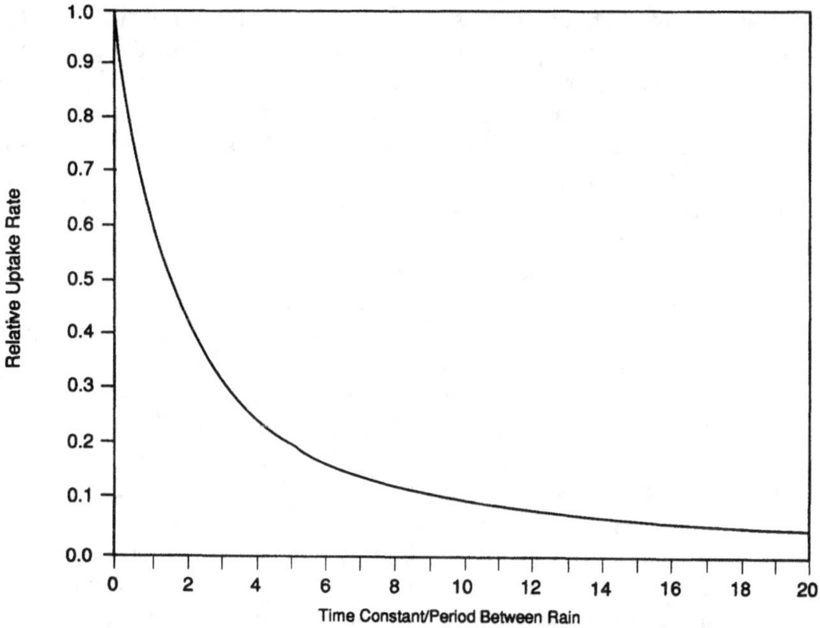

Figure 6–3. The normalized surface uptake rate for a gas, such as SO_2, which reacts with the leaf surface, assuming a first-order reaction with the surface.

and the controlling factor becomes the bulk aerodynamic canopy resistance, R_b (Chameides, 1987). Although there are no data to support the assumption, it is logical to assume that the resistance to diffusion in the forest air space is also negligible because most the HNO_3 should be deposited in the upper portions of the canopy.

The solubility of SO_2 is pH dependent, and for alkaline solutions there are indications that the aerodynamic resistance can be the controlling factor if the wetted surface is alkaline (Brimblecombe, 1978). However, because the pH of dew is typically acidic (Wisniewski, 1982), the low solubilities of SO_2, NO, and NO_2 indicate that the controlling factor to surface uptake by wet vegetation should be a surface-related resistance for these gases.

C. Deposition to the Forest Floor and Soil

Like diffusion and uptake in the forest canopy, uptake of gases by the forest floor is a combination of diffusion through the forest floor (litter) and soil matrix and sorption by sinks in the litter and soil material. Uptake of gases may occur through physical-chemical sorption or processes related to soil organisms.

Moisture content of the soil affects the processes of diffusion, chemical-physical sorption, and biological activity. At high moisture content, gaseous diffusion is limited by plugging of soil pores. At low moisture content, biological and many chemical-physical processes are limited. The forest floor is covered by a layer of partially decomposed organic matter, through which the gases must diffuse to reach the more active soil layers. The moisture content of this upper layer varies rapidly with meteorological conditions and results in varying effects on diffusion into the more active layers of the soil. The sum of these limitations results in the maximum uptake occurring at a moisture content below saturation but still high enough to promote solution and microbial activity—obviously a wide range.

The cycling of sulfur and nitrogen in soils is quite complex and has been extensively studied (see Galloway et al., 1985, for a review). It is possible for reactions in soils to be sources as well as sinks of the sulfur and nitrogen oxides. Under elevated atmospheric concentrations, soils will generally be sinks for SO_2, NO_2, and HNO_3. Because significant amounts of NO are produced in some soils, it is possible for soils to be sources of NO under a wide range of conditions.

Soils do emit sulfur compounds, but these sulfur compounds are in the reduced state and there is no soil source of SO_2 (Golden et al., 1987). It appears that SO_2 uptake is predominantly a physical-chemical process because uptake is not greatly decreased by soil sterilization. Uptake is greatest in alkaline soils because of the acidic nature of the gases. Because many forest soils are acidic, the combination of diffusion limitations by the forest floor and an acidic forest soil may cause forest soils to have relatively low uptake rates for acidic gases such as NO, NO_2, HNO_3, and SO_2. However, almost no field measurements of uptake rates have been made in forest soils.

Although detailed analytical models have been developed for gaseous diffusion with distributed sinks in soil, the data to use one of these models are rare (Lai et al., 1976). Therefore, few attempts have been made to determine effective deposition velocities for gaseous pollutants. We will use the few available sources, primarily observation from agricultural studies, to estimate soil uptake in the latter part of this chapter. It is likely that uptake by forest soils is somewhat less than that measured in agricultural soils because of the additional diffusion barrier of the forest floor litter and soil amendments that raise the pH.

D. Summary

The preceding discussion suggests that the process of dry deposition to forests as a whole and to individual components of the forest ecosystem is affected by (1) the process of turbulent atmospheric diffusion above and within the forest air space, (2) turbulent and laminar diffusion to the surface of plant elements, (3) molecular diffusion through stomata on plant leaves, (4) solution in the plant cells, (5) physiological processes that are sinks for the gases; (6) diffusion through the forest floor and soil air space, and (7) biological processes in the soil matrix. An understanding of dry deposition that could lead to dry deposition predictions based on environmental conditions, plant species, and soil type requires knowledge of the processes in all these areas. Although the present state of knowledge does not allow precise prediction, the existing information does provide a basis for crude estimates of dry deposition for whole forests.

III. Experimental Measurement of Deposition

In this section we present an overview of the methods used to measure dry deposition in forests. Our purpose is to provide the information necessary to interpret the results reviewed in the following section. Information on the details of the techniques can be found in the individual research papers. The strengths and weaknesses of each technique have been reviewed by Hicks and others (1980), Hicks and others (1986), and Davidson and Wu (see the previous chapter).

We divide the measurement techniques for dry deposition into (1) atmospheric techniques that measure the gas exchange through a plane above the forest canopy; (2) chamber techniques that measure extraction from the chamber atmosphere; and (3) techniques that measure accumulation of sorbed gas or its products, including the accumulation of tracers. Atmospheric measurements of dry deposition are normally made above, but near, the forest canopy. The methods can be broadly classified as one of two types. The profile methods depend on measuring a vertical gradient of concentration above the canopy and inferring a diffusion coefficient from the momentum transfer or energy balance of the forest. The eddy flux methods sum the net flux of gas (concentration multiplied by velocity) moving vertically past a point above the forest.

Profile methods require very precise measurements of small differences in concentration for periods of 15 minutes or greater. Eddy flux methods require very fast response measurements that are capable of resolving fluxes at the frequency of the eddies producing the transport. The sampling frequencies most often reported vary from about 1 hertz during the day to greater than 4 hertz at night. The choice of techniques is often based on the characteristics of the instrument available to measure the concentration of a particular gas. Atmospheric methods measure only the net flux through a plane above the forest. The individual processes discussed above are not investigated directly, although information about these processes can be obtained if the momentum, heat, and the water vapor flux are measured simultaneously. Analysis of the momentum flux can provide information about the atmospheric transport processes. Analysis of the heat transfer can provide information about the aerodynamics transport at the canopy forest surfaces. The analysis of the water vapor flux can do the same for stomatal diffusion (Thom, 1975). Atmospheric methods are probably the most practical means of measuring dry deposition to a particular forest vegetation type. These measurement techniques integrate the deposition to the forest some distance upwind of the measurement site. If the measurements are to be characteristic of a certain forest type, it is necessary to have the forest type be uniform for a distance of from 20 to 100 times the height of the forest upwind of the measurement site. Watershed catchment studies have estimated dry deposition for whole watersheds (Norton et al., 1988). Although estimates of deposition based on catchment elemental balances are in general agreement with the values estimated from other studies, the deposition is estimated as the residual not accounted for by measurements of the other processes and thus includes the accumulated error for all the other processes measured in the balance equation.

Chamber methods of measuring gas uptake are used in the laboratory or the field. A major problem with chamber techniques is the impossibility of enclosing vegetation in a chamber without changing the environment in the chamber. Open-top chambers minimize the chamber influence but still do not entirely do away with chamber effects (Davis, 1972). The other alternative is to accept this situation and study the characteristics of the vegetation by controlling the environment. This technique is used in many laboratory experiments in which all the factors thought to affect the plant are carefully controlled and measured. In the field, this level of control is generally not practical, and chambers are designed to maintain environmental conditions as close as possible to those outside the chamber.

One advantage of chambers is that different parts of the forest can be isolated and studied separately. The response of different species of plants or different plant components can be studied in semi-isolation. Thus it is possible to observe deposition to bark separately from deposition to leaf tissue. This has made chamber studies invaluable in determining the effect of the different processes described earlier in this chapter.

Because forests are large and often made up of many ages, species, and sizes of vegetation, only parts of a forest or tree are enclosed. Integration of chamber methods to characterize whole forests requires adequate sampling of this heterogeneous mixture of surfaces. Deposition in the chamber can be measured as the difference between gas concentration at the chamber inlet and at the outlet. It is necessary to have rapid mixing of the gas within the chamber to assure that the outlet concentration is characteristicc of the concentration in the chamber.

Other chamber techniques add gas to the chamber at a rate that maintains the inlet concentration or seals the chamber and measures the transient decrease in concentration of the gas. The deposition is then estimated as the amount of gas added during any period or the amount depleted over a set time interval. In all methods, it is necessary to take the sorption of gas onto the chamber walls into account. This requires careful selection of chamber material and tests of the sorption of the chamber system when no vegetation is present.

Measurement of the accumulation of materials in plants exposed to gases is dependent on the ability to measure change through a given period against the background concentration of the material. It is also necessary that the material be conserved in the plant or plant component under study. This high concentration of sulfur and nitrogen in plant tissue makes direct measurement of accumulation difficult. In addition, sulfur and nitrogen can be taken up from the soil through plant roots and can be quite mobile in plant tissues. For these reasons, isotopic tracers are often used to measure accumulation rates. The isotopes available include the two stable isotopes ^{34}S and ^{15}N and the radioisotope ^{35}S.

In theory, tracers can be used to measure deposition in the open atmosphere or in controlled chambers. However, tracer release in the open atmosphere requires large releases, which are generally too expensive to be practical. Chamber exposures have been used to investigate ^{35}S deposition to tree branches. This type of experiment has the potential for differentiating the site of deposition (internal versus external) and following the fate of the sulfur deposited (Garland and Branson, 1977; Dollard, 1980; Gay and Murphy, 1985). However, chamber exposures of plants to tracers have the same drawback of changing the immediate environment.

There are very few measurements of deposition of sulfur and nitrogen oxide gases to forest soils. The few measurements that have been made used chambers (Bottger et al., 1980). Chamber soil deposition measurements are very difficult because it is impossible to surround the soil volume completely in the field. Only careful design of the chambers to minimize the pressure differential between the inside and outside of the chamber can assure that gas will not be forced through the soil by mass flow. Closed, static chambers do not have the pressure differential problem but do have the problem of maintaining a representative environment under the chamber.

In discussion of the results of deposition measurements that follows, we will use the results from all the types of measurements discussed previously. The measured values will be accepted as characteristic of the system they attempted to measure. Whole forest deposition will be estimated from atmospheric measurement techniques. Deposition to individual parts of the forest is derived from chamber and accumulation techniques applied to the different surfaces. Finally, estimates of the transfer resistances based on the measurements to individual parts of the forest system will be integrated through the simple resistance model to compare the results of whole forest and individual surface measurements.

IV. Results of Measurements of Deposition

The number of measurements of dry deposition of sulfur and nitrogen oxides has steadily increased over the last decade as the recognition of the importance of dry deposition has increased and improved measurement techniques have been developed. Several reviews of dry deposition have been published (McMahon and Denison, 1979; Sehmel, 1980; Voldner et al., 1986; Davidson and Wu in this volume). We have used many of the same data sources in assembling the results presented here. In some cases, we have reported results gathered directly from the reviews when the primary publications were not available to us. The results of measurements of dry deposition to forests are shown in Table 6–1. The results are not given as an extensive list of deposition velocities or deposition rates but are what we consider best values that can be used for crude deposition estimates.

The atmospheric resistances R_a and R_{ac} are from Equattions 1 and 2 presented earlier and have been described in more detail by Davidson and Wu. Two conditions of atmospheric stability are used with gradient profile corrections of 0.5 and -1.5. These are typical values for a daylight unstable case and a nighttime stable case. These values of the stability correction are not extreme cases but provide examples of the range of the resistances that are often encountered over forests.

The leaf and stem boundary layer resistance values r_{bl} and r_{blb} are based on the

Table 6-1. Parameter estimates for deposition of acidic gases to forests.

	R_a		R_{ac}			r_{sf}		r_s				v_d Estimated[a]		v_d Measured[b]	
Gas	Unstable	Stable	Unstable	Stable	r_{bl}	Leaf	Bark	Day	Night	r_m	R_{al}	High	Low	High	Low
							Broadleaf Species								
NO	0.041	2.95	0.047	0.40	0.30			7.7	64.5	0.69	13.3	0.09	0.07		
NO_2	0.041	2.95	0.047	0.40	0.35			9.6	79.9	0.55	10.0	0.55	0.11	0.3–0.8[c]	
HNO_3	0.041	2.95	0.047	0.40	0.39	1	4	11.2	93.5	0.47	1.0	2.25	0.27		
SO_2	0.041	2.95	0.047	0.40	0.39	15	99	11.3	94.2	0.47	9.1	0.55	0.13	0.5–1.6[d]	0.0–0.3[d]
							Conifers								
NO	0.041	2.95	0.047	0.40	0.023			7.7	64.5	0.69	13.3	0.09	0.07		
NO_2	0.041	2.95	0.047	0.40	0.027			11.2	93.5	0.55	10.0	0.55	0.11		
HNO_3	0.041	2.95	0.047	0.40	0.030	0.1	4	11.2	93.5	0.47	1.0	2.26	0.27		
SO_2	0.041	2.95	0.047	0.40	0.030	15	99	11.3	94.2	0.47	9.1	0.57	0.13	0.03–1.5[e]	0.01–0.2[f]

[a] Range of values for midday conditions, wet periods excluded.

[b] Range of values for night conditions.

[c] Hicks et al. (1983).

[d] Matt et al. (1987).

[e] Hicks and Wesely (1980); Fowler and Cape (1983); Lorenz and Murphy (1985); and McMillen et al. (1987).

[f] Fowler and Cape (1983) and McMillen et al. (1987).

engineering calculations in Equations 7, 8, and 9, which can be found in text such as Monteith (1973). These calculations have been tested for broadleaf trees (Wigley and Clark, 1974; Murphy and Knoerr, 1977).

The results of Grennfelt and others (1983) suggest that surface uptake of NO and NO_2 is quite small and this path may be ignored. As stated earlier, the uptake of SO_2 can be high under some circumstances. The leaf surface resistance value used here (r_{sfl}) assumes that the uptake of the leaf surface is one third the uptake through the stomata when the stomata are open (Fowler, 1978; Hallgren et al., 1982; Gay and Murphy, 1985). The surface resistance for stem uptake, $r_{sfb} = A_b R_{sf} - r_{blb}$, is assumed to be one quarter of that for the leaves because bark is comparatively chemically inert. The results of Huebert (1983) indicate that the surface resistance for HNO_3 vapor is very low. The values used for the leaves and bark in the table are based on an estimate of the average boundary layer resistance for these surfaces. Daytime values are assumed to be greater than the night values because of lower wind speed in the canopy.

The stomatal resistance values bound the expected values for open stomata during the day and closed stomata at night. The conifer values are from the review paper of Jarvis and others (1975). The broadleaf stomatal resistances are average values based on measurements reported by Tenhunen and Gates (1975). Both stomatal resistance measurements are based on diffusion of water vapor. The resistances in the table are corrected for the molecular diffusivity of the gas in question.

The internal resistances were taken from the calculations of O'Dell and others (1977). They are liquid diffusion coefficients that were corrected for the relative molecular diffusion of the gases in water solution.

The soil resistances were modified values of those found in the literature. We assumed lowered uptake rates for in situ forest soils than the values reported because we felt that the forest floor litter would impede uptake under forest conditions. In addition, for gases reactive enough to have significant uptake, the concentraton at the soil surface would be decreased because of uptake by the canopy. The values of soil resistance were calculated for NO and NO_2 based on the deposition velocities reported by Bottger and others (1980) divided by 2.0. The value of soil resistance for SO_2 is based on an average of deposition velocities reported by Garland (1977) and Milne and others (1979) divided by 3.0. The low resistance for HNO_3 is based on Huebert's (1983) observation that HNO_3 vapor uptake appears to be limited only by diffusion.

The deposition velocities for NO, NO_2, HNO_3, and SO_2 were calculated by using the resistance network shown in Figure 6–1 and the resistances determined from measurements made on individual surfaces. Calculations were made for a broadleaf and a conifer canopy, both having a leaf area of 4 square meters per square meter of ground surface, and a bark surface area of 1 square meter per square meter of ground area. The results indicate the deposition velocities for the gases and the factors that control uptake. The ranking, in order to increasing uptake, is $NO < NO_2 < SO_2 < HNO_3$.

NO. The uptake of NO to the vegetation is limited by its low solubility. The solubility enters the deposition velocity calculation when the mesophyll resistance is divided by the Henry's law solubility. Most of the NO uptake is by the soil and is not affected by the large change in stomatal resistance between day and night conditions. However, it is quite possible that the NO deposition velocity might be even lower or that the forest soil might be a source of NO under some conditions.

NO_2. There is significant uptake of NO_2 by both the vegetation and the soil. The increased aerodynamic, canopy, and stomatal resistances decrease the uptake at night relative to day conditions. Because of the soil uptake, the deposition velocity does not drop to very low values at night.

SO_2. The deposition velocity for SO_2 is very similar to that of NO_2. However, the processes are not identical. The higher surface uptake is balanced by the increased stomatal resistance, due to the difference in molecular diffusion. The higher surface uptake is more evident at night when uptake through the stomata is very low and the deposition velocity is controlled by the vegetation surface and soil uptake.

HNO_3. The deposition of HNO_3 is assumed to be limited by the atmospheric resistance, R_a and R_{ac}, and the surface boundary layer resistance, r_{bll} and r_{blb}. Only Huebert's (1983) measurements are available at present to guide this choice.

There are a few cases in which data are available to compare the deposition velocities with measurements for whole forest using atmospheric techniques. Hicks and others (1983) have reported values of NO_x, the combination of NO and NO_2, for a deciduous forest. The range of their values of deposition velocity compare well with the range of the sum of the estimates for NO and NO_2 made by integrating the uptake of the forest surfaces using the resistance model, as shown in Table 6–1.

Several studies use atmospheric techniques to measure SO_2 deposition to conifer stands (Hicks and Wesely, 1980; Fowler and Cape, 1983; Lorenz and Murphy, 1985; McMillen et al., 1987), and one reported measurement in a broadleaf forest (Matt et al., 1987). As shown in Table 6–1, the range of deposition velocities is large. Most of the higher values are thought to result from deposition to a wet canopy, which could lead to high uptake at the foliage surfaces and deposition limited only by aerodynamic processes. The lower values are at night and may reflect more stable atmospheric conditions than those used in the calculations. When the extreme values are excluded, the remaining more frequent measurements are in the range of those calculated on the basis of the individual resistances.

In summary, the results of experiments carried out to investigate the dry deposition of SO_2 suggest that the average value of the deposition velocity and the range of values of the deposition velocity are reasonably determined. However, data on a larger variety of forest types would be desirable. The conditions are not as good for NO and NO_2, but enough understanding exists to make some tentative estimates. It also seems likely that most of the processes by which this deposition takes place have been identified. The situation is even less good for HNO_3. More investigations of the processes determining deposition, as well as the total flux by dry deposition, are needed. Although a good deal has been learned about the processes controlling deposition and the rates of deposition of the oxides of nitrogen and sulfur, more information is necessary to improve predictions of deposition to whole forest systems. This is particularly evident for deposition to soils. Table 6–1 shows that soil uptake should be significant for all of these gases; however, almost no field measurements of deposition rates have been made for forest soils.

V. Dry Deposition and the Response of Forests

Global and watershed sulfur balance studies have indicated that the dry deposition of SO_2 is a significant process in both local and global sulfur cycles (Kellogg et al., 1972; Shriner and Henderson, 1978; Norton et al., 1988). Regional and local estimates of dry deposition based on atmospheric transport processes and measured SO_2 concentrations confirm the importance of dry deposition (Garland, 1978; Meszaros and Horvath, 1984). These studies all conclude that dry deposition of sulfur to forests, predominantly by dry deposition of SO_2, can commonly account for 50% or more of the total sulfur deposited from the atmosphere.

Watershed studies suggest that as much as 70% of the atmospheric nitrogen input to a forest ecosystem is from dry deposition (Likens et al., 1977). Based on regional air concentration and atmospheric transport processes, Meszaros and Horvath (1984) concluded that HNO_3 deposition accounted for 56% of the dry deposition and 28% of the total nitrogen input from the atmosphere in the rural areas of Hungary. The corresponding figures for NO_2 were 35% of the dry deposition and 17% of the total nitrogen input.

However, the fact that dry deposition is a significant factor in the deposition of acidic gases does not determine the role of these gases in forest decline. Many forests eventually enter a stage of decline through natural processes. Senescence of stands dominated by older trees, changes in climate, and epidemics caused by pathogens are sources of forest decline. Air pollutants may be the cause of forest decline only in the sense that they are the factor that tips the balance against survival of certain species. Under these conditions, the study of air pollutant stress in isolation may not be relevant. One tool available for organizing observations of complex systems is the systems-modeling technique, which attempts to integrate the effects of all the relevant factors involved in forest dieback. The estimation of the deposition of acidic air pollutant presented in this chapter is only one part of such a model.

Acknowledgment

The information in this paper was developed during the course of work under Contract No. DE-AC09-76SR00001 with the U.S. Department of Energy.

References

Allaway, W. G., 1981. *In* P. G. Javis and T. A. Mansfield, eds. *Stomatal physiology,* Cambridge University Press, Cambridge. 71.

Baldocchi, D. D., B. B. Hicks, and P. Camara. 1987. Atmos Environ 21:91–101.

Baldocchi, D. D., and B. A. Hutchison. 1988. Boundary-Layer Meteor 42:293–311.

Bennett, J. H., and A. C. Hill. 1973. J Air Pollut Control Assoc 23:203–206.

Bennett, J. H., A. C. Hill, and D. M. Gates. 1973. J Air Pollut Control Assoc 23:957–962.

Bottger, A., D. H. Ehhalt, and G. Gravenhorst. 1980. *Atmospharische Kreislaufe von Stickstoffoxiden and Ammoniak,* Report Jul-15582. Kernforschlungsanlage Julich, West Germany.

Brimblecombe, P. 1978. Tellus 30:151–157.

Busch, N. E. 1973. *In* D. A. Hauger, ed. *Workshop on micrometeorology,* 67–100. American Meteorological Society, Boston, MA.

Businger, J. A. 1973. *In* D. A. Haugen, ed. *Workshop on micrometeorology,* 67–100. American Meteorological Society, Boston, MA.

Chameides, W. L. 1987. J Geophys Res 92:11895–11908.

Cowan, I. R. 1965. J Applied Ecology 2:221–239.

Davis, C. R. 1972. J Air Pollut Control Assoc 22:964–966.

Denmead, O. T., and E. F. Bradley. 1985. *In* B. A. Hutchison and B. B. Hicks, eds. *The forest-atmosphere interaction,* 421–442. Reidel, Dordrecht, Netherlands.

Dollard, G. J. 1980. *Wind tunnel studies on the dry deposition of* $^{35}SO_2$ *to spruce, pine and birch seedlings.* Intern Rapport IR 54/80. 37 p.

Finlayson-Pitts, B. J., and J. N. Pitts, Jr. 1986. *Atmospheric chemistry.* John Wiley and Sons, New York. 1,098 p.

Finnigan, J. J. 1985. *In* B. A. Hutchison and B. B. Hicks, eds. *The forest-atmosphere interaction,* 443–480. Reidel, Dordrecht, Netherlands.

Fowler, D., 1978. Atmos Environ 12:369–373.

Fowler, D., and J. N. Cape. 1983. *In* H. R. Pruppacher, R. G. Semonin, and W. G. N. Slinn, eds. *Precipitation scavenging, dry deposition, and resuspension,* 763–774. Elsevier, New York.

Galloway, J. N., R. J. Charlson, M. O. Andrene, and H. Rohde. 1985. *The biogeochemical cycling of sulfur and nitrogen in the remote atmosphere.* D. Reidel, Dordrecht, Netherlands. 249 p.

Garland, J. A. 1977. Proc R Soc London A 254:245–268.

Garland, J. A. 1978. Atmos Environ 12:349–362.

Garland, J. A., and J. R. Branson. 1977. Tellus 29:445–454.

Garratt, J. R., and B. B. Hicks. 1973. Quart J R Met Soc 88:680–687.

Gay, D. W., and C. E. Murphy, Jr. 1985. *Final contract report: The deposition of sulfur dioxide on forests.* Final Report EPRI Project RP1813-2.

Golden, P. D., and W. C. Kuster, D. L. Albritton, and F. C. Fehsenfeld. 1987. J Atmos Chem 5:439–467.

Grennfelt, P., C. Bengtsson, and L. Skarby. 1983. *In* H. R. Pruppacher, R. G. Semonin, and W. G. N. Slinn, eds. *Precipitation scavenging, dry deposition, and resuspension,* 753–761. Elsevier, New York.

Hallgren, J. E., S. Linder, A. Richter, E. Troeng, and L. Granat. 1982. Plant Cell Environ 5:75–83.

Heck, W. W., J. A. Dunning, and I. J. Hindawi. 1965. J Air Pollut Control Assoc 15:511–515.

Hicks, B. B., D. D. Baldocchi, T. P. Meyers, R. P. Hosker, Jr., and D. R. Matt. 1987. Water Air Soil Pollut 36:311–330.

Hicks, B. B., D. R. Matt, R. T. McMillen, J. D. Womack, and R. F. Shetter. 1983. *In* Samson, ed. *Trans meteorology of acid deposition,* an APCA specialty conference, 189–210.

Hicks, B. B., and M. L. Wesely. 1980. *In* D. S. Shriner, C. R. Richmond, and S. E. Lindberg, eds. *Atmospheric sulfur deposition.*

Hicks, B. B., M. L. Wesely, and J. L. Durham. 1980. *Critique of methods to measure dry deposition: Workshop summary.* EPA-600/9-80-050.

Hicks, B. B., M. L. Wesely, J. L. Durham, and M. A. Brown. 1982. Atmos Environ 16:2899–2903.

Hicks, B. B., M. L. Wesely, S. E. Lindberg, and S. M. Bromberg, eds. 1986. Proceedings of the NAPAP Workshop on Dry Deposition. March 25–27, Harpers Ferry WV. 77 pp.

Hill, A. C. 1971. J Air Pollut Control Assoc 21:341–346.

Hilpert, R. 1933. Warmeabgabe von geheizten Drahten und Rohren. Forsch. Gebiete Ingenieurwesen 4:220.

Huebert, B. J. 1983. *In* H. R. Pruppacher, R. G. Semonin, and W. G. N. Slinn, eds. *Precipitation scavenging, dry deposition, and resuspension,* 785–794. Elsevier, New York.

Jarvis, P. G., G. B. James, and J. J. Landsberg. 1976. *In* J. L. Monteith, ed. *Vegetation and the atmosphere,* vol. 2, 171–238. Academic Press, London.

Jarvis, P. G. and T. A. Mansfield. 1981. *Stomatal Physiology.* Cambridge University Press, New York.

Johansson, C., A. Richter, and L. Granat. 1983. *In* H. R. Pruppacher, R. G. Semonin, and W. G. N. Slinn, eds. *Precipitation scavenging, dry deposition, and resuspension,* 763–774. Elsevier, New York.

Judeikis, H. S., and A. G. Wrenn. 1978. Atmos Environ 12:2315–2319.

Kellogg, W. W., R. D. Cadle, E. R. Allen, A. L. Lazrus, and E. A. Martell. 1972. Science 175:587–596.

Lai, S., J. M. Tiedje, and A. E. Erickson. 1976. J Soil Sci Soc Am 40:3–6.

Likens, G. E., F. H. Bormann, R. S. Pierce, J. S. Eaton, and N. M. Johnson. 1977. *Biogeochemistry of a forested ecosystem.* Springer-Verlag, New York. 146 p.

Lorenz, R., and C. E. Murphy, Jr. 1985. Atmos Environ 19:797–802.

MacRobbie, E. A. C. 1981. *In* P. G. Javis and T. A. Mansfield, eds. *Stomatal physiology,* 51, Cambridge University Press, Cambridge.

Majernik, O., and T. A. Mansfield. 1970. Nature 227:377–378.

Matt, D. R., R. T. McMillen, J. D. Womack, and B. B. Hicks. 1987. Water Air Soil Pollut 36:331–347.

McMahon, T. A., and P. J. Denison. 1979. Atmos Environ 13:571–584.

McMillen, R. T., D. R. Matt, B. B. Hicks, and J. D. Womack. 1987. *Dry deposition measurements of sulfur dioxide to a spruce-fir forest in the Black Forest: A data report.* NOAA Technical Memorandum ERL ARL-152, Atmospheric Turbulence and Diffusion Division, Oak Ridge, TN.

Meszaros, E., and L. Horvath. 1984. Atmos Environ 18:1725–1730.

Meyers, T. P. 1987. Water Air Soil Pollut 35:261–278.

Milne, J. W., D. B. Roberts, and D. J. Williams. 1979. Atmos Environ 13:373–379.

Monteith, J. L. 1965. In *The state and movements of water in living organisms,* 19th Symp Soc Expl Biol:205.

Monteith, J. L. 1973. *Principles of environmental physics.* American Elsevier, New York. 241 p.

Murphy, C. E., Jr. 1985. Carbon dioxide exchange and growth of a pine plantation. Forest Ecology and Management 11:203–224.

Murphy, C. E., Jr., and D. R. Knoerr. 1977. Boundary-Layer Met 11:223–241.

Murphy, C. E., Jr., T. R. Sinclair, and K. R. Knoerr. 1977. Environ Quality 6:388–396.

Norton, S. A., J. S. Kahl, D. F. Brakke, G. F. Brewer, T. A. Haines, and S. C. Nodvin. 1988. Sci Total Environ 72:183–196.

O'Dell, R. A., M. Taheri, and R. L. Kabel. 1977. J Air Pollution Control Assoc 27:1104–1109.

Owen, P. R., and W. R. Thompson. 1963. J Fluid Mech 15:321–334.

Pierson, W. R., W. W. Brachaczek, R. A. Gorse, Jr., S. M. Japar, and J. M Norbeck. 1986. J Geophys Res 91:4083–4096.

Sehmel, G. A. 1980. *In* D. S. Shriner, C. R. Richmond, and S. E. Lindberg, eds. *Atmospheric sulfur deposition,* 223–236. Ann Arbor Science, Ann Arbor, MI.

Shaw, R. H. 1977. J Appl Meteorol 16:514–521.

Shreffler, J. H. 1978. Atmos Environ 16:1497–1503.

Shriner, D. S., and G. S. Henderson. 1978. J Environ Qual 7:392–397.

Sinclair, T. R., C. E. Murphy, Jr., and K. R. Knoerr. 1976. J Applied Ecology 13:813–829.

Spittlehouse, D. L., and T. A. Black. 1981. Water Resources Res 17:1651–1656.

Stewart, J. B., and A. S. Thom. 1973. Quart J Royal Meteorol Soc 99:154–170.

Temple, P. J., C. H. Fa, and O. C. Taylor. 1985. Environ Pollut 37 (Series A):267–279.

Tenhunen, J. D., and D. M. Gates. 1975. *In* D. M. Gates and R. B. Schmerl, eds. *Perspectives of biophysical ecology,* 213–226. Springer-Verlag, Berlin.

Thom, A. S. 1975. *In* J. L. Monteith, ed. *Vegetation and the atmosphere,* 57–109. Academic Press, London.

Thomas, M. D., R. H. Hendricks, and G. R. Hill. 1944. Some chemical reactions of sulphur dioxide after absorption by alfalfa and sugar beets. Plant Phys 19:212–226.

Unsworth, M. H. 1981. *In* J. Grace, E. D. Ford, and P. G. Jarvis, eds. *Plants and their atmospheric environment,* 111–138. Blackwell, Oxford.

Unsworth, M. H., P. V. Biscoe, and H. R. Pinckney. 1972. Nature 239:458–459.

Unsworth, M. H., and V. J. Black. 1981. *In* P. G. Javis and T. A. Mansfield, eds. *Stomatal physiology,* 187. Cambridge University Press, Cambridge.

Urone, P., and W. H. Schroeder. 1978. *In* J. O. Nriagu, ed. *Sulfur in the environment, Part I: The atmospheric cycle,* 297–324. John Wiley & Sons, New York.

Voldner, E. C., L. A. Barrie, and A. Sirois. 1986. Atmos Environ 20:2101–2123.

Wesely, M. L., and B. B. Hicks. 1977. J Air Pollut Control Assoc 27:1110–1116.

Wigley, G., and J. A. Clark. 1974. Boundary-Layer Meteorol 7:139–150.

Wisniewski, J. 1982. Water Air Soil Pollut 17:361–377.

Throughfall Chemistry and Canopy Processing Mechanisms

Douglas A. Schaefer* and William A. Reiners†

Abstract

Forest canopies receive chemical inputs from the atmosphere by rainfall, cloud droplet capture, and the accumulation of particles and vapors by dry deposition. These chemical inputs interact with surfaces in the canopy and are released to the forest floor primarily as throughfall (TF). A quantitative understanding of chemical fluxes in TF requires examination of the mechanisms of TF processing by forest canopy components. One such mechanism involves interactions between anthropogenic acidity inputs and the forest canopy, which may increase chemical fluxes in TF. The processes that control the inorganic chemistry of coniferous forest TF include atmospheric inputs from wet and dry deposition as well as physical, chemical, and biological processes that occur on forest canopy surfaces. A review of these processes suggests that dry deposition washoff, diffusion, uptake, and cation exchange control TF chemistry. In this chapter we use these processes to generate hypothetical patterns for ion-specific TF chemical fluxes. We compare these hypotheses to short-term sequential samples of the net ionic fluxes in TF and two coniferous forests that differ in atmospheric inputs. Substantial progress has been made toward the development of a general model of TF chemistry. Experiments are proposed to address the questions that remain for the development of such a model.

I. Introduction

Forest canopies receive chemical inputs from the atmosphere by rainfall, cloud droplet capture, and the accumulation of particles and vapors by dry deposition. These chemical inputs interact with surfaces in the canopy and are released to the forest floor primarily as throughfall (TF).

*Syracuse University, Department of Civil Engineering, 220 Hinds Hall, Syracuse, NY 13244, USA.

†Department of Botany, University of Wyoming, Laramie, WY 82071, USA.

Compared to the substantial body of literature on TF chemistry (cf. Parker, 1983), relatively few studies address the mechanisms by which forest canopies may alter the chemistry of TF during its descent to the forest floor (Eaton et al., 1973, 1978; Luxmoore et al., 1978; Hoffman et al., 1980a; Olson et al., 1981; Parker, 1983; Cronan and Reiners, 1983; Lovett et al., 1984; Miller, 1984; Lovett and Lindberg, 1984; Lovett et al., 1985; Lindberg et al., 1986).

The advance from empirical to mechanistic understanding of TF chemistry depends on quantitative examinations of the mechanisms of TF processing by forest canopy components. An understanding of these mechanisms is necessary for the development of models that provide predictive power over a range of environmental conditions. For example, increases in ionic loss from forest canopies exposed to acidic inputs have led several authors to speculate on the involvement of anthropogenic acidity in this process (Abrahamsen et al., 1980; Rehfuess et al., 1982; Mollitor and Raynal, 1983; Ulrich, 1983; Evans, 1984; Jacobson, 1984; Miller, 1984; Johnson and Richter, 1984; Landolt and Keller, 1985; Lovett et al., 1985; McLaughlin, 1985; Schutt and Cowling, 1985). One potential pathway is through direct interaction between surface water acidity and the canopy components themselves (Evans, 1982a, 1982b).

If the acidity of atmospheric inputs affects TF processing, then canopy chemical fluxes may vary with the pH of atmospheric inputs (Hornvedt et al., 1980; Evans et al., 1981; Lindberg and Harriss, 1981; Keever and Jacobson, 1983a, 1983b; Richter et al., 1983; Ulrich, 1983; Lovett et al., 1985; Kreutzer and Bittersohl, 1986; Skiba et al., 1986; Parker et al., 1987; Joslin et al., 1988). Manipulations of input acidity in small-scale field experiments have supported the theory that acidic inputs increase ionic losses from plant canopies (Wood and Bormann, 1975; Lee and Weber, 1982; Scherbatskoy and Klein, 1983).

Although chemical concentrations of stemflow (SF) are generally much higher than those in TF, SF fluxes represent only a small percent of TF fluxes in coniferous forests (Nihlgard, 1970; Mahendrappa and Ogden, 1973; Olson et al., 1981; Freedman and Praeger, 1986; but see Miller, 1984, for an exception). For this reason, our review specifically addresses processes controlling TF chemistry. However, it should be recognized that many of these mechanisms are important for SF as well.

The objective of this review is to examine the processes that control the chemistry of TF in forest canopies. Our major focus is the inorganic ions in coniferous forest TF, but organic compounds and deciduous canopies are considered in certain topics as well. Our review uses two prototype coniferous forests situated in strongly contrasting environments to assess TF chemistry. First we compare the atmospheric inputs to which these two forests are exposed; then we examine in detail processes controlling TF chemistry that occur in forest canopies in general. These processes are then used to predict ion-specific hypothetical TF chemistry patterns, which are compared to short-term sequential samples of the net ionic fluxes in TF in these two forests. We conclude by considering the progress that has been made and the questions that remain to be answered to develop a general model of TF chemistry.

II. Processes That Control Throughfall Chemistry

A. Atmospheric Inputs

Atmospheric inputs to forest canopies include rain and cloud water during the growing season and their solid-phase counterparts, snow and rime ice, during winter. Dry deposition of aerosol particles and vapors also takes place throughout the year.

This review does not focus on the mechanisms by which forest canopies acquire chemical inputs from the atmosphere. These mechanisms are treated elsewhere in this series (Davidson and Wu, 1988). Instead, we consider the types of processing these inputs undergo during their passage through the canopy as TF. To discuss processing, however, we first must present information on inputs to our two prototype canopies from rain and snow, cloud and rime, and dry deposition of particles and vapors. Dry deposition inputs depend both on mean air concentrations and the deposition velocities of the species involved. These deposition velocities, which are not well known, may show substantial intersite differences. The comparison of inputs is therefore limited to mean air concentrations.

1. Rain and Snow

The balsam fir (BF) forest has an estimated annual precipitation of 180 cm (Dingman, 1981; Reiners et al., 1984). Only 35 to 40 cm falls as rain during the growing season (Cronan, 1980; Foster, 1984): The rest occurs primarily as snow. The most appropriate snow chemistry data available are those of Hornbeck and others (1977), which were collected 30 km away at the Hubbard Brook Experimental Forest, New Hampshire. For the BF site, the annual chemical flux is derived by applying Cronan's data (1980) to an annual rainfall of 50 cm rain and that of Hornbeck and others (1977) to an annual snowfall of 130 cm. Volume-weighted concentrations and annual fluxes are presented in Table 7–1.

All wet-deposition inputs to the loblolly pine (LP) forest site (elevation 230 m) in Tennessee have been collected since January 1986 (Lindberg et al., 1988a, 1988b). At this site wet deposition is dominated by 98 cm of rain annually, with only 3 cm of snow. Volume-weighted concentrations and annual fluxes are also presented in Table 7–1.

2. Cloud Water and Rime Ice

Cloud deposition at the BF site has been estimated as 840 mm/year (Lovett et al., 1982). To model cloud deposition, Lovett assumed a cloud liquid water content (LWC) of 0.4 g m^{-3}, based on literature values. We used Lovett's (1981) passive collector capture efficiency and 55 h of additional cloud data to derive a time-weighted modal LWC value of 0.23 g m^{-3}. This value was determined during mixed cloud and rain events, but the cloud deposition estimates of Lovett and others (1982) were based on cloud-only events. With estimates of cloud chemistry obtained during both cloud-only events and mixed cloud and rain events

Table 7-1. Volume-weighted mean concentrations and annual fluxes in rain (R) and snow (S) to the balsam fir[a] and loblolly pine[b] forests described in the Appendix. The pH values are determined from volume-weighted mean concentrations.

	pH	H^+	SO_4^{2-}	NO_3^-	Cl^-	Ca^{2+}	K^+	Mg^{2+}	Na^+	NH_4^+
Balsam fir										
R μmol(c) L^{-1} (50 cm)	4.08	83	75	21	7	9	3	3	4	13
R mmol(c) m^{-2}		42	38	11	3.5	4.5	1.5	1.5	2	6.5
S μmol(c) L^{-1} (130 cm)	4.26	55	42	23	7	9	3	3	4	13
S mmol(c) m^{-2}		71	55	30	9.1	12	3.9	3.9	5.2	17
R + S mmol(c) m^{-2}	4.20	113	92	40	13	16	5.4	5.4	7.2	23
Loblolly pine										
R μmol(c) L^{-1} (97.5 cm)	4.36	44	41	18	5.9	7.2	0.47	2.4	3.6	11
R mmol(c) m^{-2}		43	40	17	5.8	7.0	0.46	2.3	3.5	11
S μmol(c) L^{-1} (2.6 cm)	4.39	41	27	29	5.6	8.0	0.51	2.1	2.0	9.4
S mmol(c) m^{-2}		1	0.7	0.77	0.15	0.21	0.01	0.05	0.05	0.25
R + S mmol(c) m^{-2}	4.36	44	42	18	6.0	7.2	0.47	2.4	3.5	11

[a] From Hornbeck et al. (1977), Cronan (1980), and Foster (1984).

[b] From Lindberg et al. (1988a, 1988b).

and with new data on cloud LWC and duration of canopy immersion, we recalculated cloud water and chemical deposition to the BF canopy.

To calculate cloud deposition fluxes to the BF forest, we followed Lovett and others (1982) for rime deposition, assuming 40% cloud immersion time during the winter. We used the Lovett and others (1982) cloud chemistry data for rain-free cloud events. We used Foster's (1984) estimates for relative humidity and wind speed throughout the year and for duration of cloud events during the summer. Data for volume-weighted mean cloud LWC and chemical concentrations for mixed rain and cloud events are from Schaefer (1986). This revised estimate of 40.5 cm year^{-1} is presented in Table 7–2. Although these deposition estimates are only half those of Lovett and others (1982), cloud deposition still appears to be a major chemical input to the BF canopy. Mount Moosilauke is now a site in the Mountain Cloud Chemistry Program (although those samples are collected at lower elevation), and it is anticipated that cloud water input estimates for that site should improve in the near future.

Fog water inputs at the LP site result from nocturnal adiabatic cooling fogs formed under calm wind conditions. Approximately 20 such events have been sampled with an active collector, and the resulting volume-weighted mean chemical concentrations are presented in Table 7–2 (Lindberg et al., 1988a). Due to the low wind speeds (and resultant low impaction) during these events and their very intermittent nature (fog less than 10% of the time), we estimate that deposition to this canopy is on the order of 0.8 cm year^{-1} (Lindberg et al., 1988b).

3. Vapors and Particles

Concentrations of sulfur- and nitrogen-containing atmospheric vapors and particles have not been published for Mt. Moosilauke, but measurements made at Whiteface Mountain, New York (200 km away), at the Hubbard Brook Experiment Forest, New Hampshire (10 km away), and at other sites in the northeastern United States (Altschuller, 1983) are presented in Table 7–3.

Concentrations of gaseous, aerosol, and fine particulate species have been measured intermittently at the LP site (approximately 40% of the total dry hours have been sampled) since January 1986 (Lindberg et al., 1988a, 1988b). Time-weighted concentrations are presented in Table 7–3. It appears that gaseous and particulate concentrations of N and S are two to ten times higher in the pine forest than in the fir forest.

4. Summary

At the BF site, approximately equal amounts of H^+ are deposited by rain/snow and cloud/rime. Cloud/rime inputs of SO_4^{2-}, NH_4^+, NO_3^-, and K^+ exceed those from rain/snow, even with the cloud/rime inputs reduced relative to the estimate of Lovett and others (1982). At the LP site, much higher air concentrations suggest that a relatively large proportion of H^+, SO_4^{2-}, and NO_3^- deposition results from particles and vapors. Fog water inputs at LP are generally insignificant and approach 20% of total inputs only for NH_4^+.

Table 7-2. Volume-weighted mean concentrations and annual fluxes in cloud (C) and fog (F) water to the balsam fir[a] and loblolly pine[b] forests described in the text. The pH values are determined from volume-weighted mean concentrations.

	pH	H^+	SO_4^{2-}	NO_3^-	Cl^-	Ca^{2+}	K^+	Mg^{2+}	Na^+	NH_4^+
Balsam fir										
C μmol(c) L^{-1} (40.5 cm)	3.56	270	330	180	9	10	11	19	32	102
C mmol(c) m^{-2}		110	130	74	3.7	4.1	4.3	7.6	13	42
Loblolly pine										
F μmol(c) L^{-1} (0.8 cm)	4.72	19	170	52	44	88	40	19	64	206
F mmol(c) m^{-2}		0.2	2	0.5	0.4	0.8	0.4	0.2	0.6	2

[a] From Lovett (1981) and Schaefer (1986).

[b] From Lindberg et al. (1988a, 1988b).

Table 7-3. Time-weighted mean concentrations of vapors (V) and aerosols (A) at the balsam fir[a] and loblolly pine[b] forests described in the text. The vapor phase concentrations are those of SO_2 and HNO_3. The BF values are estimated from values determined at Hubbard Brook Experimental Forest, New Hampshire, and at other remote sites in the northeastern United States.

	SO_4^{2-}	NO_3^-	Cl^-	Ca^{2+}	K^+	Mg^{2+}	Na^+	NH_4^+
Balsam fir								
V $\mu g\ m^{-3}$	3	1						
A $\mu g\ m^{-3}$	0.5	1						1
Loblolly pine								
V $\mu g\ m^{-3}$	12	2.8						
A $\mu g\ m^{-3}$	7.9	0.39	0.10	0.53	0.25	0.06	0.11	1.94

[a]From Altschuller (1983).

[b]From Lindberg et al. (1988a, 1988b).

It is clear that these two sites differ markedly in their atmospheric chemical inputs. Conventional wet deposition inputs are proportionately similar at these sites. In broad terms, cloud deposition at the BF site and dry deposition at the LP site may be the analogous major inputs, resulting from differences in air chemistry and in geographic and topographic setting.

Wet and cloud deposition may undergo liquid phase reactions (i.e., diffusion and cation exchange, discussed later), but particle deposition may simply accumulate on canopy surfaces until a liquid phase is present. A liquid phase is always present for substomatally deposited vapors. These contrasting patterns of atmospheric inputs are expected to cause differences in TF chemical processing.

B. Physical Processes

1. Dry Deposition

As shown above, dry deposition may contribute a much larger fraction of total atmospheric chemical input to the LP canopy. Although dry deposition per se is not a major topic of this review, we summarize here major processes that are thought to be involved. Mechanisms of particle capture by vegetation are reviewed by Chamberlain and Little (1981) and Davidson and Wu (1988); gaseous exchange is reviewed by Fowler (1981), Unsworth (1981), and Davidson and Wu (1988). Hosker and Lindberg (1982) review deposition and assimilation of both particles and gases.

For coarse particles (greater than 10 μm), sedimentation is the primary deposition mechanism. For the smaller particles (1 to 10 μm), sedimentation and impaction are both important mechanisms. For particles between 0.1 and 1.0 μm, deposition to surfaces is very inefficient, as neither mechanism is effective for the penetration of particles in this size range through the laminar boundary layer. For the smallest of the particles (less than 0.1 μm) and gases, Brownian diffusion is the major mechanism. The deposition of gases soluble in water is greater to a wet forest canopy than to a dry one, other factors being equal (Fowler, 1980, 1981; Granat, 1983).

Coniferous forest canopies have been found to be more efficient collectors of atmospheric deposition than broad-leaved canopies (White et al., 1971; Mayer and Ulrich, 1978; Likens et al., 1977; Gosz, 1980; Dochinger, 1980). Coniferous needles have smaller effective diameters than broad-leaved foliage (Wedding et al., 1975, 1977; Schuepp, 1982). A small effective diameter means that particles in a moving airstream are close to the underlying surface when they enter the laminar boundary layer; their momentum can drive them to the surface. A thinner boundary layer may also increase the capture of gases and submicron particles by diffusion (Chamberlain, 1975). Coniferous canopies such as BF and LP are thus relatively efficient at removing particles and gases from the atmosphere. Not all canopy surface particulates are entirely water soluble (Lindberg and Harriss, 1981; Dutkiewicz and Halstead, 1983; Lawson and Winchester, 1979; Ferek and Lazarus, 1983; Munger and Eisenreich, 1983). This means that canopy washoff occurs partially as dissolved ions and partially as particles. However, the sulfate-containing particles of anthropogenic origin are quite soluble (Ferek and Lazarus, 1983; Lindberg and Harriss, 1981).

Sulfate concentration increases in TF have been attributed to deposition of particulate sulfate and gaseous SO_2 (Bache, 1977; Eaton et al., 1973; Garland and Branson, 1977; Mayer and Ulrich, 1978; Miller, 1984; Lindberg et al., 1986; Garten et al., 1988), and to diffusion from the foliar apoplast (Olson et al., 1985; Lovett and Lindberg, 1984; Reiners and Olson, 1984). In any particular forested ecosystem, the proportional contributions of these mechanisms depend on both atmospheric SO_2 concentrations and the availability of sulfate to the plants from soil solution. Although it is probably an important factor at the LP site, SO_2 deposition (because of both lower air concentrations and additional sulfate inputs from cloud water) appears to be a relatively minor contributor of sulfate to TF at the BF site.

When dry-deposited gaseous SO_2 is absorbed by the canopy and assimilated into canopy tissues, it may undergo subsequent release in TF as sulfate in solution. This sulfate is not easily distinguished from apoplastic sulfate derived from root uptake (Galloway and Parker, 1980; Parker, 1983; Johnson, 1984).

2. Particle Resuspension

Although particle attachment to forest canopy surfaces is not generally a reversible process, some particle reentrainment has been measured (Beauford et al., 1975, 1977; Crozat, 1974; Curtin et al., 1974). Indirect evidence for particle release comes from negative deposition velocity measurements derived by the flux gradient method (Droppo, 1980). In this technique, negative deposition velocities result from negative concentration gradients of particulates above the canopy.

3. Evaporation

Water descending the canopy during storms is subject to evaporation. This evaporation can occur whenever a saturation deficit and net radiation are available (Monteith, 1975). Because only pure water can distill by this process, concentrations of particles and ions may increase by this process whereas fluxes in TF do not (Parker, 1983).

4. Passive Excretion of Ions

Plants may passively excrete ions to or near their external surfaces (Stenlid, 1958; Reiners et al., 1986b). This excretion may result from diffusion of ions into the cuticle or from ions left behind by cuticular transpiration. Morgan (1963) reported carbonate crystals forming on the leaves of greenhouse plants, and Rehfuess and others (1982) noted the appearance of calcium sulfate crystals in and around antestomatal cavaties in spruce in the field. An attempt to quantify this process in BF (Reiners et al., 1986b) showed that it is measurable ultramicroscopically. However, excretion increases the surface loadings of ions by only about 1% per day for potassium and sulfate and even less for the other ions measured. These rates are quite low compared to field estimates of dry deposition to foliage (Reiners et al., 1986b).

5. Summary

Dry deposition to forest canopies is a primary source of the inorganic ions in TF. The "dose" of a particular dry-deposited compound to the canopy is the product of its air concentration, deposition velocity, and the duration of the dry interstorm period.

Canopy washoff of dry deposition is a regionally important component of net TF (TF minus rain) fluxes, especially for elements with anthropogenic sources (e.g., N and S; Lindberg and Harriss, 1981; Lindberg and Lovett, 1986; Lindberg et al., 1986). This seems especially likely for sites exhibiting high concentrations of atmospheric particles and vapors and/or relatively low precipitation inputs.

Field rinsing of branches suggests that dry-deposition inputs are removed quite rapidly (Lindberg and Lovett, 1985). The hypothetical patterns of TF chemical fluxes (described in section III) consider both the dry-deposition dose to the canopy and the kinetics of its removal by TF. Specifically, it predicts the following for dry-deposition washoff:

1. It will be proportional to "dose" as described above.
2. It will influence TF fluxes only during the early part of a storm.

C. Chemical Processes

Chemical processing of TF refers to passive diffusion and ion exchange between the surface water and underlying apoplast of canopy tissues. In this section we provide evidence that passive diffusion is a major cause of elevated anionic concentrations in net TF and that diffusion and ion exchange both contribute to cationic concentrations in net TF.

Contact with water causes a loss of ions from plant tissues. Two early reviews of this phenonenon concentrated on ions important to plant nutrition (Arens, 1934; Stenlid, 1958). This field of study grew rapidly with the availability of radioisotopic tracers (Tukey et al., 1958). Reviews by Tukey (1966, 1970, 1971, 1980) chart the growth of understanding of this process.

Contact with water can also result in ionic uptake. Possible mechanisms of foliar uptake were presented by Wittwer and Teubner (1959). A more modern treatment of this subject by Kannan (1980) concentrated on plant species of agronomic importance. There have also been recent attempts to quantify foliar uptake of herbicides with radioisotopically labeled tracer compounds (Cutler et al., 1982).

1. Isolated Foliar Cuticles

The plant cuticle was quite early recognized as the major barrier to chemical exchange between the interior of the plant and the external environment (e.g., Dybing and Currier, 1961). The experiments described below have been performed on cuticles isolated from deciduous plant leaves, but chemical analyses suggest that coniferous and deciduous cuticles should be quite similar (Hunneman and Eglinton, 1972).

Isolation of the cuticle has been performed by both chemical (Holloway and Baker, 1968; Martin and Juniper, 1970) and enzymatic means (Orgell, 1957; Norris and Bukovac, 1968). Both techniques decompose the pectic polysaccharides that attach the cuticle to the underlying epidermal cells, while presumably leaving the cuticle itself intact (Cutler et al., 1982). However, the removal of pectins and underlying epidermal cell walls may alter both ionic diffusion pathways and the ion exchange capacity of the isolate. This drawback remains to extrapolating results from flux experiments on isolated cuticles to whole-plant fluxes.

A greater abundance of ionic binding sites was found on the internal than on the external surfaces of isolated cuticles, and this asymmetrical distribution of binding sites was correlated with rates of ionic flux outward greater than those inward through isolated cuticles (Yamada et al., 1964b, 1966). Presence of cuticular cationic exchange sites was demonstrated by Yamada and others (1964a), who found that $^{45}Ca^{2+}$ accumulation would saturate in time and that a large proportion of those accumulated ions were retained after rinsing with distilled H_2O. However, $^{35}SO_4^{2-}$ accumulation in cuticles did not saturate, and these ions were completely removed by distilled H_2O rinses.

Schonherr and Bukovac (1973) characterized chemical properties of chemically isolated, dried, ground tomato fruit cuticle. They found that cuticle had a density of 1.1 g cm^{-3} and a cation exchange capacity of 0.1 to 1.0 mmol(c) g^{-1}. They also determined by titration that this cuticular material had three pK_a values: 3 (attributed to pectins), 6 (hydroxy fatty acids), and 12 (phenolic hydroxyls).

Schonherr (1976b) measured the flux of tritiated water through isolated cuticles. He concluded that water moved by a combination of diffusion and viscous flow in dewaxed cuticle through pores with an average radius of 4.5×10^{-8} cm. The number of these pores (but not their size) increased with rising pH values (buffered with organic acids) from 5×10^{10} pores cm^{-2} at pH 3 to 11×10^{10} pores cm^{-2} at pH 9. Schonherr attributed these values to increased deprotonation of carboxyl groups within the pores at higher pH values, making the pores more permeable to water molecules. In a similar experiment not preceded by wax removal, cuticular resistance to water flux was much higher and less dependent on pH (Schonherr, 1976a). The waxed resistance was 320 times higher than the dewaxed resistance at pH 3 and 550 times higher at pH 9.

McFarlane and Berry (1974) also measured cation permeabilities of isolated cuticles at different pH values and found that cation penetration was hindered relative to that of water at low pH. They found 10^{12} to 10^{13} cationic binding sites cm^{-2} cuticle and substantial variation among cuticles in permeability to ionic flux.

The findings described above suggest that that heterogeneity of plant waxy cuticles as barriers to ionic fluxes results from the distribution of polar pores and cationic exchange sites, the presence and development of surface waxes, and the chemistry of solutions to which the cuticles are exposed.

2. Foliage with Unacidified Water

For intact leaves, the presence of stomata raises questions as to their involvement in chemical fluxes. Dybing and Currier (1961) showed that surfactants increased dye penetration of stomatal apertures. Schonherr and Bukovac (1972) theoretically demonstrated that droplets of dilute aqueous solutions (with surface tensions greater than 70 dyn cm^{-1}) could not physically penetrate stomata and that the penetration threshold was 25 to 35 dyn cm^{-1}. This surface tension is obtained routinely with surfactants added to foliar herbicides and other sprays.

The effect of altered water surface tension on stomatal penetration was demonstrated empirically in the field by Grieve and Pitman (1978). In their study, Norfolk Island pines' stomata (which are plugged with wax, as is generally true for *Pinus* and *Abies*) were demonstrated to be impermeable to synthetic salt solutions unless solution surface tension was reduced by about 32 dyn cm^{-1}. Detergents released from a newly constructed sewage outfall lowered the surface tension of sea-salt spray at this site, enabled stomatal penetration by the salts, and led to a sudden decline of the trees. These two theoretical and empirical studies show that under natural circumstances stomatal penetration by aqueous solutions should be only a minor influence on TF chemical fluxes.

In the absence of stomatal penetration, intact leaf surfaces have been treated experimentally as homogenous. Kylin (1960a, 1960b) used $^{35}SO_4^{2-}$ to trace sulfate flux into immersed leaves. He found flux rates directly proportional to external sulfate concentration. Jyung and Wittwer (1964) also observed this linear dependence on external concentration for $^{86}Rb^+$ and $H_2^{32}PO_4^-$. Following isotopic influx, they rinsed the then-labeled leaves in distilled water and monitored efflux into distilled water without radioisotopic labeling. This efflux was very slight (5% of the previous influx), but unfortunately the initial distilled water rinse was not assayed. This apparently unidirectional flux—coupled with evidence from light, temperature, and metabolic inhibitor experiments—led the authors to conclude that, although uptake involves crossing a passive diffusion barrier, the overall process was an active (metabolically driven) one.

Levi (1970) found that for intact leaves (as had been shown for isolated cuticles) cation penetration was inversely related to the hydrated ionic radii. Clement and others (1972) rinsed leaves with distilled H_2O and found that the concentration of released K^+ fell rapidly in sequential rinses. This observation led Clement and others to conclude that several light rains would result in more ionic losses than a single heavy one. Zamierowski (1975) effected substantial foliar losses of K^+, Ca^{2+} and Mg^{2+} in the same manner, even from tropical species with thick, waxy cuticles. For coniferous species that retain foliage for more than one year, cuticular weathering reduced cuticular thickness through time (Chabot and Chabot, 1977). Branch rinsing experiments on BF show that newly expanded foliage exhibits lower ionic fluxes than does foliage formed in previous years (Reiners and Olson, 1984).

The linear dependence of uptake on external concentrations and the rapid decay of ionic release in sequential rinses suggest a single mechanism: diffusion between water on leaf surfaces and an aqueous pool of ions within the leaf, external to cell membranes (the apoplast). This diffusion is controlled by the (ion-specific) resistance of the cuticle and epicuticular wax and by the concentration gradient between leaf surface water and apoplast. This mechanism was invoked by Reiners and Olson (1984), Olson and others (1985), and Schaefer and others (1988) to explain observed patterns of TF chemistry; it was explicitly modeled by Lovett and others (1984).

Turkey (1970), summarizing a voluminous literature on chemical losses from aboveground plant tissues, concluded that ion exchange, diffusion, and mass flow were the primary pathways. The experiments cited here show that in spite of low ionic permeability, ionic fluxes do occur through foliar surfaces. Results showing similar flux behavior of intact leaves and isolated cuticles argue against an active metabolic component in this process. These fluxes are controlled by concentration gradients across the cuticle and by chemical properties of specific ions (as expected for diffusion). Cationic fluxes are also affected by cuticular cation exchange.

3. Foliage with Acidified Water

Natural and simulated acidic rain or mist causes increased foliar losses of cations in both herbaceous tree species (Fairfax and Lepp, 1975; Wood and Bormann, 1975; Hindawi et al., 1980; Evans et al., 1981; Reed and Tukey, 1978a, 1978b), in hardwood trees (Eaton et al., 1973; Wood and Bormann, 1975; Hoffman et al., 1980a, 1986b; Cronan and Reiners, 1983; Lovett et al., 1985; Parker et al., 1987), and in coniferous tree species (Abrahamsen et al., 1977, 1980; Cronan and Reiners, 1983; Scherbatskoy and Klein, 1983; Schaefer et al., 1988).

Schonherr demonstrated that the resistance of isolated cuticles increased with increases in external acidity in solutions buffered with organic acids (Schonherr, 1976b) but also that this effect was far less pronounced for cuticles with intact epicuticlar wax (Schonherr, 1976a). Nevertheless, cuticular resistance of intact leaves appears to decrease with increases in strong inorganic external acidity, both for cations and anions (Reed and Tukey, 1978a, 1978b; Evans et al., 1981; Evans, 1984).

The mechanisms by which low external pH may reduce the liquid-phase diffusive resistances of forest canopy components are not yet understood, and several possibilities remain to be examined. Starting from the surfaces of forest canopy tissues, the first possibility is that interactions between external water and canopy surfaces are altered in some way. Increases in foliar surface wettability following exposure to acidifying substances have been observed (Cape, 1983), and erosion of epicuticular waxes and/or cutin has been demonstrated under similar circumstances (Shriner, 1977; Hoffman et al., 1980b; Haines et al., 1985; Huttunen and Soikkeli, 1984; Paparozzi and Tukey, 1983; Magel and Ziegler, 1986; Rinallo et al., 1986). Such erosion is suspected to increase foliar diffusion of ions (Evans, 1982a, 1984; Skeffington and Roberts, 1985; Rinallo et al., 1986).

As discussed earlier, functional groups in cuticular pores may become more charged at higher pH (Schonherr, 1976a, 1976b), leading to increased permeability. This finding apparently contradicts the experiments with intact plants described above. Schonherr's experiments used chemically isolated cuticles exposed to organic acid buffers and may therefore not apply to field situations. Another possibility is that some of the ester linkages in the cutin itself may reversibly depolymerize upon exposure to strong inorganic acidity (Roberts and Caserio, 1965). If this depolymerization occurs, it would explain reversible increases in diffusive resistances at low pH, but the possibility remains to be tested. Complete depolymerization would make cuticle monomers available for dissolution into the TF stream. Hoffman and others (1980b) found long-chain fatty acids in TF and proposed cuticular weathering as a source.

Cell walls may be "loosened" by acidity, a factor important for plant cell elongation (Salisbury and Ross, 1978). Whether externally applied acidity can reduce the resistance of foliar cell walls to ionic fluxes has apparently not been considered.

Cell membranes may lose diffusive resistance when exposed to reduced pH in the apoplast (Huttunen and Soikkeli, 1984; Landolt and Keller, 1985). This loss of diffusive resistance would increase potential transfer of symplastic ions to the apoplast. Cell membranes have also been described as susceptible to damage from foliar injury in general (Tukey and Morgan, 1963). These mechanisms suggest that effects of external pH on diffusive resistances may be quite complex and may take place under conditions less severe than those shown to cause macroscopic foliar damage (e.g., Haines et al., 1980; Evans, 1982b; Bell, 1986).

Diffusion of ions out of the foliar apoplast differs from hydrogen ion exchange for cations held on discrete cuticular exchange sites, however. Mecklenburg and others (1966) suggested that cations were removed from intact leaves by ionic exchange, with carbonic acid derived from atmospheric CO_2 supplying the protons. The cuticular structure studies of Kollatakudy (1980, 1981) and the isotopic flux measurements of Yamada and others (1964a) suggest that cuticular anion exchange sites are far fewer in number than cation exchange sites. Cations (specifically those abundant in the foliar apoplast) are released from cuticular ionic binding sites in exchange for hydrogen ions retained by the foliage (Keppel, 1967; Eaton et al., 1973; Hoffman et al., 1980a, 1980b; Rehfuess et al., 1982; Cronan and Reiners, 1983) and may eventually occupy exchange sites. To our knowledge, however, no studies have dealt explicitly with implications of the lyotropic exchange series or with the actual abundance and saturability during storms of cation exchange sites in forest canopies. Because the number of exchange sites must be finite, these sites may become saturated after long exposures. This phenomenon was noted in the experimental observations of Tukey (1970, 1980), Parker (1983), and Reiners and Olson (1984).

To set an upper limit on the contribution to net TF cation flux by exchange from fixed sites in the foliar cuticle, we estimate the number of cuticular cation exchange sites in the BF forest. This estimate is a product of a foliar surface area of 6×10^8 cm^2 ha^{-1} (Lovett, 1981) and a cation exchange site density of 10^{12} cm^{-2} for isolated cuticles (McFarlane and Berry, 1974). This estimate suggests that the BF canopy possesses 6×10^{20} exchange sites ha^{-1}. Because 1 mol of ions occupies 6×10^{23} exchange sites, only 10^{-3} mol(c) ha^{-1} can be released (assuming full saturation and no recharge) from such sites in exchange for protons retained by the canopy. A different estimate is derived from Schonherr and Huber's (1977) work with isolated fruit cuticle, which indicated 0.1 to 1 mmol g^{-1} cation exchange capacity, a value leading to a much larger estimate of canopy cation exchange: 30 to 300 mol(c) ha^{-1}. The fact that two different methods (intact plant tissue versus cuticular isolate) yield such different estimates of canopy cation exchange (10^{-3} to about 10^2 mol(c) ha^{-1}) suggests that the potential involvement of fixed foliar cationic exchange sites in foliage has not yet been resolved. Three possible methods of quantifying these exchange sites (and their degrees of saturation) are mentioned in section IV. We consider the first estimate with intact foliage to be more realistic for the BF site. We therefore conclude that exchange at fixed foliar sites is not necessarily the major source of cations from the canopy, even when atmospheric inputs provide a more than ample supply of protons for exchange.

Cations can also be released from foliage along with weak organic acid anions (Hoffman et al., 1980a, 1980b; Cronan and Reiners, 1983; Lovett et al., 1985) or along with inorganic anions (Cronan and Reiners, 1983; Lovett et al., 1985). Lovett and others (1985) used dry deposition estimates and charge balance in throughfall to estimate that approximately half of the cations released from three deciduous canopies were associated with weak acidity, with the remainder presumed to be lost as a result of cation exchange in the canopies. Further measurements of this sort should prove useful, especially when coupled with direct measurements of canopy cation exchange capacity.

Coniferous and deciduous forest canopies show distinctly different behaviors with respect to hydrogen ions. When exposed to acidic precipitation, deciduous canopies generally raise the pH of TF, whereas coniferous canopies cause decreases or only small increases in the pH of TF (Cronan and Reiners, 1983; Eaton et al., 1973; Hoffman et al., 1980a; Mayer and Ulrich, 1978; Mahendrappa, 1983). The causes of this difference are not understood, but recent experiments show that weak acidity in coniferous TF is as least as concentrated as that measured in TF from nearby deciduous stands (Lovett et al., 1985; Schaefer and Lindberg, 1988).

Of the sources of TF cations discussed by Cronan and Reiners (1983), diffusion, accompanied by either inorganic or organic acid anions, may predominate over exchange for protons in the foliar cuticle. The studies cited above indicate that diffusion and (hydrogen) ion exchange are both involved for cations lost from the canopy and that anion loss is primarily controlled by diffusion.

4. Foliage with Ozone

Ozone exposure may lead to erosion of cuticular waxes (Fowler et al., 1980; Karhu and Huttunen, 1986). Cape (1983) found that contact angles formed by water droplets on foliar surfaces would decrease after O_3 exposure. Other researchers have noted that ionic losses from foliage can increase after O_3 exposure (Prinz et al., 1984; Hinrichsen, 1986). Wax morphology of current year (but not older) needles were observed to change following both O_3 and acidic exposure (Magel and Ziegler, 1986). Ionic efflux from foliage has also been experimentally determined to increase following both O_3 and acidic exposure (Parker et al., 1987). Cuticular and epicuticular wax structure may also be modified by other atmospheric gases such as SO_2 (Percy and Riding, 1978; Riding and Percy, 1985). It appears that ozone (and other anthropogenic gases) may alter the structure of foliar surfaces. Whether subsequent increases in foliar fluxes result from increased surface wetting, from physical removal of part of the wax-cuticle barrier, or from increased permeability of underlying living membranes has not been determined.

5. Branches and Boles

The external surfaces of branches and boles are analogous in structure to those of foliage, albeit with less resistance to liquid-phase diffusion and with greater wettability (Martin and Juniper, 1970; Kollatakudy, 1980). Both diffusion between surface water and apoplast and cation exchange on extracellular suberin and cell wall pectins are mechanisms available for chemical processing (Carlisle et al., 1967; Mayer and Ulrich, 1980; Leonardi and Fluckinger, 1987). Reduced contact between external water and the apoplast (resulting from layers of dead cells) and much smaller surface areas of these components (compared to those of foliage) both reduce their potential effect on TF chemical processing.

6. Lichens and Other Epibiota

Chemical interactions between lichens and TF water are likely to be more rapid than those involving foliage, both because of a lack of a strong (cuticular) barrier to diffusion and the direct physical contact between TF water and the lichen "apoplast" (Benzing, 1984; Lawrey, 1984).

Cation uptake by lichens appears to take place by adsorption to ion exchange sites external to cell membranes. Uptake experiments reveal that cations are exchanged competitively for protons or other cations (Wainwright and Beckett, 1975; Nieboer et al., 1978) and that exchange for different cations follows the lyotropic series (Nieboer et al., 1976). Farrar (1976a, 1976b, 1976c) found phosphate uptake to be modulated by metabolic inhibitors, light, and temperature. For ammonium and nutrient anions, uptake also appears to be metabolically driven. Active uptake by epiphytes is discussed in section II, D, 5.

As cell wall composition varies among microepiphytes potentially present on forest canopies (Lehninger, 1970; Lawrey, 1984), the potential for cell wall ion exchange also varies. In spite of differences in detail, all of these cell wall polymers expose cation-exchanging carboxyl and hydroxyl groups to the exterior, as do higher plant cell walls. The phosphate groups present in bacterial cell walls seem to increase their cation exchange capacity. The amino groups found in both fungal and bacterial cell walls may inpart an anion exchange capacity to these cells. In spite of the chemical differences, we are aware of no studies that address comparative ionic exchange capacities among these microorganisms.

Evidence from studies of microbial physiology suggests that uptake of nutrient ions from the surrounding environment is an active biological process (Lehninger, 1970). (Microbial uptake is discussed in section II, D, 5.) It is also probable that some ionic uptake is preceded by adsorption to cell wall ionic exchange sites in the epibiotic cell walls, as is the case in plant roots (Epstein, 1972).

7. Summary

Passive diffusion out of foliage, branches, lichens, and other epiphytes appears to be the dominant chemical processing mechanism for anions in TF. For cations, both diffusion and ion exchange occur. Foliage and epiphytes are the major tissues involved in that they make up the bulk of canopy surface area. Epiphytes exhibit metabolically active processing of certain ions.

Diffusional losses and cation exchange from canopy surfaces will increase TF chemical fluxes. The hypothetical patterns described in section III incorporate both diffusion and cation exchange. For ions removed from the canopy in net TF (TF fluxes minus rain fluxes), we hypothesize that canopy apoplastic pools of ions are depleted during a storm, as did Reiners and others (1986a) and Schaefer and others (1988). This leads to several predictions:

1. As these apoplastic pools are depleted, their influence on net TF will decline through storms.
2. Storms with greater precipitation depth will deplete the apoplastic pools to a greater extent than will smaller storms.
3. Varying degrees of depletion of apoplastic pools during storms can influence net TF fluxes in subsequent storms.
4. If ionic diffusion out of apoplastic pools is the rate-limiting step in net TF

fluxes, storms of lower intensity (where the residence time of water in the canopy is longer) will lead to greater net TF fluxes.

The evidence presented above concerning the effects of acidity on ionic fluxes suggests that deposition of acidity to the canopy may increase cation exchange and diffusion out of apoplastic pools. More specifically, the following conclusions can be drawn:

1. Increased acidic inputs to the canopy will increase inorganic cation removal from the canopy (see Lovett et al., 1985).
2. Cation exchange sites will be depleted through storms, but increased rainfall acidity during a storm may remove additional cations from the canopy.

D. Biological Processes

The term *biological processes* refers here to all fluxes of ions between living tissue and surface water that require metabolic energy. In discussing decomposition, we treat all canopy components together because all tissues in the canopy decompose into the same array of inorganic ions.

1. Active Uptake by Higher Plant Foliage

Experimental descriptions of active foliar uptake are based on the response of uptake rates to light, metabolic uncouplers or inhibitors (Epstein, 1972; Wittwer and Teubner, 1959; Jyung and Wittwer, 1964) or to temperature with a Q_{10} value (the ratio of rates measured at two temperatures differing by 10°C) of 2 or greater (Nobel, 1974; Luttge and Higinbotham, 1979).

There are two mechanisms by which active foliar uptake may occur. First, active uptake from apoplast into symplast (cytoplasm or tonoplast) across a membrane surface (or surfaces) could lower the apoplastic concentration of a given chemical species, establishing (or increasing) the concentration gradient between the apoplast and the external solution. This would "drive" passive diffusion into the apoplast, retaining the first nonbiological step and allowing modulation by temperature, light, and externally applied antimetabolites. This mechanism is consistent with the experimental results described earlier and suggests a reason by externally applied adenosine triphosphate (ATP) or urea in some cases facilitates foliar uptake (Kannan, 1980). The metabolic stimulation afforded leaf cells by these treatments may lead to an increase in ionic uptake from the apoplast.

Second, metabolic conversion within the symplast may follow active ionic uptake. For example, sulfate ions could be esterified to the adenosine phosphate carrier (Schiff, 1983) or nitrate reduced to an amine (Smirnoff et al., 1984). These processes would again lead to reduced apoplastic ion concentrations, maintaining the gradient from higher concentrations in external (TF) water to lower concentrations in the apoplast. Several experiments demonstrate conversion of foliar-applied dissolved inorganic ions (Kylin and Hylmo, 1957) and gases (Garsed and Read, 1977) into intracellular catabolic products.

It should also be mentioned that ionized species, strictly speaking, flow along an electropotential gradient, as opposed to a simple concentration gradient (Luttge and Higinbotham, 1979). These gradients are not equal when an active membrane potential is exposed to the aqueous medium. The fixed negative charges of the cuticle, as they appear to be electrically balanced by adsorbed protons or metallic cations, do not alter the effect of the concentration gradient for anions between the exterior and the apoplast (Ritchie and Larkum, 1982).

2. Active Excretion by Aboveground Plant Tissues

Active excretion of ions to foliar surfaces occurs in plants bearing hydathodes, nectaries, salt glands, or other such specialized structures (Arena, 1934; Esau, 1962). These are not generally found in conifers and are absent from *Pinus* and *Abies* in particular (Napp-Zinn, 1966). Various root tissues excrete protons in the process of ion uptake (Epstein, 1972; Lauchli, 1984), but to our knowledge no such process has been reported in foliage.

Branches and boles appear to be similarly devoid of any active excretion mechanism, save sap exudation in response to wounding (Kramer and Kozlowski, 1977).

3. Foliar H_2S Emission

Foliar surfaces may also release gases; the best known example is hydrogen sulfide. Spaleny (1977) measured H_2S release from spruce seedlings under excess sulfur nutrition, and Hallgren and Fredrikkson (1982) found it to occur in pines fumigated with SO_2. Garten (1988) inferred the emission of this gas from mass balance experiments on LP and maple trees labeled with $^{35}SO_4^{2-}$. Rennenberg (1984) found H_2S emission to be common for most plants. Foliar emission of reduced sulfur gases may reduce the amount of SO_4^{2-} that appears in TF, but the influence of this process on SO_4^{2-} fluxes in TF is not yet clear. Whether H_2S is emitted by BF is not known.

4. Arthropod Conversions

Arthropod feeding converts plant canopy tissues into microparticulate and soluble forms (Mattson and Addy, 1975), but the involvement of these compounds in TF chemistry has received scant attention (e.g., Reichle et al., 1973; Crossley and Seastedt, 1981; Seastedt and Crossley, 1984). Arthropods release nutrients in the excreta (Kimmins, 1972; Mattson and Addy, 1975) and by damage to canopy tissues may increase nutrient losses directly from those tissues (Tukey, 1970; Preece and Dickinson, 1971).

In certain coniferous forests, nutrient cycling by canopy arthropods is a continuous, endemic process (Carroll, 1980; Voegtlin, 1982). Canopy arthropod activity is characterized by occasional epidemic outbreaks in BF (Martineau, 1984) and in LP (Baker, 1972). In the absence of such outbreaks in these forests, we expect canopy arthropods to have little influence on TF fluxes.

5. Active Uptake by Lichens and Microepiphytes

Lichens and microepiphytes are treated together because they share a common mode of nutrient acquisition: the active uptake through cell surfaces from external water. This has been demonstrated for lichens (Rasmussen and Johnsen, 1976; Garraway and Evans, 1984; Farrar, 1976c; Nieboer et al., 1978; Pike, 1978) and in particular for epiphytic lichens of the BF canopy (Lang et al., 1976). Active uptake of ions through cell surfaces is the reason that algae and bacteria can be routinely cultured in media far more dilute in nutrient ions than is their cytoplasm (Bowen, 1966; Lehninger, 1970). Lang and others (1976) showed that for epiphytic lichens common in the BF canopy, ammonium and nitrate uptake and cationic loss rates tended to increase with temperature. The Q_{10} of these processes suggested to the authors that active uptake and loss were involved.

Epiphytic algal cells (either free-living or in fungal symbioses) are suspected to excrete sulfonated polysaccharides (J. A. Schiff, personal communication), as has been noted for other algae (Roberts et al., 1980; Percival and McDowell, 1981). Bentley and Carpenter (1984) used $^{15}N_2$ labeling to demonstrate direct (and presumably active) transfer of fixed atmospheric nitrogen from epiphyllous microbes to their host leaves. Other microbial excretions are well known in laboratory culture (Garraway and Evans, 1984) but have not been investigated in the forest canopy environment. Epilithic lichens are known for their ability to excrete organic acids involved in mineral weathering, and epiphytic lichens seem to possess this ability as well (Nieboer et al., 1978). Unfortunately, canopy microbiology remains poorly understood, and further knowledge is essential for detailed treatment of TF processing by forest canopies possessing abundant epibiota.

6. Release of Dissolved Substances by Decomposition

Following decomposition in the canopy, dissolved substances are washed from the canopy in the SF-TF stream. The magnitude of this flux is dependent on (1) the pool of decomposing organic matter in the canopy and (2) the strongly seasonal mineralization, immobilization, and particulate release rates of this matter.

Lichens, foliage, branches, and boles all may die and decompose above the forest floor, thus releasing ions to the TF stream. In Table 7–4, we estimate the maximum necromass pools available in the BF canopy to contribute to ionic flux by mineralization. As fir foliage is shed and falls to the forest floor upon (or before) senescence (Foster, 1984), it does not represent a decomposable pool in the BF canopy. It is therefore not estimated in Table 7–4. For nitrogen, mineralization of even a small fraction of the necromass pool of 150 mmol m^{-2} (Table 7–4) would constitute a significant flux. This finding suggests that the BF canopy uptake of ammonium and nitrate observed by Olson and others (1981) may actually be underestimated. Because of low tissue S content and relatively high precipitation SO_4^{2-}, in-canopy mineralization appears to be a minor source of sulfate to the TF stream. The cations K^+, Mg^{2+}, and Ca^{2+} may show large fluxes by this pathway, but most should appear as SF down standing dead boles.

Table 7-4. Estimated canopy necromass pools in the balsam fir[a] forest described in the text.

Tissue	N	P	K	Ca	Mg	Na	S
Standing boles	130	4.7	110	100	40	3.3	47
Branches	17	0.7	2	4.3	2	0.03	2.7
Lichens	4.3	0.2	0.1	0.03	0.06	0.06	0.08
Total pools [mmol(c) m^{-2}]	150	5.6	110	100	42	3.4	50

[a]From Schaefer (1986).

With some exceptions (e.g., Carlisle et al., 1967; Carroll, 1980), microparticulates are intentionally (by filtration) or unintentionally (by settling during storage) excluded from chemical analyses of TF, even in cases where canopy microepiphytes are suspected to play an important role in nutrient cycling (e.g., Jordan et al., 1980). Although macroparticulate litterfall (greater than 0.1 cm) is routinely collected and analyzed in most nutrient cycling studies (Cole and Rapp, 1981; Vitousek, 1984), microparticulates (for which canopy washoff during storms is presumably a major transfer pathway to forest floor) are not as carefully quantified. In cases where microepiphyte turnover and other sources of microparticulates constitute significant nutrient fluxes, disregarding this pathway weakens the conclusions drawn in studies of TF chemical cycling.

7. Summary

Biological processing of TF increases the concentrations of organic and particulate compounds in TF and reduces the concentrations of inorganic ions, mainly those of nitrogen (there is no evidence concerning other ions). Studies with individual canopy components (Lang et al., 1976; Reiners and Olson, 1984) show that this retention capacity becomes saturated during an individual rain event. Whole-canopy studies suggest that relatively more inorganic nitrogen can be retained when TF water remains on canopy surfaces longer (Reiners et al., 1986a; Schaefer et al., 1988). The hypothesized TF chemical fluxes in section III are based on the observations that the ability of the canopy to retain NO_3^- and especially NH_4^+ is time-limited and that the canopy's retention capacity becomes saturated during a storm.

III. Hypothesized and Observed Throughfall Fluxes

A. Introduction

In the previous section, we concluded that dry deposition, cation exchange, and ion retention and release by diffusion all affect TF fluxes (and do so in rather distinct ways). Descriptions of canopy processing based on these mechanisms are as yet uncommon (Abrahamsen et al., 1977; Olson et al., 1981; Cronan and

Reiners, 1983; Parker, 1983; Lovett and Lindberg, 1984; Olson et al., 1985). Notably lacking are descriptions that take into account canopy ionic uptake (potentially important for N and P) and release of particulates and dissolved organics. The biologically active mechanisms of canopy processing are only infrequently included (but see Lang et al., 1976; Pike, 1978; Carroll, 1980; Olson et al., 1981, 1985; Lovett and Lindberg, 1984).

In this section, we hypothesize how dry-deposition washoff, cation exchange, ion diffusion, and active uptake influence TF chemical fluxes. We compare the hypothesized fluxes to those observed in sequentially sampled TF and rainfall during individual rain events at the BF and LP forests. This methods provides fine temporal resolution of the processes taking place during these events (Parker, 1983; Olson et al., 1985; Reiners et al., 1986a; Schaefer et al., 1988). Site descriptions and methods used in the sequential sampling of TF are presented in the Appendix.

B. Patterns for Ions and Groups of Ions

1. Nitrate

Because nitric acid vapor deposition at the LP site is the primary source of NO_3^- in net TF (Lindberg et al., 1988a), this ion is dominated by dry-deposition washoff at LP. Dry-deposition washoff should result in rapid declines of net TF fluxes for those ions with a large dry-deposition component, as well as a strong relationship with canopy loading during the preceding interstorm period (Lindberg and Lovett, 1983; Lovett and Lindberg, 1984). After washoff of dry-deposited nitrate, net TF fluxes would be negative because of canopy uptake. Canopy uptake would be the dominant process throughout the event in environments with less dry deposition of nitrate. Figure 7–1 shows these two contrasting hypothetical patterns, and Figure 7–2 shows net NO_3^- fluxes in TF measured at LP and BF. Note that at LP (Figure 7–2A) the washoff curve declines steeply to the zero line and the net TF fluxes increase with the interstorm dry deposition dose in the order of sequential storms 17, 11, and 12 (indicated as LP 17, LP 11, and LP 12; with 17, 61, and 285 antecedent dry h, respectively). During events 2 and 5 (BF 2 and BF 5) at the BF site, however, there is very little indication of dry deposition washoff early in the storms, and net TF apparently remains negative throughout the storm (Figure 7–2B), as hypothesized in Figure 7–1. Canopy nitrate retention at BF was attributed to epiphytic lichens by Lang and others (1976).

2. Sulfate

For sulfate, we expect that both dry-deposition washoff and diffusion from apoplastic pools can contribute to net TF fluxes (Olson et al., 1981; Parker, 1983; Garten et al., 1988). Figure 7–3 shows the hypothetical pattern, and Figure 7–4 shows net SO_4^{2-} fluxes in TF measured at LP and BF. For this ion a simple washoff curve may occur (as in LP 10), but it may not decline to zero (as in LP 11). In some cases, the initial fluxes may be exceeded by those later in storms (as in LP 12 and

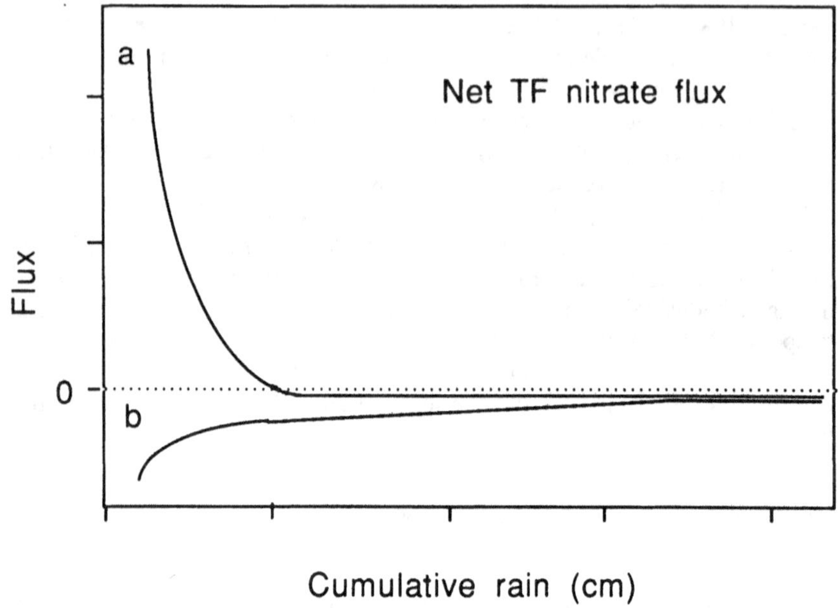

Figure 7–1. Hypothetical pattern of nitrate net throughfall (TF) flux (net TF; flux in TF minus flux in rain) during storms of high (curve a) and low (curve b) antecedent doses of HNO_3 by dry deposition.

BF 3), a finding we, along with Olson and others (1981), attribute to initial low-intensity rain that incompletely wets canopy surfaces.

3. Potassium, Ammonium, and Chloride

We hypothesize that the net TF fluxes of potassium are dominated by diffusion from apoplastic pools. It is important to note that this source also yields exponential decay curves because of the depletion of canopy pools. However, the potassium curve differs from the nitrate curve in two ways: First, the potassium curve decays more slowly and stays positive (hypothetical model shown in Figure 7–5A); second, there can be a strong intensity effect on net TF potassium fluxes. Figure 7–5B shows net fluxes of potassium at LP. A pattern of increases and decreases (see discussion of residence time below) is superimposed on the familiar decay curve. The sequential storms analyzed from BF generally support these patterns for potassium in that storms with the longest residence time show the largest net TF fluxes.

Figure 7–6A illustrates the hypothesized pattern of NH_4^+ net TF flux. This figure suggests that the canopy's capacity to retain NH_4^+ reaches a limit and that greater canopy water residence time increases this retention. Figure 7–6B shows net fluxes of NH_4^+ and residence time in BF 5.

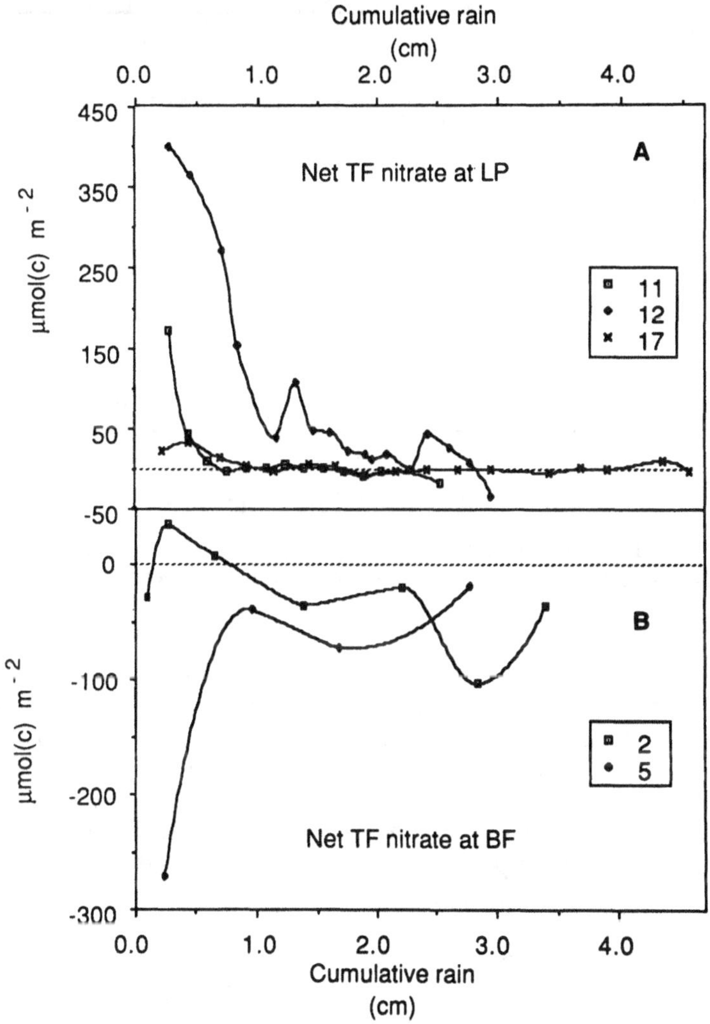

Figure 7–2. (A) Net throughfall (TF) nitrate (net TF; flux in TF minus flux in rain) at the loblolly pine (LP) site in μmol(c) m^{-2}. Storm 17 was preceded by 17 dry h, storm 11 by 61 dry h, and storm 12 by 285 dry h. (B) Net TF nitrate flux at the balsam fir (BF) site in μmol(c) m^{-2}.

Figure 7–7A shows net TF fluxes of potassium and ammonium in LP 13. Note that beyond 0.8 cm, ammonium behaves in essentially an opposite manner to potassium, being retained when potassium is released. Figure 7–7B is a logarithmic plot of the residence time (the inverse of intensity) of rain in LP 13. After about 0.8 cm rain (which may include deposition washoff), more potassium is released (and more ammonium retained) at greater water residence times. An interesting

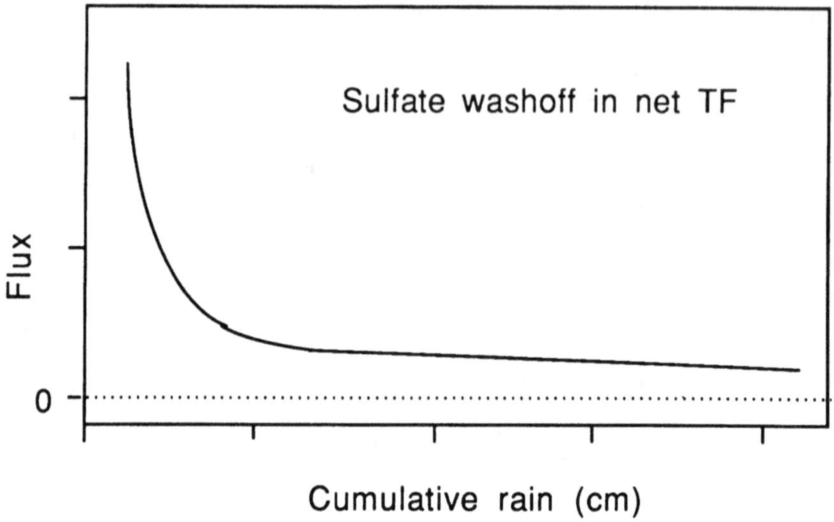

Figure 7–3. Hypothetical pattern of net throughfall (TF) sulfate flux (net TF; flux in TF minus flux in rain). Note that the flux remains positive following a rapid decline.

secondary feature can be seen at 1.4 cm, where residence time has already declined, but potassium release and ammonium retention in net TF remain high. We interpret this feature as the intrinsic time lag required for TF water of a particular composition to be removed from the canopy. This effect has already been noted by Reiners and others (1986a) in their graphical analysis of TF fluxes. The relationship between potassium release and ammonium retention in LP 18 is shown in Figure 7–8A, and the responses of these elements to water residence time are shown in Figure 7–8B.

LP 12 and LP 18 (Figure 7–9A and 7–9B) show very strong correspondences between the net TF fluxes of potassium and chloride, despite varying storm intensities. This suggests that these two ions may have similar canopy sources and that chloride fluxes may be modeled in much the same way as those of potassium. It also suggests that cation exchange may not be an important mechanism controlling net TF potassium fluxes.

4. Calcium

Net TF fluxes of Ca^{2+} should consist of both dry-deposition washoff and cation exchange (cf. Lovett and Lindberg, 1984; hypothetical pattern shown in Figure 7–10A), but the kinetics of the two sources may be quite similar. To distinguish these two processes, we need data from a storm with high rainfall acidity after the initial TF washoff. In such an event, we would predict greater net TF calcium fluxes (resulting from exchange due to the increased rainfall acidity), but such an event has not been sampled. Figure 7–10B shows Ca^{2+} fluxes in LP 13 and LP 18.

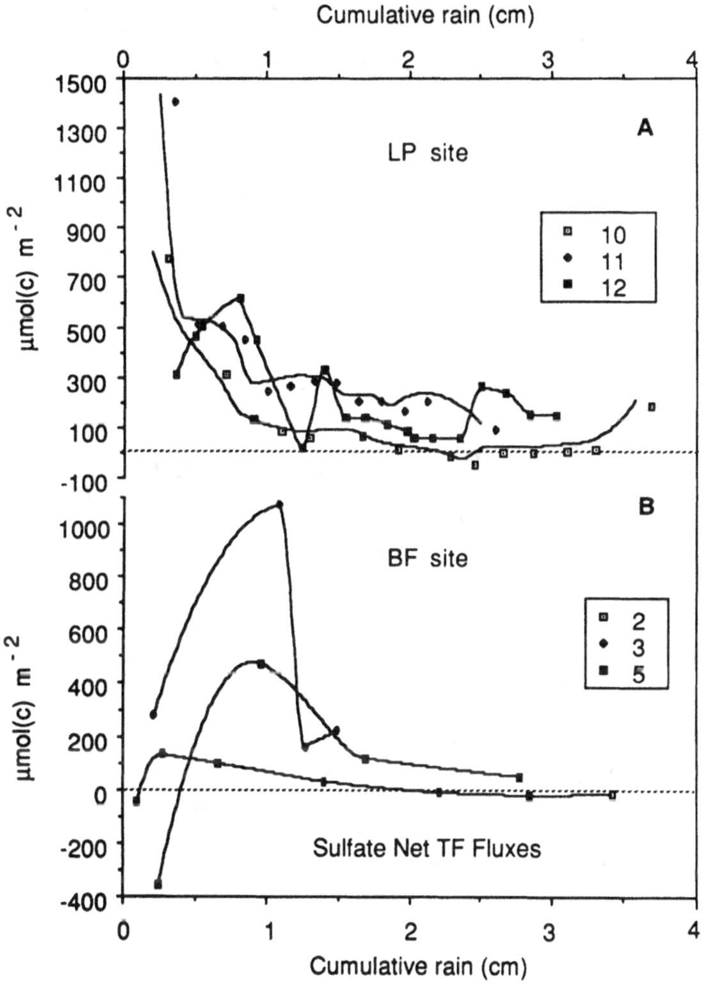

Figure 7–4. (A) Net throughfall (TF) sulfate (net TF; flux in TF minus flux in rain) at the loblolly pine (LP) site (in μmol(c) m^{-2}). (B) Net TF sulfate flux at the balsam fir (BF) site in μmol(c) m^{-2}.

In these events, Ca^{2+} continues to be released by the canopy (in very small amounts) after the initial "pulse" at a rate that is not appreciably influenced by canopy and water retention time. This implies that calcium fluxes in net TF are not time-limited, as are those for potassium, ammonium, and chloride. Unfortunately, neither Ca^{2+} nor Mg^{2+} was analyzed in the BF sequential storms.

One implication of the depletion of canopy apoplastic pools is that storms of differing depth deplete these pools to a different degree. A long storm (relative to

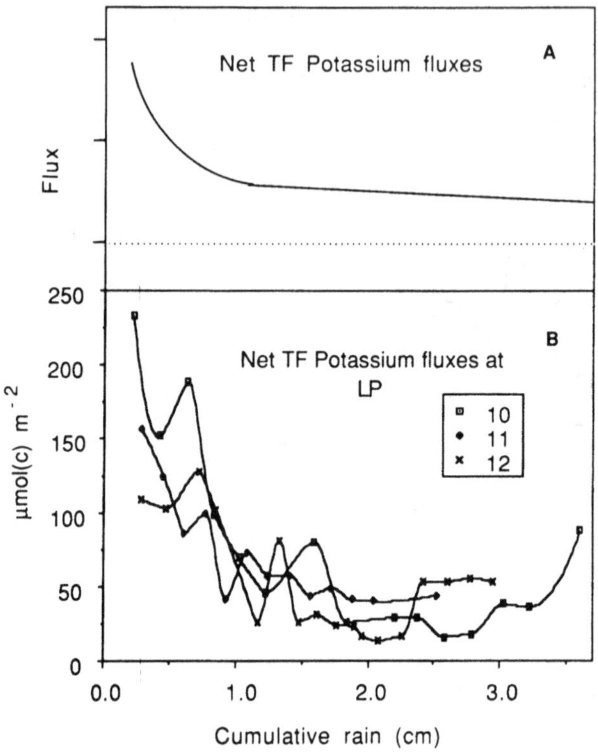

Figure 7–5. (A) Hypothetical pattern of net throughfall (TF) potassium flux (net TF; flux in TF minus flux in rain). Note that the flux remains positive and relatively large following a moderate decline. (B) Net TF potassium flux at the loblolly pine (LP) site in μmol(c) m^{-2}. Note the discontinuous nature of the decline.

the "washout curve" of the ion in question) completely depletes these pools, but a very short one does not. Calcium net TF fluxes for two such LP events are compared in Figure 7–10B. Both of these storms were preceded by a 60-hour (relatively short) dry period, which would minimize the interstorm loading. Although the storm before LP 13 was only 1.3 mm (approximately equal to the canopy interception and not enough to "wash" the canopy), the storm before LP 18 was 10.7 mm.

C. Patterns for Storms in Succession

Ion-specific predictions of TF chemistry can also be made for storms in succession. Positive net TF fluxes for a series of three storms that differ in rainfall duration and intensity are depicted in Figure 7–11A. The interstorm intervals also differ in duration and dry deposition rates. The horizontal axis of the plot can be

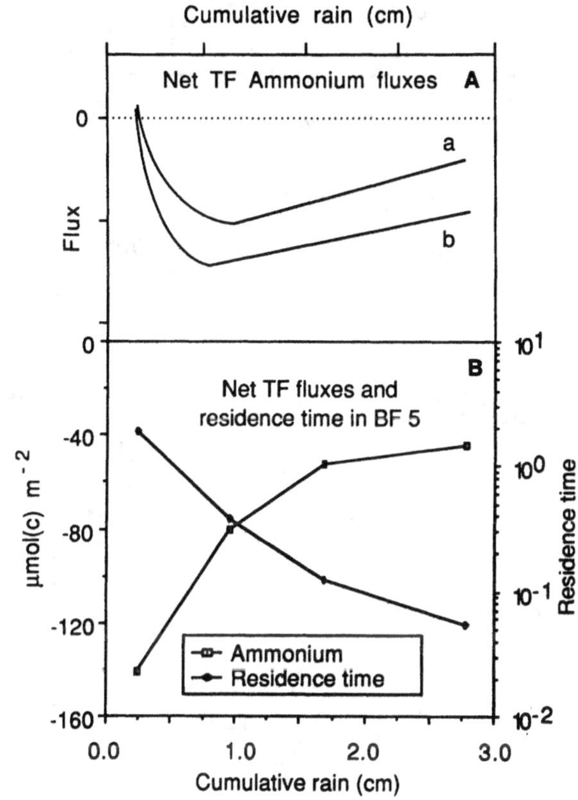

Figure 7–6. (A) Hypothetical pattern of net throughfall (TF) ammonium flux (net TF; flux in TF minus flux in rain) in storms of high (curve a) and low (curve b) rain intensity. Note that the canopy retains ammonium following a rapid intial period of release. (B) Net TF potassium flux in μmol(c) m^{-2} and water residence time (in h/mm) during loblolly pine (LP) sequential storm 18. Note that ammonium retention by the canopy (negative net TF) and the residence times respond inversely during the storm.

regarded as hours (between storms) and as centimeters of rain (during storms). The vertical axis of the plot combined the canopy surface loading from dry deposition and the apoplastic pools available for diffusion of ions from the canopy. The height of the curve declines as ions appear in net TF. In the first storm, the total availability of ions in the canopy is high, but the storm is of limited duration. The second storm is of much longer duration and essentially depletes net TF ions from the canopy. In the third storm, available pools are low, and even long duration does not result in much net TF. Where net TF is negative (as can occur for ammonium and nitrate), the curve would be inverted. This framework shows how interstorm dry deposition loading can be coupled with canopy behavior during storms to yield a complete picture of TF processes.

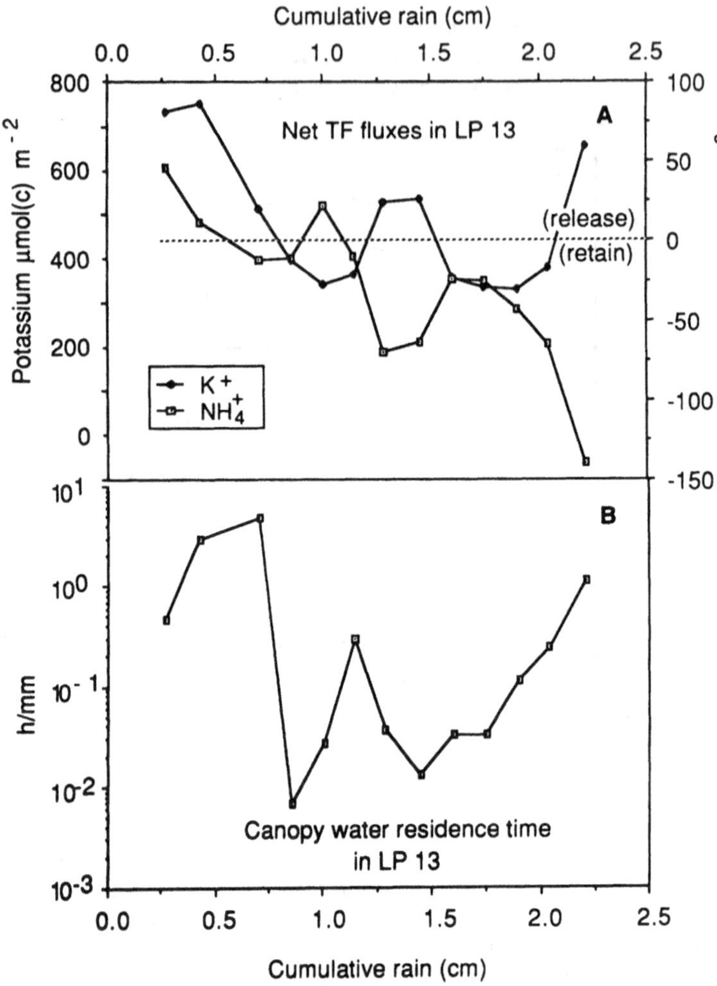

Figure 7–7. (A) Net throughfall (TF) potassium and ammonium (net TF; flux in TF minus flux in rain) in loblolly pine (LP) sequential storm 13 in μmol(c) m^{-2}. Note that ammonium is retained by the canopy (negative net TF) through almost the entire storm and that potassium and ammonium respond inversely during the storm. (B) The inverse of rain rate (i.e., h/mm) plotted on a logarithmic scale. This represents the residence time of water in the canopy during loblolly pine (LP) sequential storm 13. Note that in panel A both potassium release and ammonium retention increase with residence time.

IV. Future Research

A. Throughfall Chemistry Models

1. Introduction

In the previous section, hypothetical TF chemical flux patterns based on TF processing mechanisms were described and compared with field observations. This section synthesizes this material with a quantitative model of chemical fluxes

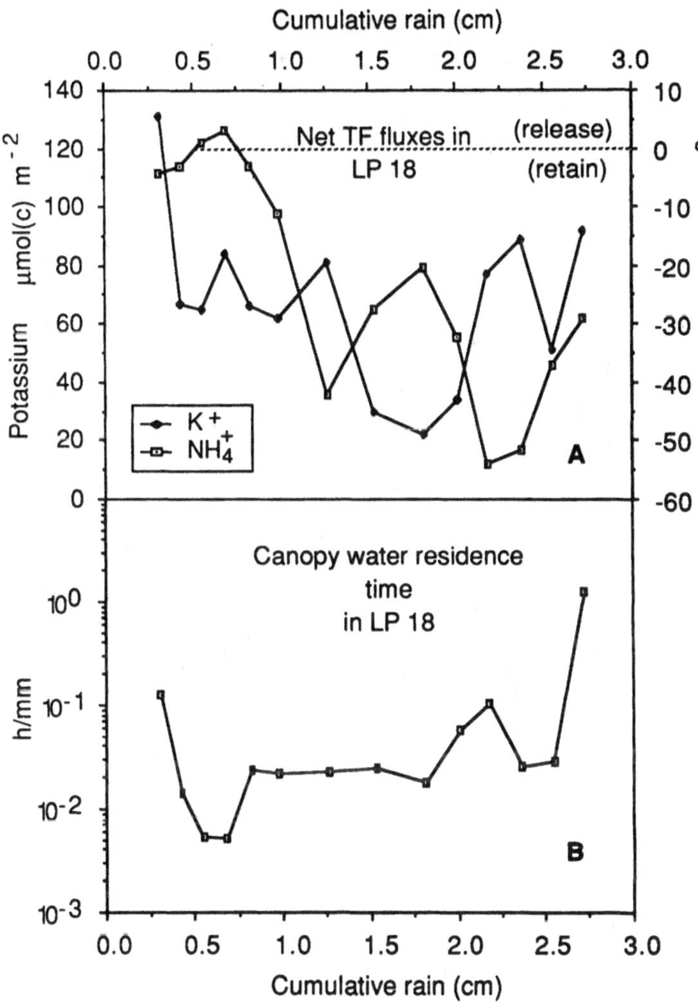

Figure 7–8. (A) Net throughfall (TF) potassium and ammonium (net TF; flux in TF minus flux in rain) in loblolly pine (LP) sequential storm 18 in μmol(c) m^{-2}. Note that ammonium is retained by the canopy (negative net TF) through almost the entire storm and that potassium and ammonium respond inversely during the storm. (B) The inverse of rain rate (i.e., h/mm) plotted on a logarithmic scale. This represents the residence time of water in the canopy during loblolly pine (LP) sequential storm 18. Note that increased residence time (in B) corresponds with both potassium release and ammonium retention (in A).

based on those patterns. Quantitative throughfall chemistry models serve several useful purposes: They link interacting processes in a common framework; they generate testable predictions over a range of time scales and input conditions by simulation modeling; and they help to illuminate research needs.

To date, few predictive canopy processing models have been presented. Luxmoore and others (1978) modeled passive diffusion of ions out of a canopy and

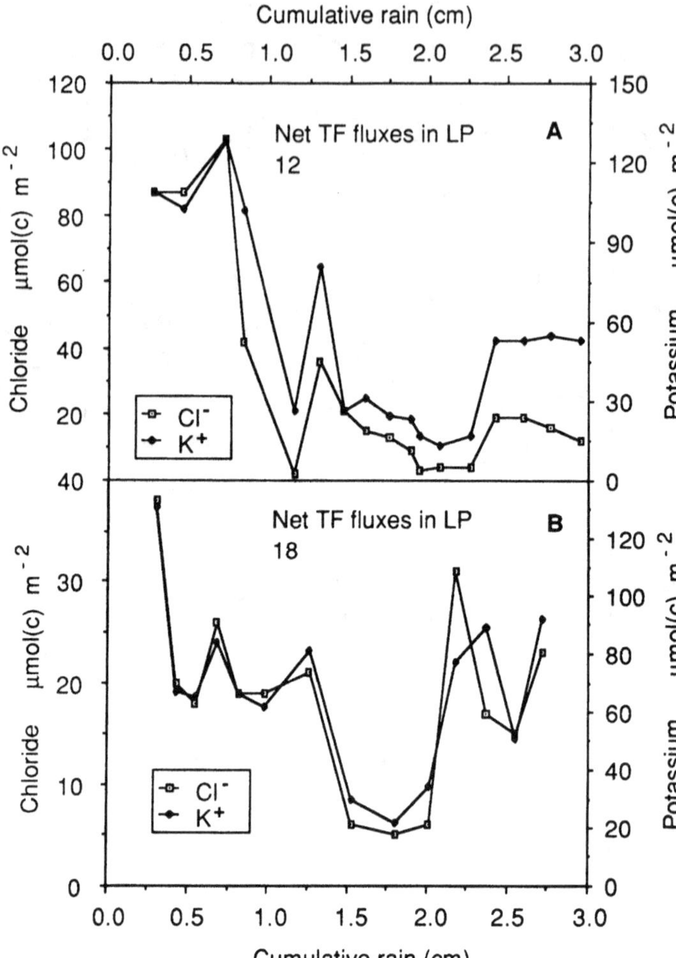

Figure 7–9. (A) Net throughfall (net TF; flux in TF minus flux in rain) potassium and chloride during loblolly pine (LP) sequential storm 12 in μmol(c) m^{-2}. Although they are plotted on different scales, these net ionic fluxes track each other very closely. (B) Net TF potassium and chloride fluxes during loblolly pine (LP) sequential storm 18 in μmol(c) m^{-2}. Although they are plotted on different scales, these net ionic fluxes track each other very closely.

treated the entire canopy as a single layer of foliage. In a single-layer model, canopy tissues are exposed only to surface water with the chemistry of incident precipitation. This type of model implicitly ignores dry-deposition washoff and the effects of changing concentration gradients with canopy depth. Lovett and others (1984) used results from foliage- and branch-washing experiments (Reiners and Olson, 1984) to predict TF chemistry to potassium and ammonium with a multilayer diffusion model that included both diffusion and intrastorm apoplastic recharge. Olson and others (1985) used time-course studies of TF composition

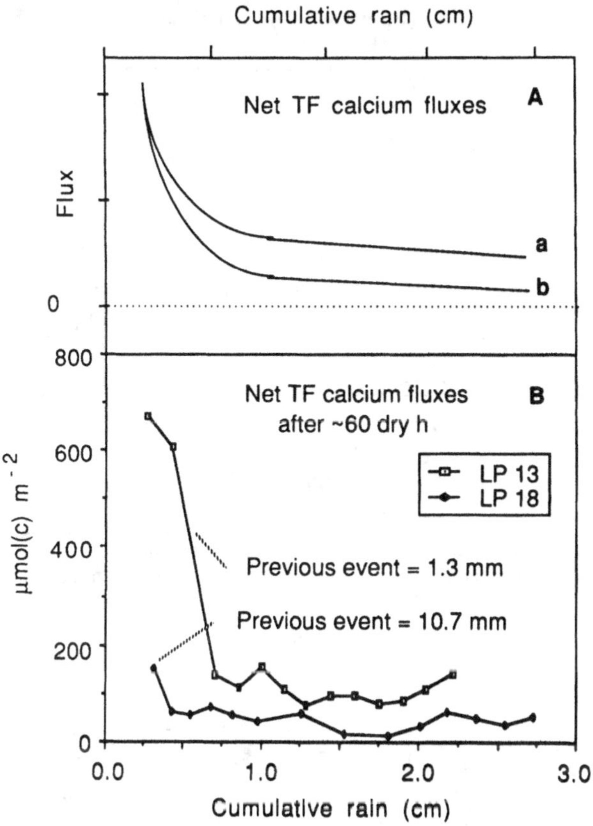

Figure 7–10. (A) Hypothetical pattern of net throughfall (TF) calcium flux (net TF; flux in TF minus flux in rain) during storms with lower (curve a) and higher (curve b) rainfall activity. Note that the flux remains positive following a rapid decline. (B) Net TF calcium flux in µmol(c) m^{-2} during two loblolly pine (LP) storms that differ in the depth of the previous storm. Note that even though the antecedent dry periods were of comparable lengths, the initial release of calcium was much greater from the event that was preceded by a smaller storm.

during storms to show empirically distinct behaviors of groups of ions, and Schaefer and others (1988) developed a series of linear regressions between TF chemistry and storm depth, intensity, chemistry, and acidity.

In this section, we propose a framework for such a quantitative model of chemical fluxes, describe its operation, and discuss the information required to parameterize it.

2. Operation

Proposed model operation is shown diagrammatically in Figure 7–11B. During dry interstorm conditions, canopy surfaces accumulate dry deposition, based on air concentrations and specific deposition velocities. Canopy apoplastic and

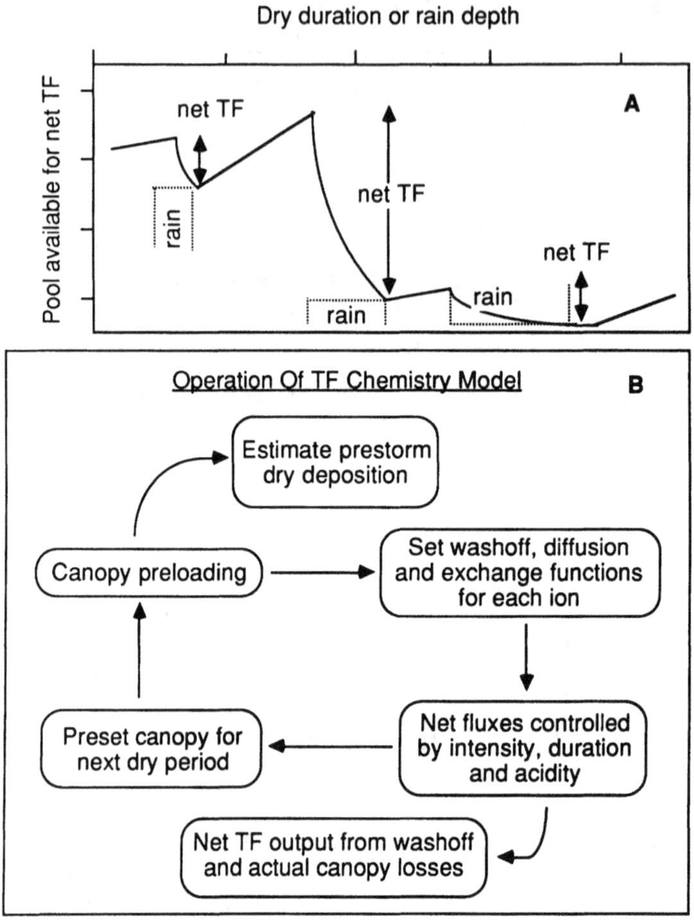

Figure 7–11. (A) Hypothetical pattern of net throughfall (TF) flux (net TF; flux in TF minus flux in rain) for an ion lost from the canopy. The horizontal axis represents interstorm dry hours or rain (in centimeters). The vertical axis represents the ion pool available for release to net TF (both from dry deposition and canopy sources). This ion pool is depleted (through the release of net TF) during each period of rain. Net TF released in each storm is a function both of storm characteristics (i.e., depth and intensity) and of the available ion pool. For ions retained by the canopy the curve would be inverted. (B) Schematic framework for the TF chemistry model as described in the text. Between storms, canopy surfaces accumulate dry deposition, and apoplastic and exchangeable cation pools are recharged. These factors set the net TF curve for each ion. During the storm, intensity, duration, and acidity of the incident precipitation set net TF rates for specific ions. Net TF is released, and canopy apoplastic pools are depleted. The TF water left in the canopy evaporates, and the ions left behind combine with dry deposition for preloading prior to the next storm.

exchangeable cation pools (depleted during the previous rain) are recharged, based on mass flow on the transpiration stream. The combination of these factors sets the height of a "decay function" for each ion. This function consists of a dry deposition curve and a canopy sources (diffusion and ion exchange) curve, in different proportions for each ion. During the storm, flux rates are calculated on ion-specific washoff, cation exchange, and diffusion rates based on rain intensity and concentration gradients between diffusible pools and the external water (Reiners et al., 1986a; Schaefer et al., 1988). The intensity, duration, and acidity of the incident precipitation may all affect net TF fluxes. As the storm progresses, dry deposition is washed off and canopy apoplastic pools are depleted (for ions with positive net TF fluxes), or retention capacity is saturated (for ions with negative net TF fluxes), resulting in exponential decay curves. After the storm ends, TF water left in the canopy (equal to the interception loss) is evaporated, and the ions left behind contribute to the surface deposit. The canopy preloading proceeds during the dry period.

3. Parameterization and Testing

The first need is to quantify the washoff of particles and vapors from dry deposition. Dry deposition measurements are nearing the point where realistic estimates for the amount of material available for canopy washoff can be provided for individual rain events (Lovett and Lindberg, 1984; Hicks et al., 1987). For sites receiving substantial anthropogenic deposition, this may constitute the major portion of the N and S compounds in net TF (Lindberg et al., 1986).

We do not yet know how much cation exchange occurs in forest canopies and to what extent this process differs between coniferous and deciduous canopies. Experimental manipulations on intact foliage are needed. Foliar rinsing with acidified and nonacidified water is underway to determine acidic effects on diffusion and cation exchange (Schaefer and Lindberg, unpublished results). Cations on fixed exchange sites can potentially be quantified by displacement with dilute NH_4Cl solutions. Foliage rinsing in the presence of ion exchange resin beads has been suggested as another means to quantify foliar cation exchange capacity (D. Binkley, personal communication).

The canopy "pool" of ions available for loss by diffusion is certainly much less than the total foliar pools, but the size of this diffusible pool is not well known. Indirect determinations from concentration decay curves are obtained in rinsing experiments (Lindberg and Lovett, 1983; Reiners and Olson, 1984). This diffusible pool may be measured directly by immersing tissue into a solution of a substance not normally found in the plant and determining the solution concentration before immersion and after equilibration. This would be analogous to "free-space" determinations in roots (Epstein, 1972) or stems (Cosgrove and Cleland, 1983). Chemical removal of the external wax barrier may be required to make this feasible. The dynamics of this pool has also been approached indirectly with considerations of "recharge time" (e.g., Reiners and Olson, 1984) between storms, and sequential TF sampling (as presented here) clearly demonstrates that the diffusible pools of certain ions are depleted during storms. Intensity effects for different ions can be determined from a subset of the sequentially sampled storms.

With a parameterized model, the next step is to test the outputs with whole-event or sequential TF chemistry at BF and LP (with samples not used for parameterization). For some other sites, measurement of atmospheric deposition and of apoplastic and exchangeable ion pools (or at least reasonable estimates) are available. For these sites a modified version of this model may be directly testable. For other sites, spatial and temporal simplifications would be required, but a simplified version could still be used to link interacting processes, generate testable predictions for various input conditions, and suggest which of the many unknowns may be most profitably pursued.

B. Throughfall Processes in an Ecosystem Perspective

We conceive three general research goals as vital steps to understanding both how canopies influence TF fluxes and the significance of these fluxes to forest ecosystems. The first goal in the study of TF processing mechanisms is the separation of internal and external sources of chemicals in TF. This will help to clarify wet and dry deposition fluxes and net TF fluxes relative to plant nutritional requirements. The second goal is to determine whether atmospheric inputs of acids or other compounds can alter chemical fluxes from internal canopy pools. This goal will have implications for plant nutrition if foliar nutrient deficiencies result from atmospheric inputs. The third goal is to resolve whether such canopy chemical losses significantly alter the physiology of the trees themselves from the standpoint of carbon balance or plant nutrition.

Substantial progress has been made toward the first goal. Comparisons of TF chemistry among environments that differ greatly in anthropogenic deposition of sulfur compounds show striking differences (Parker, 1983). Statistical techniques to separate the washoff of dry deposited compounds from canopy exchange and release have become increasingly sophisticated (Ulrich, 1983; Lakhani and Miller, 1980; Miller and Miller, 1980; Lovett and Lindberg, 1984). As illustrated earlier in this chapter, studies of the chemical dynamics of individual TF events can separate dry deposition washoff from canopy internal sources by release rates in some cases. Direct (Matt et al., 1987) and inferential (Lindberg et al., 1986; Hicks et al., 1987; Garten et al., 1988) methods to determine dry deposition have become practical only recently. We expect further progress toward this first goal to be made by the application of TF chemistry models to forest canopies in environments with known differences in anthropogenic deposition.

Observations of TF chemistry in environments known to be affected by anthropogenic inputs of acids and oxidants (Rehfuess et al., 1982; Prinz et al., 1982, 1984; Ulrich, 1983; Lovett et al., 1985; Joslin et al., 1988) have been the source of inferences concerning the second goal. Additional progress has resulted from experimental manipulations of canopy components (Reiners and Olson, 1984; Parker et al., 1987) and of intact trees (as discussed in section II). The physiological processes underlying acidic and oxidant effects on canopy ionic fluxes are not yet clear. Foliar chemical losses resulting from these physiological processes currently appear to be minor, except in environments where other stresses on the forest (e.g., climate, disease, or soil chemistry) also appear to be important. Accomplishing the second goal, therefore, involves quantifying canopy chemical losses in environments where forests are known to be stressed. The difficulties here are perhaps symptomatic of those faced by acidic deposition research in general: the separation of many interacting effects.

The third goal concerns whether these TF processes affect plant physiology. There are several ways that this might occur. As previously noted, a substantial amount of inorganic nitrogen deposition may be retained by the canopy and not appear in TF at all (Olson et al., 1981; Lindberg and Lovett, 1986). It is important to realize that because of canopy uptake nitrogen fluxes per se in TF cannot quantify atmospheric deposition. This "fertilization" may have adverse effects on foliar physiology (Freidland et al., 1985; Nihlgard, 1985). That organic nitrogen compounds are released from canopies is also known (Carlisle et al., 1967; Carroll, 1980; Cronan, 1980). In the BF canopy considered here, releases of other forms of nitrogen in TF appear comparable to inorganic uptake by the canopy (Schaefer et al., unpublished data). In the LP canopy, such nitrogen releases are relatively small (Lindberg et al., 1988a). The effects of these nitrogenous fluxes on canopies and the forests floors on which they deposit are not known as yet.

Whereas attaining the first two goals would quantify the effects of acidic deposition on ionic losses from forest canopies, the third goal is concerned with the physiological effects of these losses. Clearly, if ions lost from foliage are replaced by passive root uptake, then the net effect is only to increase the rate of ionic cycling in TF (Mecklenburg and Tukey, 1964; Tukey and Mecklenburg, 1964). If the supply of these ions in the rooting zone becomes limited or if metabolic energy is expended to reacquire them, however, then physiological effects need to be considered. The metabolic costs of replacing inorganic ions lost in TF have been estimated to be minor (Amthor, 1986; Kennedy, 1987), but these estimates are based on crop plants grown hydroponically. We are not aware of studies concerning root uptake metabolism for trees growing in soils of low base status. Other compounds (such as weak organic acids) are also released from canopies in TF (Hoffman et al., 1980b; Lovett et al., 1985; Hantschel and Klemm, 1987; Schaefer et al., 1989). If excess sulfate deposition is reduced and released from the canopy in the form of H_2S (Garten, 1988), sulfate reduction may represent an additional metabolic drain. An understanding of the metabolic consequences of these losses and of the organic nitrogen compounds mentioned above awaits future research.

Although temperate coniferous forests are the focus of this review, TF processing mechanisms may have important implications for forests of other regions as well. As an example, tropical forests on nutrient-poor soils may be among the most dependent on atmospheric inputs for fertility (Vitousek, 1984), yet these forests have received scant attention concerning anthropogenic deposition effects (Whelan, 1986). Future industrial development in these regions will almost certainly be accompanied by regional increases in anthropogenic emissions. The evidence from temperate forests suggests that initial positive effects from nitrogen and sulfur fertilization may later be counteracted by negative effects of acidity, oxidants, and other potential stresses. An understanding of the influence of anthropogenic deposition on TF processes may help to mitigate these effects as they relate to forest nutrient cycles.

Achieving the first two goals depends on the development of TF chemical processing models based on the release of dry deposited compounds and on diffusion and cation exchange from internal canopy pools. Both the second and third goals depend on controlled experiments into the effects of acids and oxidants on the permeability of cuticular and membrane barriers to these fluxes and the physiological implications of these organic and inorganic fluxes to individual plants and forest ecosystems. Many questions remain, but when these are answered, a clearer understanding of TF processing mechanisms will emerge.

Appendix: Sequential Throughfall Sampling Methods

Balsam Fir Site

The balsam fir (BF) site was located at 1,220 m in the subalpine zone of Mt. Moosilauke, New Hampshire. Rainfall for chemical analysis was collected using three 20-cm-diameter polyethylene funnels mounted above the canopy on a 13-m tower at the study site.

Throughfall (TF) collectors were located in the 10-by-10-m plot used by Olson and others (1981, 1985) for stemflow (SF) and TF chemical flux measurements and by Lovett (1981) for SF/TF hydrological flux measurements. These collectors were placed 1 m above the ground to be above the herb layer and to preclude splash effects from forest floor or bare soil.

The five sampled storms occurred during August 1982 and August 1983 ($n = 10$ TF collectors) and during July 1984 ($n = 15$ TF collectors). All rainfall and TF samples were collected manually within a 15-to-20-minute period at the end of the sampling intervals. The lengths of these sampling intervals were proportional to precipitation rates. For storms with very light rain at onset, the individual collectors were composited when the 60-mL minimum volume for analyses was obtained. At greater input rates, individual samples were collected when each TF collector obtained $\geqslant$60 mL. In practice, the sampling intervals varied in length from 20 minutes to 12 hours. Sample volumes were measured in the field, and a 60-mL subsample of each was returned to the laboratory for chemical analyses. The TF water flux was calculated by dividing funnel collection volume by the summed funnel collection areas. Rainfall flux was measured with the tipping bucket gauge in a nearby clearing. Precipitation from five storms was collected in this fashion, with a total of 25 sampling intervals.

After a storm, samples were transported to the laboratory and brought to room temperature ($\simeq$22°C) within 12 hours. Measurements of pH (by Beckman Altex with separate pH and reference electrodes) and conductivity (by Wheatstone conductivity bridge and 1-cm cell with platinum electrodes) were made on unfiltered samples at that time.

The remainder of each sample was filtered through Whatman No. 1 filter paper. The filtrate was stored at 2°C for 1 to 16 weeks prior to inorganic ion analyses. Ammonium, nitrate, and sulfate were measured by automated colorimetric tests with the Technicon Autoanalyzer II. Potassium was measured by flame photometry.

Loblolly Pine Site

The loblolly pine (LP) site was located in a valley site (elevation 230 m) near Oak Ridge, Tennessee. Rain was collected in a canopy clearing with a HASL-type wet-only collector (Volchok et al., 1974). The TF was similarly collected, with a single collector placed under a particularly dense section of LP canopy. Water was routed from the funnels in both collectors into a computer-controlled carousel-style fraction collector. This device operated under the control of a Rockwell AIM 65 microcomputer, with the sample bottles being advanced in unison when a tipping bucket rain gauge in the clearing indicated that 1.4 mm of rain had fallen.

After a storm, samples were transported to the laboratory and brought to room temperature ($\simeq$22°C) within 24 hours. Measurements of pH (by Beckman pHI 21 with Orion "Ross" combination pH electrode) and conductivity (by YSI 35 conductivity meter and 1-cm cell with platinum electrodes) were made on unfiltered samples at that time.

Ion chromatography was used for the inorganic anions: ammonium by Technicon Autoanalyzer, sodium and potassium by flame photometry, and calcium and magnesium by atomic absorption. Samples were stored at 2°C but not filtered prior to analyses. The eight LP storms presented here were collected from October 1986 through July 1987. In contrast to that at the BF site, the winter is warm at the LP site (soil does not freeze), and seasonal effects on TF chemistry there are not very strong (Schaefer and Lindberg, unpublished results)

The ionic flux in rain and TF are calculated as concentrations multiplied by water depth (expressed as μmol(c) m^{-2}). Net TF fluxes are the differences between TF and rain fluxes.

Acknowledgments

Drs. George E. Taylor, Jr., Charles T. Garten, Stephen E. Herbes, and Steven E. Lindberg provided helpful reviews of the manuscript. Richard Olson and James Owens provided substantial field and laboratory assistance. Research was funded by National Science Foundation Grants DEB-7907346 and BSR-8306228 (at the balsam fir site) and by the Electric Power Research Institute's Integrated Forest Study Program (RP-2621) and the Office of Health and Environmental Research, U.S. Department of Energy, under Contract DE-AC05-84OR21400 with Martin

Marietta Energy Systems, Inc. (at the loblolly pine site). The experimental work was conducted in part on the Oak Ridge National Environmental Research Park, Publication No. 3193, Environmental Sciences Division, Oak Ridge National Laboratory.

References

Abrahamsen, G. 1984. Philos Trans Roy Soc Lond Ser B 305:369–382.

Abrahamsen, G., K. Bjor, R. Hornvedt, and B. Tveite. 1980. *In* D. Drablos and A. Tollan, eds. *Proceedings of the international conference on the ecological impact of acid precipitation,* 58–63. SNSF Project, Swedish National Science Foundation, Oslo, Norway.

Abrahamsen, G., R. Hornvedt, and B. Tveite. 1977. Water Air Soil Pollut 8:57–73.

Altschuller, E. B. 1983. *In The acidic deposition phenomenon and its effects,* Vol. 1: *Atmospheric sciences,* 5.1–5.103. EPA-600/8-83-016A. EPA, Washington, DC.

Amthor, J. S. 1986. New Phytol 102:359–364.

Arens, K. 1934. Jahrb Wiss Bot 80:248–300.

Bache, D. H. 1977. J Appl Ecol 14:881–895.

Baker, W. L. 1972. *Eastern forest insects.* US Forest Service Misc Publ 1175. USDA Forest Service, Washington, DC.

Beauford, W., J Barber, and A. R. Barringer. 1975. Nature 256:35–37.

Beauford, W. J. Barber, and A. R. Barringer. 1977. Science 195:571–573.

Bell, J. N. B. 1986. Experientia 42:363–371.

Bentley, B. L., and E. J. Carpenter. 1984. Oecologia 63:52–56.

Benzing, D. H. 1984. *In* G. Medina, H. A. Mooney, and C. Vasquez-Yanes, eds. *Physiological ecology of plants of the wet tropics,* 155–171. Junk, Dordrecht, Netherlands.

Bowen, H. J. M. 1966. *Trace elements in biochemistry.* Academic Press, London. 241 p.

Cape, J. N. 1983. New Phytol 93:293–299.

Carlisle, A., A. H. F. Brown, and E. J. White. 1967. J Ecol 54:87–98.

Carroll, G. C. 1980. *In* R. H. Waring, ed. *Forests: Fresh perspectives from ecosystems analysis* (OSU Biology Colloquium), 87–107. Oregon State University Press, Corvallis.

Chabot, J., and B. F. Chabot. 1977. Can J Bot 55:1064–1075.

Chamberlain, A. C. 1975. *In* J. L. Monteith, ed. *Vegetation and the atmosphere,* 155–203. Academic Press, London.

Chamberlain, A. C., and P. Little. 1981. *In* J. Grace, E. D. Ford, and P. G. Jarvis, eds. *Plants and their atmospheric environment* (Twenty-first symposium of the British Ecological Society), 147–174. Blackwell Scientific, Oxford, UK.

Clement, C. R., C. H. P. Jones, and M. J. Hopper. 1972. J Appl Ecol 9:249–260.

Cole, D. W., and M. Rapp. 1981. *In* D. E. Reichle, ed. *Dynamic properties of forest ecosystems,* 341–409. Cambridge University Press, New York.

Cosgrove, D. J., and R. E. Cleland. 1983. Plant Physiol 72:326–331.

Cronan, C. S. 1980. Oikos 34:272–281.

Cronan, C. S., and W. A. Reiners. 1983. Oecologia 59:216–223.

Crossley, D. A., Jr., and T. R. Seastedt. 1981. Bull Ecol Soc Am 62:106.

Crozat, G. 1974. Tellus 31:52–57.

Curtin, G. C., H. D. King, and E. L. Mosier. 1974. J Geophys Explor 3:245–263.

Cutler, D. F., K. L. Alvin, and C. E. Price. 1982. *The plant cuticle*. Academic Press, London. 461 p.

Davidson, C. I., Jr., and Y.-L. Wu. 1988. *In* S. E. Lindberg, A. L. Page, and S. A. Norton, eds. *Acidic Precipitation, volume 3. Sources, depositions, and canopy interactions* (Advances in Environmental Science Series). Springer-Verlag, New York. (In press.)

Dingman, S. L. 1981. Hydrol Sci Bull 26:399–413.

Dochinger, L. S. 1980. J Environ Qual 9:265–268.

Droppo, J. G. 1980. *In* D. S. Shriner, C. R. Richmond, and S. E. Lindberg, eds. *Atmospheric sulphur deposition*, 209–221. Ann Arbor Science, Ann Arbor, MI.

Dutkiewicz, V. A., and J. A. Halstead. 1983. Atmos Environ 17:1475–1482.

Dybing, C. D., and H. B. Currier. 1961. Plant Physiol 36:169–174.

Eaton, J. S., G. E. Likens, and F. H. Bormann. 1973. J Ecol 61:495–508.

Eaton, J. S., G. E. Likens, and F. H. Bormann. 1978. Tellus 30:546–551.

Epstein, E. 1972. *Mineral nutrition of plants: Principles and perspectives*. John Wiley and Sons, New York. 412 p.

Esau, C. 1962. *Plant anatomy*, 2d ed. John Wiley and Sons, New York. 767 p.

Evans, L. S. 1982a. Water Air Soil Pollut 18:395–403.

Evans, L. S. 1982b. Environ Exp Bot 22:155–169.

Evans, L. S. 1984. Bot Rev 50:449–490.

Evans, L. S., T. M. Curry, and K. F. Lewin. 1981. New Phytol 88:403–420.

Fairfax, J. A. W., and N. W. Lepp. 1975. Nature 255:324–325.

Farrar, J. F. 1976a. New Phytol 77:93–103.

Farrar, J. F. 1976b. New Phytol 77:105–113.

Farrar, J. F. 1976c. New Phytol 77:127–134.

Ferek, R. J., and A. L. Lazarus. 1983. Atmos Environ 17:1545–1561.

Foster, J. R. 1984. Ph.D. thesis, Dartmouth College, Hanover, NH. 384 p.

Fowler, D. 1980. *In* D. Drablos and A. Tollan, eds. *Ecological impact of acid precipitation*, 22–32. SNSF Project. Swedish National Science Foundation, Oslo, Norway.

Fowler, D. 1981. *In* J. Grace, E. D. Ford, and P. G. Jarvis, eds. *Plants and their atmospheric environment* (Twenty-First Symposium of the British Ecological Society), 139–146. Blackwell Scientific, Oxford, UK.

Fowler, D., J. N. Cape, I. A. Nicholson, W. J. Kinnaird, and I. S. Patterson. 1980. *In* D. Drablos and A. Tollan. eds. *Ecological impact of acid precipitation* (Proceedings of an international conference), 146. SNSF Project. Swedish National Science Foundation, Oslo, Norway.

Freedman, B., and U. Praeger. 1986. Can J For Res 16:854–860.

Friedland, A. J., R. A. Gregory, L. Karenlampi, and A. H. Johnson. 1985. Can J For Res 14:963–965.

Galloway, J. N., and G. G. Parker. 1980. *In* T. C. Hutchinson and H. Havas, eds. *Effects of acid precipitation on terrestrial ecosystems*, 57–68. Plenum Press, New York.

Garland, J. R., and J. R. Branson. 1977. Tellus 29:445–454.

Garraway, M. O., and R. C. Evans. 1984. *Fungal nutrition and physiology*. John Wiley and Sons, New York. 401 p.

Garsed, S. G., and D. J. Read. 1977. New Phytol 99:583–592.

Garten, C. T., Jr. 1988. Oecologia 76:43–50.

Garten, C. T., Jr., E. A. Bondietti, and R. D. Lomax. 1988. Atmos Environ 22: 1425–1432.

Gosz, J. R. 1980. Ecology 61:515–521.

Granat, L. 1983. *In* B. Ulrich and J. Pankrath, eds. *Effects of accumulation of air pollution in forest ecosystems*, 83–89. D. Reidel, Dordrecht, Netherlands.

Grieve, A. M., and M. G. Pitman. 1978. Aust J Plant Physiol 5:397–413.
Haines, B., J. A. Jernstedt, and H. S. Neufeld. 1985. New Phytol 99:407–416.
Haines, B. M. Stefani, and F. Hendrix. 1980. Water Air Soil Pollut 14:403–407.
Hallgren, J. E., and S. A. Fredrikkson. 1982. Plant Physiol 70:456–459.
Hantschel, R., O. Klemm. 1987. Tellus 39B:354–362.
Hicks, B. B., D. D. Baldocchi, T. P. Meyers, R. P. Hosker, Jr., and D. R. Matt. 1987. Water Air Soil Pollut 36:311–330.
Hindawi, I. J., J. A. Rea, and W. L. Griffiths. 1980. Am J Bot 67:168–172.
Hinrichsen, D. 1986. Ambio 15:258–265.
Hoffman, W. A., Jr., S. E. Lindberg, and R. R. Turner. 1980a. J Environ Qual 9:95–100.
Hoffman, W. A., Jr., S. E. Lindberg, and R. R. Turner. 1980b. Environ Sci Technol 14:999–1002.
Holloway, P. J., and E. A. Baker. 1968. Plant Physiol 43:1878–1879.
Hornbeck, J. W., G. E. Likens, and J. S. Eaton. 1977. Water Air Soil Pollut 7:355–365.
Hornvedt, R., G. J. Dollard, and E. Joranger. 1980. *In* D. Drablos and A. Tollan, eds. *Ecological impact of acid precipitation,* 192–193. SNSF Project. Swedish National Science Foundation, Oslo, Norway.
Hosker, R. P., and S. E. Lindberg. 1982. Atmos Environ 16:889–910.
Hunneman, D. H., and G. Eglinton. 1972. Phytochem 11:1989–2001.
Huttunen, S., and S. Soikkeli. 1984. *In* M. J. Koziol and F. R. Whatley, eds. *Gaseous air pollutants and plant metabolism,* 117–127. Butterworths, London.
Jacobson, J. S. 1984. Philos Trans Roy Soc Lond Ser B 305:327–338.
Johnson, D. W. 1984. Biogeochem 1:29–43.
Johnson, D. W., and D. D. Richter. 1984. TAPPI J 67:82–83.
Jordan, C., F. Golley, J. Hall, and J. Hall. 1980. Biotropica 12:61–66.
Joslin, J. D., C. McDuffie, and P. F. Brewer. 1988. Water Air Soil Pollut 39:355–363.
Jyung, W. H., and S. H. Wittwer. 1964. Am J Bot 51:437–444.
Kannan, S. 1980. J Plant Nutr 2:717–735.
Karhu, M., and S. Huttunen. 1986. Water Air Soil Pollut 31:417–423.
Keever, G. J., and J. S. Jacobson. 1983a. J Am Soc Hortic Sci 108:80–83.
Keever, G. J., and J. S. Jacobson. 1983b. Field Crops Res 6:241–250.
Kennedy, I. R. 1987. *Acid soil and acid rain.* John Wiley and Sons, Hertfordshire, UK. 610 p.
Keppel, J. 1967. *Isotopes in plant nutrition and physiology,* 324–346. International Atomic Energy Agency, Vienna, Austria.
Kimmins, J. P. 1972. Oikos 23:226–234.
Kolattakudy, P. E. 1980. Science 208:990–1000.
Kolattakudy, P. E. 1981. Annu Rev Plant Physiol 32:539–567.
Kramer, P. J., and T. T. Kozlowski. 1977. *Physiology of woody plants.* Academic Press, New York. 811 p.
Kreutzer, K., and J. Bittersohl. 1986. Forstwiss Centralbl 105:357–363.
Kylin, A. 1960a. Physiol Plant 13:148–154.
Kylin, A. 1960b. Physiol Plant 13:366–379.
Kylin, A., and B. Hylmo. 1957. Physiol Plant 10:467–484.
Lakhani, K. H., and H. G. Miller. 1980. *In* T. C. Hutchinson, and M. Havas, eds. *Effects of acid precipitation on terrestrial ecosystems,* 161–172. Plenum Press, New York.
Landolt, W., and T. Keller. 1985. Experientia 41:301–310.
Lang, G. E., W. A. Reiners, and R. K. Heier. 1976. Oecologia 25:224–241.
Lauchli, A. 1984. *In* D. A. Kral, ed. *Roots, nutrient and water influx and plant growth* (ASA Special Pub 4), 1–25. American Society of Agronomy, Madison, WI.

Lawrey, J. D. 1984. *The biology of lichenized fungi*. Praeger Publishers, New York. 407 p.
Lawson, D. R., and J. W. Winchester. 1979. J Geophys Res 84(C7):3723–3727.
Lee, J. J., and D. I. Weber. 1982. J Environ Qual 11:57–65.
Lehninger, R. A. 1970. *Biochemistry*. Worth Publishers, New York. 833 p.
Leonardi, S., and W. Fluckinger. 1987. Tree Physiol 3:137–145.
Levi, E. 1970. Physiol Plantarum 23:811–819.
Likens, G. E., F. H. Bormann, R. W. Pierce, J. S. Eaton, and N. M. Johnson. 1977. *Biogeochemistry of a forested ecosystem*. Springer-Verlag, New York. 146 p.
Lindberg, S. E., and R. C. Harriss. 1981. Water Air Soil Pollut 16:13–31.
Lindberg, S. E., and G. M. Lovett. 1983. *In* H. R. Pruppacher, R. G. Semonin, and W. G. N. Slinn, eds. *Precipitation scavenging, dry deposition, and resuspension,* 837–848. Elsevier, New York.
Lindberg, S. E., and G. M. Lovett. 1985. Environ Sci Technol 19:238–244.
Lindberg, S. E., and G. M. Lovett. 1986. Biogeochem 2:137–148.
Lindberg, S. E., G. M. Lovett, D. D. Richter, and D. W. Johnson. 1986. Science 231:141–145.
Lindberg, S. E., G. M. Lovett, D. A. Schaefer, M. Mitchell, D. Cole, W. Swank, N. Foster, and K. Knoerr. 1988a. *In* S. E. Lindberg and D. Johnson, eds. *Third annual progress report on the integrated forest study*. Electric Power Research Institute, Palo Alto, CA.
Lindberg, S. E., D. Silsbee, D. A. Schaefer, J. G. Owens, and W. Petty. 1988b. *In* M. H. Unsworth and D. Fowler, eds. *Proceedings of the NATO workshop on deposition at high-elevation sites*. Edinburgh, Scotland. pp. 321–344.
Lovett, G. M. 1981. Ph.D. thesis. Dartmouth College, Hanover, NH. 225 p.
Lovett, G. M. and S. E. Lindberg. 1984. J Appl Ecol 21:1013–1027.
Lovett, G. M., S. E. Lindberg, D. D. Richter, and D. W. Johnson. 1985. Can J For Res 15:1055–1060.
Lovett, G. M., R. K. Olson, W. A. Reiners, and D. A. Schaefer. 1984. Bull Ecol Soc Am 65:238.
Lovett, G. M., W. A. Reiners, and R. K. Olson. 1982. Science 218:1303–1304.
Luttge, U., and N. Higinbotham. 1979. *Transport in plants*. Springer-Verlag, New York. 468 p.
Luxmoore, R. J., C. L. Begovich, and K. R. Dixon. 1978. Ecol Modell 5:137–171.
Magel, E., and H. Ziegler 1986. Forstwiss Centralbl 105:234–238.
Mahendrappa, M. K. 1983. Can J For Res 13:948–955.
Mahendrappa, M. K., and E. D. Ogden. 1973. Can J For Res 4:1–7.
Martin, J. T., and B. E. Juniper. 1970. *The cuticles of plants*. E. Arnold Publishers Limited, Edinburgh, Scotland. 347 p.
Martineau, R. 1984. *Insects harmful to forest trees*. Multiscience Publications Ltd., Ottawa, Canada. 261 p.
Matt, D. R., R. T. McMillan, J. D. Womack, and B. B. Hicks. 1987. Water Air Soil Pollut 36:331–347.
Mattson, W. J., and N. D. Addy. 1975. Science 190:515–522.
Mayer, R., and B. Ulrich. 1978. Atmos Environ 12:375–377.
Mayer, R., and B. Ulrich. 1980. *In* T. C. Hutchinson and M. Havas, eds. *Effects of acid precipitation on terrestrial ecosystems,* 173–182. Plenum Press, New York.
McFarlane, J. C., and W. L. Berry, 1974. Plant Physiol 53:723–727.
McLaughlin, S. B. 1985. J Air Pollut Control Assoc 35:512–534.
Mecklenburg, R. A., and H. B. Tukey. 1964. Plant Physiol 39:533–536.
Mecklenburg, R. A., H. B. Tukey, Jr, and J. V. Morgan. 1966. Plant Physiol 41:610–613.

Miller, H. G. 1984. Philos Trans Roy Soc Lond Ser B 305:339–352.
Miller, H. G., and J. D. Miller. 1980. *In* D. Drablos and A. Tollan, eds. *Ecological impact of acid precipitation,* 33–40. SNSF Project. Swedish National Science Foundation, Oslo, Norway.
Mollitor, A. W., and D. J. Raynal. 1983. J Air Pollut Control Assoc 33:1032–1036.
Monteith, F. L. 1975. *Vegetation and the atmosphere*. Academic Press, London.
Morgan, J. V. 1963. M. S. thesis. Cornell University, Ithaca, NY.
Munger, J. W., and S. J. Eisenreich. 1983. Environ Sci Technol 17:32a–42a.
Napp-Zinn, K. 1966. *Anatomie des blattes I. blattanatomie der gymnospermen*. Gebruden Borntraeger, Berlin. 369 p.
Nieboer, E. P., P. Lavoie, R. L. P. Sasreville, K. J. Puckett, and D. H. S. Richardson. 1976. Can J Bot 54:720–723.
Nieboer, E. P., D. H. S. Richardson, and F. D. Tomassini. 1978. Bryologist 81:226–246.
Nihlgard, B. 1970. Oikos 21:208–217.
Nihlgard, B. 1985. Ambio 14:2–8.
Nobel, P. S. 1974. *Introduction to biophysical plant physiology*. W. H. Freeman and Company, New York. 488 p.
Norris, R. F., and M. J. Bukovac. 1968. Am J Bot 55:975–983.
Olson, R. K., W. A. Reiners, C. S. Cronan, and G. E. Lang. 1981. Holarctic Ecol 4:291–300.
Olson, R. K., W. A. Reiners, and G. M. Lovett. 1985. Biogeochem 1:361–373.
Orgell, W. H. 1957. Proc Iowa Acad Sci 64:189–198.
Paparozzi, E. T., and H. B. Tukey, Jr. 1983. J Am Soc Hortic Sci 108:890–898.
Parker, G. G. 1983. Adv Ecol Res 13:57–133.
Parker, G. G., G. M. Lovett, and R. Mickler. 1987. Bull Ecol Soc Am 68:383.
Percival, G., and R. N. McDowell. 1981. *In* W. Tanner and F. A. Loewus, eds. *Plant carbohydrates II. Extracellular carbohydrates*. Encyclopedia of plant physiology, new series, (vol. 13B), 277–316. Springer-Verlag, Berlin.
Percy, K. E., and R. T. Riding. 1978. Can J For Res 8:474–477.
Pike, L. H. 1978. Bryologist 81:247–257.
Preece, T. F., and C. H. Dickinson. 1971. *Ecology of leaf-surface microorganisms*. Academic Press, New York. 640 p.
Prinz, B., K. D. Jung, and G. H. M. Krause. 1984. *Forest effects in West Germany* (Symposium on air pollution and the productivity of the forest). Izaak Walton League of America, Arlington, VA.
Prinz, B., G. H. M. Krause, and H. Stratmann. 1982. *Forest damage in the Federal Republic of Germany* (LIS Report No. 28). LIS, Essen, FRG.
Rasmussen, L., and I. Johnsen. 1976. Oikos 27:483–487.
Reed, W. D., and H. B. Tukey, Jr. 1978a. J Am Soc Hortic Sci 103:336–340.
Reed, W. D., and H. B. Tukey, Jr. 1978b. J Am Soc Hortic Sci 103:815–817.
Rehfuess, K. E., C. Bosch, and E. Pfannkuch. 1982. *Nutrient imbalances in coniferous stands in southern Germany* (International workshop on growth disturbances of forest trees, 10–13 October, 1982). I.F.F.R.J., Jvaskyla, Finland.
Reichle, D. E., R. A. Goldstein, R. I. Van Hook, Jr., and G. J. Dodson. 1973. Ecology 54:1076–1084.
Reiners, W. A., D. J. Hollinger, and G. E. Lang. 1984. Arct and Alp Res 16:31–36.
Reiners, W. A., G. M. Lovett, and R. K. Olson. 1986a. *Proceedings of the forest/atmosphere interaction workshop,* 111–146. DOE, Office of Energy Research, Washington, DC.

Reiners, W. A., and R. K. Olson. 1984. Oecologia 63:320–330.
Reiners, W. A., R. K. Olson, L. Howard, and D. A. Schaefer. 1986b. Environ Exp Bot 26:227–231.
Rennenberg, H. 1984. Annu Rev Plant Physiol 135:121–153.
Richter, D. D., D. W. Johnson, and D. E. Todd. 1983. J Environ Qual 12:263–270.
Riding, R. T., and K. E. Percy. 1985. New Phytol 99:555–563.
Rinallo, C., P. Raddi, R. Gellini, and V. DiLeonardo. 1986. Eur J For Pathol 16:440–446.
Ritchie, R. J., and A. W. D. Larkum. 1982. J Exp Bot 33:140–153.
Roberts, J. D., and M. C. Caserio. 1965. *Basic principles of organic chemistry*. W. A. Benjamin, New York. 1315 p.
Roberts, K., M. R. Gay, and G. J. Halls. 1980. Physiol Plant 49:421–424.
Salisbury, F. B., and C. W. Ross. 1978. *Plant physiology,* 2d ed. Wadsworth, Belmont, CA. 436 p.
Schaefer, D. A. 1986. Ph.D. thesis. Dartmouth College, Hanover, NH. 369 p.
Schaefer, D. A., S. E. Lindberg, and W. A. Hoffman, Jr. 1988. Tellus. (In press.)
Schaefer, D. A., W. A. Reiners, and R. K. Olson. 1988. Environ Exp Bot 28:175–189.
Scherbatskoy, T., and R. M. Klein. 1983. J Environ Qual 12:189–193.
Schiff, J. A. 1983. *In* A. Lauchli and R. L. Bieleski, eds. *Inorganic plant nutrition,* vol. 15, 401–421. Springer-Verlag, Berlin.
Schonherr, J. 1976a. Planta 128:113–126.
Schonherr, J. 1976b. Planta 131:159–164.
Schonherr, J., and M. J. Bukovac. 1972. Plant Physiol 49:813–819.
Schonherr, J., and M. J. Bukovac. 1973. Planta 109:73–93.
Schonherr, J., and R. Huber. 1977. Plant Physiol 59:145–150.
Schuepp, P. H. 1982. Boundary Layer Meteorol 24:464–480.
Schutt, P., and E. B. Cowling. 1985. Plant Dis 69:548–558.
Seastedt, T. R., and D. A. Crossley. 1984. Bioscience 34:157–161.
Shriner, D. S. 1977. Water Air Soil Pollut 8:9–14.
Skeffington, R. A., and T. M. Roberts. 1985. Oecologia 65:201–206.
Skiba, U., T. J. Peirson-Smith, and M. S. Cresser. 1986. Environ Pollut Ser B 11:255–270.
Smirnoff, N., P. Todd, and G. R. Stewart. 1984. Annu Bot London 54:363–374.
Spaleny, J. 1977. Plant Soil 48:557–563.
Stenlid, G. 1958. *In* W. Ruhland, ed. *Encyclopedia of plant physiology*. vol. IV, *Mineral nutrition of plants,* 615–637. Springer-Verlag, Berlin.
Tukey, H. B., and J. V. Morgan. 1963. Physiol Plantarum 16:557–564.
Tukey, H. B., Jr. 1966. Bull Torrey Bot Club 93:385–401.
Tukey, H. B., Jr. 1970. Annu Rev Plant Physiol 21:305–324.
Tukey, H. B., Jr. 1971. *In* T. F. Preece and C. H. Dickinson, eds. *Ecology of leaf surface microorganisms,* 67–80. Academic Press, London.
Tukey, H. B., Jr. 1980. *In* T. C. Hutchinson and M. Havas, eds. *Effects of acid precipitation on terrestrial ecosystems,* 141–150. Plenum Press, New York.
Tukey, H. B., Jr., and R. A. Mecklenburg. 1964. Am J Bot 51:737–742.
Tukey, H. B., Jr., H. B. Tukey, and S. H. Wittwer. 1958. Proc Am Soc Hortic Sci 71:496–506.
Ulrich, B. 1983. *In* H. Ulrich and J. Pankranth, eds. *Effects of air pollutants on forest ecosystems,* 33–45. D. Reidel Publishing Company, Dordrecht, Netherlands.
Unsworth, M. H. 1981. *In* J. Grace, E. D. Ford, and P. G. Jarvis, eds. *Plants and their atmospheric environment* (Twenty-first symposium of the British Ecological Society), 263–272. Blackwell Scientific, Oxford, UK.

Vitousek, P. M. 1984. Ecology 65:285–298.

Voegtlin, D. J. 1982. *Invertebrates of the H.J. Andrews Experimental Forest, Western Cascade Mountains, Oregon: A survey of arthropods associated with the canopy of old growth Pseudotsuga menszezii,* Spec. Publ. 4. School of Forestry Oregon State University, Corvallis. 31 p.

Volchok, H. L., L. E. Toonkel, and M. Schonberg. 1974. *Trace metals: Fallout in New York City II.* U.S. Atomic Energy Commission Rep HASL-281. U.S. Atomic Energy Commission, Health and Safety Laboratory, New York.

Wainwright, J. J., and P. J. Beckett. 1975. New Phytol 75:91–98.

Wedding, J. B., R. W. Carlson, J. J. Stukel, and F. A. Bazzazz. 1975. Environ Sci Technol 9:151–153.

Wedding, J. B., R. W. Carlson, J. J. Stukel, and F. A. Bazzazz. 1977. Water Air Soil Pollut 7:545–550.

Whelan, T. 1986. Ambio 15:252.

White, E. J., R. S. Starkey, and M. J. Saunders. 1971. J Appl Ecol 8:743–749.

Wittwer, S. H., and F. G. Teubner. 1959. Annu Rev Plant Physiol 10:13–32.

Wood, T., and F. H. Bormann. 1975. Ambio 4:169–171.

Yamada, Y., M. J. Bukovac, and S. H. Wittwer. 1964a. Plant Physiol 39:28–32.

Yamada, Y., H. P. Rasmussen, M. J. Bukovac, and S. H. Wittwer. 1966. Am J Bot 53:170–172.

Yamada, Y., S. H. Wittwer, and M. J. Bukovac. 1964b. Plant Physiol 39:28–32.

Zamierowski, E. E. 1975. J Ecol 63:679–687.

Abatement of Atmospheric Emissions in North America: Progress to Date and Promise for the Future

E.C. Ellis,* R.E. Erbes,† and J.K. Grott‡

Abstract

Much progress has been made in acidic rain abatement in North America. This chapter focuses on man-made SO_2 and NO_x emissions that contribute to acidic deposition and the potential for atmospheric NH_3 and Ca-containing minerals to neutralize SO_2 and NO_x emissions. A review of U.S. historical trends of SO_2 and NO_x emissions since 1900 and projections of future emissions through the end of this century show emissions of SO_2 decreasing from a peak in 1970 of 29 Tg yr^{-1} to about 26 Tg yr^{-1}, but NO_x emissions continuing an upward trend to about 25 Tg yr^{-1}. In Canada, SO_2 and NO_x emissions are less than 20% of those in the United States, and the trends are similar, with SO_2 showing future decreases and NO_x continuing to increase. Future industry in North America is expected to emit much lower levels of SO_2 and NO_x. Technology is also available to limit NO_x emissions from future motor vehicles. Recent acidic deposition legislation in the U.S. Congress to reduce electric utility and industrial emissions of SO_2 by 9 to 13 Tg yr^{-1} is reviewed. The estimates of the cost to implement the proposals range from $2 billion to $23 billion over a 5-year period. Retrofitting existing utility and industrial boilers for maximum SO_2 and NO_x reduction carries the highest price tag. Several environmental policy options are explored for preventing emission increases and also promoting decreases in future emissions of SO_2 and NO_x in North America. Focus on NO_x emissions may be critical because population growth could cause significant increases of NO_x from motor vehicle use. Environmental policy will need to address this concern.

*Southern California Edison Company, System Planning & Research, P.O. Box 800, Rosemead, CA 91770, USA.

†Geoscience Consultants, Ltd., 500 Copper Avenue NW, Suite 200, Albuquerque, NM 87102, USA.

‡Southern Company Services, P.O. Box 2625, Birmingham, AL 35202, USA.

I. Introduction

Much progress has been made in acidic rain abatement in North America. In this chapter, we examine the progress and focus on man-made emissions of sulfur dioxide (SO_2) and nitrogen oxides (NO_x)* that contribute to acidic deposition. Historical trends of SO_2 and NO_x emissions from 1900 through 1980 with projections to the end of the century are summarized with the associated uncertainties. We also explore the potential for alkaline material such as NH_3 and Ca-containing minerals to neutralize acidic precursors over the continent. Current knowledge on the technological feasibility of further SO_2 and NO_x emission reductions and the associated cost of various technologies, especially retrofitting existing industry, is examined. In the past few years, a number of SO_2 and NO_x emission reduction proposals have been debated in the U.S. Congress. Both industry and government have estimated the costs for many of these proposals. This expense is summarized to gain perspective on the aggregate economic cost of further acidic rain abatement. Finally, we deviate from the pure science of acidic deposition precursors and explore policy development and the related future expectations for reductions in SO_2 and NO_x emissions. This information is organized into six sections:

1. Acidic rain as a political and technical issue
2. Emission patterns
3. Emission reduction approaches for SO_2 and NO_x
4. Forecasting future emission trends
5. Cost of emission reductions
6. Policy options for reducing emissions

II. Acidic Rain as a Political and Technical Issue

Concern for acidic deposition in North America evolved in the early 1970s to become the major environmental issue of the 1980s. As in Europe, early issues of air pollution in North America dealt with the elimination of black smoke from home chimneys and factories in primarily urban or heavily industrialized settings. With the Scandinavian studies of the 1950s and 1960s, increasing evidence pointed to acidic precipitation as a key factor in the deterioration of fisheries in remote lakes. Sulfate and nitrate from air pollution were blamed for the rainfall's acidity (Oden, 1968; Drablos and Tollan, 1980). The pollution in Scandinavia was thought to be from distant sources. In short, the concern for local air quality rapidly evolved into international debates of one country's air pollution affecting other countries downwind (UN, 1972).

The Scandinavian studies sparked parallel efforts by scientists in North America

*In this chapter, NO_x includes both NO and NO_2, but NO_x emission quantities are expressed as NO_2.

(Hutchinson and Havas, 1980). Evidence suggesting increased acidity in the rainfall of eastern North America prompted further study of water quality and terrestrial conditions in parts of Canada and the northeastern United States. By the early 1980s, there was a widespread call for more stringent controls of SO_2 emissions in the eastern United States to prevent further harm to acid-sensitive ecosystems in eastern North America. The Canadians argued that transboundary air pollution from the United States was damaging lakes and forests in their eastern provinces. With only limited North American data available on the cause-and-effect relation of SO_2 deposition and ecosystem distress, the U.S. government expanded research through the National Acid Precipitation Assessment Program (NAPAP) (U.S.–Canada, 1983). Although favoring immediate SO_2 emission controls on industry in the United States, the Canadian government also joined in the expanded research effort.

Intense debate continues about the efficacy of further reductions of emissions beyond those of the U.S. Clean Air Act and its amendments (CAA, 1977). Proposals to reduce SO_2 by 50% in the eastern United States have cost projections of $10 billion annually. These estimated costs are comparable to the total expense of implementing the Clean Air Act and 1977 Amendments (CAA, 1977). In addition to SO_2 controls, later proposals also include NO_x controls, which push the cost as high as $25 billion. At such a high societal cost, there is a need for great certainty in the effectiveness of any emission control strategy adopted and its benefits.

Even after the NAPAP Interim Assessment of 1986 (NAPAP, 1987), we are still not confident in our ability to project reductions in acidic deposition from SO_2 and NO_x emission changes. Research into source–receptor relationships continues with the expectation that the 1990 NAPAP results will provide the necessary confidence to act. With highly uncertain benefits, opinions remain divided on control strategies. Despite many years of national debate within the United States, and of the intensive international negotiations with Canada (Lewis and Davis, 1986), no steps to control emissions have been agreed on.

Because most concern has focused on eastern North America, our discussion highlights this region. Aside from health issues of urban air pollution, Mexico and the Central American countries have voiced little concern for ecological changes due to acidic deposition. The NAPAP research focuses primarily on the eastern United States and eastern Canada with less emphasis on the other half of both countries (U.S. EPA, 1984). However, research is progressing in western North America, especially at regional and state or provincial levels. Although sensitive ecosystems have been identified (Roth et al., 1985), analyses have shown less cause for concern for these regions (Hidy et al., 1986; Young et al., 1988).

III. Emission Patterns

In North America, the primary precursors of acidic deposition are largely emissions of SO_2 and NO_x from human activities. Natural emissions are estimated to be only 6% to 10% of the total emissions of SO_2 and NO_x in the United States

Table 8-1. Major emission source categories in North America.

Source	SO_2	NO_x
Electric utilities	X	X
Industrial boilers and space heaters	X	X
Commercial and residential uses	X	X
Metal smelters (nonferrous)	X	
Transportation		X
All others	X	X

and Canada (U.S. EPA, 1984). Emissions of primary sulfate, HCl, and HF are, on average, more than an order of magnitude less than the emissions of SO_2 and NO_x (U.S. EPA, 1986) and contribute little to acidity in the atmosphere on a regional scale. On a local scale, primary sulfate in particular could be as important a source of S deposition as SO_2, especially in areas where oil is a major combustion fuel. Table 8–1 summarizes the major sources of SO_2 and NO_x emissions from man-made origins. Emissions of SO_2 in the United States are dominated principally by fossil fuel combustion in stationary sources—primarily electric utilities. In Canada, nonferrous metal smelting causes most SO_2 emissions, with a lesser contribution from fuel combustion. Transportation and fossil fuel combustion in industrial or electric utility boilers are the major sources of NO_x in both the United States and Canada. Few data are available to estimate past SO_2 and NO_x emissions in Mexico. In general, NO_x emissions have increased with population growth and urbanization. Industrial development has led to SO_2 emission increases. During the 1980s, the expansion of nonferrous smelting operations in northwestern Mexico has increased SO_2 emissions in the region nearly fivefold (as discussed below).

A. Historical SO_2 and NO_x Emissions

Gschwandtner and others (1985, 1986) compiled an inventory of SO_2 and NO_x emissions for the United States from 1900 to 1980. Emissions were estimated for 5-year intervals for each of the 48 contiguous states. The historical compilation is broken into two periods—1900 to 1945 and 1950 to 1980— reflective of the quality and completeness of information. They compiled historic SO_2 and NO_x emission data primarily by source category and relevant fuel consumption, emission factors, and fuel S content. The fuel types include coal, petroleum distillates, natural gas, gasoline and diesel fuel, and wood. Uncertainty estimates for emissions during the 1950 to 1980 period range from $\pm 15\%$ to $\pm 20\%$. Although many estimates were necessary to compile the earlier 1900-to-1945 period, the assumptions used are reasonable, and the compilation method is consistent with the later period.

In the 1986 National Academy of Sciences report on acidic deposition (NAS, 1986), Husar compiled SO_2 and NO_x emission trends for eastern North America from 1880 through 1980. Husar's trend estimates for both SO_2 and NO_x agree qualitatively with Gschwandtner and others (1985). Both approaches are similar, and the differences between the two trends lie within the range of uncertainty of both analyses.

Figure 8–1 shows a trend plot of total SO_2 emissions for the United States. The century begins with an SO_2 emission level of about 8 Tg yr^{-1}, increasing to 20 Tg yr^{-1} by the mid-1920s, decreasing sharply during the Great Depression, peaking at 24 Tg yr^{-1} from 1940 to 1945 and again around 29 Tg yr^{-1} by 1970, and declining to 25 Tg yr^{-1} by 1980. The overall SO_2 trend is parallel to the trend in coal consumption through 1970. Thereafter, the SO_2 emissions decreased substantially due primarily to the reduced S content of coal in use as well as the increased use of various emission control technologies by industry. Figure 8–1 also depicts the SO_2 emissions by source category. Prior to 1940, the major SO_2 emission source categories were industrial boilers and others, including steel furnaces, coke ovens, smelters, and many smaller sources using coal as fuel. After 1950, the contribution of electric utilities to SO_2 emissions increased sharply to 30% by 1960 and reached 70% by the 1970s.

Annual NO_x emissions have increased steadily from about 3 Tg yr^{-1} in 1900 to about 21 Tg yr^{-1} in 1980 (Figure 8–2). Increased use of natural gas and oil as a major energy source, especially since 1950, has led to this NO_x increase. Transportation, the largest source category, accounted for over a third of NO_x emissions by 1980. In contrast to their dominant contribution to SO_2 emissions, electric utilities contribute about 30% to NO_x emissions. Although metropolitan areas have continued to grow since 1950, urban SO_2 emissions have decreased significantly while urban NO_x emissions have increased. The increase in NO_x

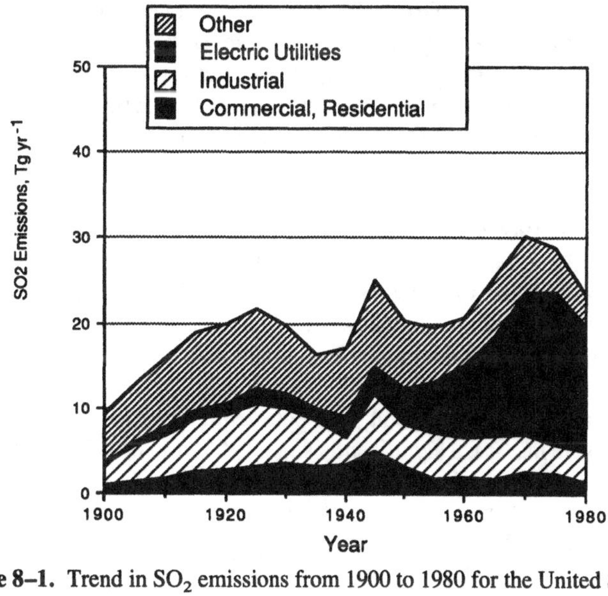

Figure 8–1. Trend in SO_2 emissions from 1900 to 1980 for the United States.

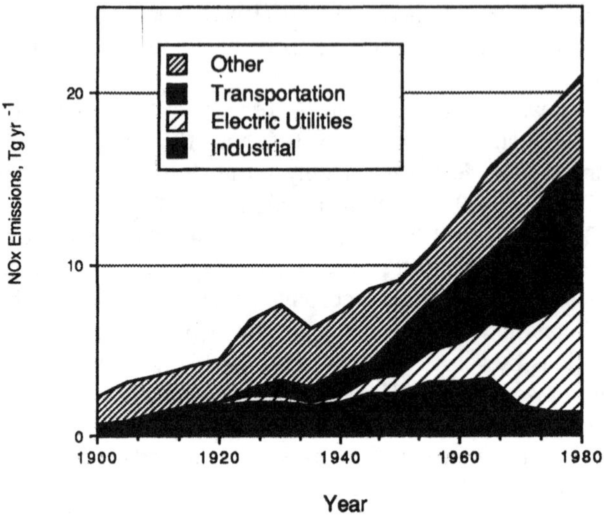

Figure 8–2. Trend in NO_x emissions from 1900 to 1980 for the United States.

emissions can be related to automobile use from population growth and technology changes.

Canadian SO_2 and NO_x emission estimates are listed in Table 8–2 from 1955 through 1980 (U.S.–Canada, 1982; U.S. EPA, 1984; U.S. EPA, 1986). The 1955 and 1965 estimates were based on fuel use records (U.S.–Canada, 1982); actual emissions data were available only for nonferrous smelters and some power plants. Data from 1976 represent a more accurate estimate because they were compiled from a nationwide inventory of SO_2 and NO_x sources (U.S. EPA, 1984). The 1980 Canadian emission estimate should be of comparable quality to the 1980 estimate for the United States (U.S. EPA, 1986) because similar compilation methods were used for both.

Table 8–2 shows that the SO_2 emissions grew from 4.5 Tg yr^{-1} in 1955 to a peak of 6.6 Tg yr^{-1} in 1965 and then decreased to 4.7 Tg yr^{-1} by 1980. The fluctuations in SO_2 emissions were due primarily to changes in production and emission control methods of the smelting industry in eastern Canada. Smelters contributed 60% of the SO_2 emissions in 1980—down from 68% in 1970. The emissions from fossil fuel power plants have grown from a little over 1% of the emissions in 1955 to about 15% in 1980. The "other" SO_2 emissions category includes iron ore processing, natural gas processing, and petroleum refining. Total NO_x emissions for Canada have increased significantly since 1955 from 0.63 Tg yr^{-1} to about 1.7 Tg yr^{-1} in 1980. Increases in transportation and in electric power generation from fossil fuels are the primary reasons for the increases. Motor vehicle NO_x emissions have increased by a factor of three and now generate about 65% of Canadian NO_x emissions.

Table 8-2. Historical emissions of SO_2 and NO_x in Canada in Tg yr^{-1}.

	1955		1965		1976		1980	
Sector	SO_2	NO_x[a]	SO_2	NO_x[a]	SO_2	NO_x[a]	SO_2	NO_x[a]
Nonferrous metal smelters	2.89	—	3.90	—	2.60	—	2.72	—
Electric utilities	0.06	0.01	0.26	0.06	0.61	0.21	0.73	0.20
Other combustion[b]	1.21	0.23	1.13	0.25	0.88	0.44	0.60	0.30
Transportation	0.08	0.32	0.05	0.51	0.08	1.02	0.10	1.10
Others	0.30	0.07	1.25	0.03	1.13	0.19	0.52	0.10
Total	4.54	0.63	6.59	0.85	5.30	1.86	4.67	1.70

From U.S.-Canada (1982); U.S. EPA (1984); U.S. EPA (1986).

[a] NO_x expressed as NO_2.

[b] Includes residential, commercial, industrial, and fuelwood combustion.

B. Emission of Alkaline Material

A significant amount of atmospheric NH_3 is biogenic in origin. Robinson (U.S. EPA, 1984) concluded that soil processes and animal wastes are the dominant natural sources of NH_3. He estimates an emission rate of 1.3 Tg yr^{-1} of NH_3 from biogenic sources for the continental United States. Given the large uncertainty in the biogenic NH_3 data, this rate is certainly a very qualitative estimate. Animal wastes from cattle feedlots and agricultural application of ammonia-based fertilizers are the dominant sources of NH_3 emissions from human activities. Anthropogenic NH_3 emissions in the United States are estimated to be about 3.4 Tg yr^{-1} by Harriss and Michaels (1982). Similarly, the NAPAP 1980 emission inventory (U.S. EPA, 1986) estimates U.S. NH_3 emissions from human activities at 3.8 Tg yr^{-1}. From very sparse data, the NAPAP 1980 inventory lists NH_3 emissions in Canada at only 0.02 Tg yr^{-1} from man-made sources. This is likely a significant underestimate, given the large farming and ranching activities in the country.

Another potential atmospheric neutralizing material is Ca from crustal sources. Emission estimates for Ca from soil dust have recently become available (Gatz et al., 1985; Stensland et al., 1985). The prevailing argument has been that in humid regions of the continent, where soils are predominantly acidic and well covered by vegetation, Ca minerals could play an insignificant role as an atmospheric neutralizer. Gatz and others (1985, 1986) and Stensland and others (1985) estimated Ca dust emissions from unpaved roads at 4.1 Tg yr^{-1} for the United States east of the Mississippi River. Calcium minerals may be even more significant in areas with highly alkaline soils, especially in the arid deserts of the southwestern United States. Hidy and others (1986) and Young and others (1988) have shown that Ca minerals make a significant contribution to the cations in precipitation from the region and even correlate quite well with rain $SO_4^=$ concentrations. One possible explanation for such a correlation is the reaction between $CaCO_3$ from soil dust with acid sulfate in atmospheric aerosols or in precipitation. A recent study in the southwestern United States of arid soils subject to wind erosion supports the importance of Ca in the region's precipitation chemistry (Schlesinger and Peterjohn, 1988). In the United States, both man-made NH_3 emissions (3.4 Tg yr^{-1} and 3.8 Tg yr^{-1}) and biogenic emissions estimated at 1.3 Tg yr^{-1} contribute to the neutralization of SO_2 and NO_x. The NH_3 emissions have the potential to neutralize as much as a fourth of the combined SO_2 and NO_x (Gatz et al., 1986). Better estimates of all natural or man-made alkaline, airborne material are essential to our understanding of the net effect of the increases in SO_2 and NO_x emissions on the deposition of free acidity throughout this century.

C. Geographical Distributions of SO_2, NO_x, and NH_3 Emissions

The geographical distributions of SO_2, NO_x, and NH_3 emissions in the United States and Canada are seen in Figures 8–3, 8–4, and 8–5. Annual emissions are aggregated on a state and provincial political boundary level for the year 1980 (U.S. EPA, 1986). These figures have the shortcoming of exaggerating the emission contribution from very large states or provinces that have a few very large emission sources, such as Ontario and Quebec. A plot of emissions density by uniform area would be preferred, but such information is not readily available. Figures 8–3 and 8–4 appear qualitatively similar to plots for 1982 SO_2 and NO_x emissions densities in the United States by Hidy and others (1986) and Young and others (1988). In 1980, SO_2 emissions in the United States were five times greater than in Canada—24.5 Tg yr^{-1} as compared to 4.5 Tg yr^{-1}. For NO_x emissions, the United States emitted over ten times more than Canada—21 Tg yr^{-1} compared

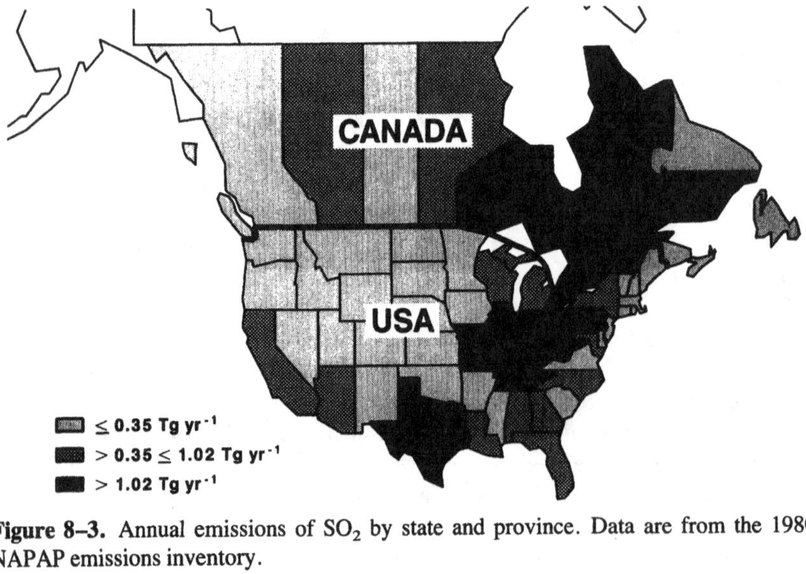

Figure 8–3. Annual emissions of SO_2 by state and province. Data are from the 1980 NAPAP emissions inventory.

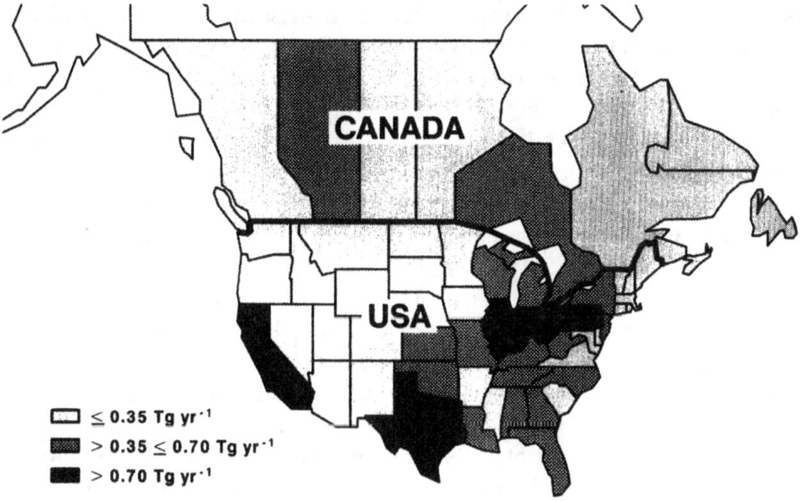

Figure 8–4. Annual emissions of NO_x by state and province. Data are from the 1980 NAPAP emissions inventory.

to 1.7 Tg yr^{-1}. On a per capita basis, however, Canada emitted about twice the SO_2 and comparable levels of NO_x, compared to the United States.

The region east of the Mississippi River in the United States has historically emitted the largest fraction of SO_2 and NO_x for the United States. By 1980, the SO_2 and NO_x emissions in this area represented 74% and 57%, respectively, of

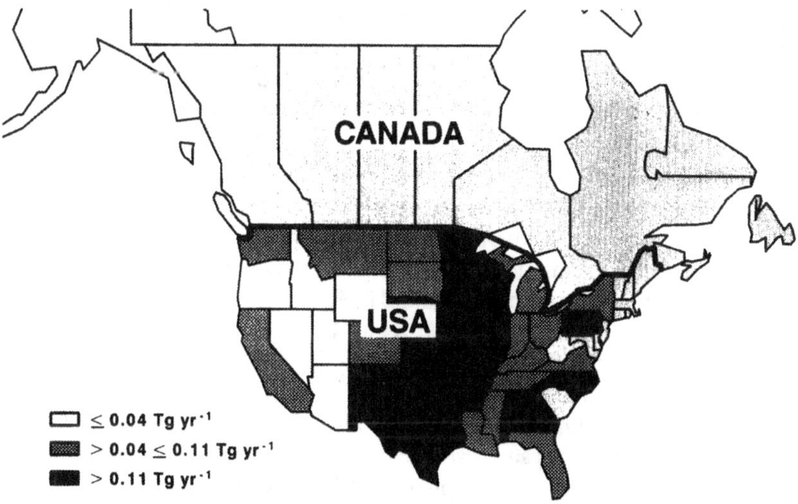

Figure 8–5. Annual emissions of NH_3 by state and province. Data are from the 1980 NAPAP emissions inventory.

total U.S. emissions. Currently, the area of highest SO_2 emissions borders the Ohio River and is bound by Pennsylvania on the east and Missouri on the west. This area emitted almost half of the 1980 U.S. emissions, with 2.4 Tg yr^{-1} in Ohio alone. Figure 8–4 shows that the regions of highest NO_x emissions in the United States are the Midwest, Gulf Coast, and California. Husar (NAS, 1986) noted that U.S. SO_2 emissions north of the Ohio River have increased about 33% since the 1920s. In contrast, emissions south of the Ohio River show a threefold increase since the 1930s. By 1980, the emission densities by unit area were comparable in the two regions for both SO_2 and NO_x emissions.

The far western United States is a minor contributor of SO_2 and NO_x emissions, accounting for no more than 10% of the SO_2 and 15% of the NO_x (Hidy et al., 1986; Young et al., 1988). Electric utilities emit about a fourth of western SO_2; smelters emit about a third. Transportation activities represent close to 50% of NO_x emissions in the region. Recent growth in the far west has been offset by the use of coal with low-S content, a decline in smelter operations, and widespread use of emission control technology. Also, substantial control of NO_x from transportation and electric utilities has occurred, especially in western urban areas such as Los Angeles.

The relative proportions of emissions from various source types are significantly different in the eastern and western United States, and the distribution of emissions is significantly different (Heisler et al., 1985). In the west, the stationary sources are spatially scattered but individually emit a relatively large amount of SO_2. Although there are very large point source emissions in the east also, their spatial separation is not nearly as great. This spatial distribution, which is often ignored in acidic rain control strategy evaluations, leads to enhanced dispersion and reduced ambient concentration levels in areas away from sources.

Historically, the Canadian provinces east of Saskatchewan have emitted the major portion of both SO_2 and NO_x emissions—about 80% and 60%, respectively. Southern Ontario is the area of highest SO_2 emissions in Canada (1.8 Tg yr^{-1}), primarily due to smelting operations. The Great Lakes region is the area of highest NO_x emissions. A shift in industrial activity and population to western Canada reduced the eastern NO_x contribution from 71% in 1955 to 61% by 1975. The SO_2 emissions in Alberta and British Columbia from natural gas processing are expected to increase over the next 10 to 15 years.

The midwestern and southern agricultural regions of the United States are the areas of highest NH_3 emissions. Warm, moist, heavily vegetated areas of North America are expected to have high natural NH_3 emissions. This would most likely be the southern, central, and eastern regions of the United States. Comparing Figure 8–5 with Figure 8–3, and taking into account the potential for biogenic NH_3, one could conclude that in the areas of highest SO_2 emissions, there is the strong potential for acidic neutralization by NH_3. The combined work of Gatz and others (1986) and Munger and Eisenreich (1983) shows that the neutralization of acids by alkaline materials in precipitation (NH_3 as well as crustal elements) is an important factor in the deposition of free acidity in the United States and Canada.

D. Tall Stacks, Distant Transport, and Receptor Impacts

The SO_2 and NO_x trend study by Gschwandtner and others (1986) also includes information for emission release heights. In the early 1900s, SO_2 emissions were from short stacks, but since 1945 the release of SO_2 has been primarily from stacks taller than 75 m (with $\simeq$30% of SO_2 emissions from stacks taller than 150 m by 1980). These stacks are primarily for electric utilities and smelters. A dramatic example of the use of tall stacks is the 381-m smelter chimney at Sudbury, Ontario, which in 1980 emitted $\simeq$0.9 Tg SO_2—about 50% of Ontario's total SO_2 emissions. In contrast, the height of NO_x emission releases has changed little with time, because NO_x is released predominantly at ground level from transportation sources and commercial and residential space heating.

The move to taller stacks for industrial emission releases developed as a benefit for minimizing local pollution. The practice has proved successful in many industrialized areas in both the United States and Canada, bringing them into compliance with ambient air quality standards, especially for SO_2. The trade-off has been that material emitted from tall stacks can travel farther downwind than material released at ground level. However, available analyses are unable to detect increasing ambient pollutant levels in air or precipitation far distant from industrial sources since tall stacks have been introduced (Hidy, 1984). This may result from a rapid mixing of emissions at higher elevations over large geographical regions (NAS, 1983).

Earlier in this chapter, we noted that SO_2 emissions increased throughout this century but peaked in the 1970s, whereas NO_x emissions increased very sharply after 1950 and were trending upward in 1980. It is beyond the scope of this chapter to relate changes in SO_2 or NO_x emissions to changes in ambient pollutant deposition. The critical question is the proportionality of emissions to deposition and any consequent ecological effects. Although the National Academy of Sciences (NAS, 1983) concluded a linear proportionality and suggested that SO_2 emission reductions would result in proportionate reductions in S deposition downwind of source areas, the question remains a topic of active debate (Hidy, 1984; Oppenheimer et al., 1985; Science, 1986). The NAPAP research continues to focus on the proportionality question (NAPAP, 1987).

IV. Emission Reduction Approaches for SO_2 and NO_x

The new source performance standards (NSPS) of the U.S. Clean Air Act (CAA, 1977) restrict the SO_2 emissions that future industry and electric power stations built in the United States can produce, especially when coal is the fuel. To the extent that effective methods for NO_x emission controls are also available, a similar requirement holds for limiting NO_x emissions. Table 8–3 summarizes the status of individual control technologies that are now available or near commercial operation to reduce emissions from industrial and electric utility sources. Below,

Table 8-3. Control technology summary.

Technology	Commercial status	Applicability	SO_2 and NO_x reduction potential	Critical issues	Retrofit cost-effectiveness (\$/1,000 kg SO_2 removed)[a]
Precombustion control					
Physical coal cleaning	Commercial	Most economically applied to coals >1% S	10% to 30% SO_2 reduction typical Up to 60% SO_2 reduction on easy-to-clean coals	SO_2 reduction limited by pyritic S content Benefits of cleaning need quantification Potential heat rate penalty due to increased moisture content and fines loss	350–1,000[b]
Coal switching	Commercial	Cyclone and wet bottom boilers require low ash fusion temperature coal	SO_2 reduction dependent on sulfur content of original and alternative coals	Potential boiler derating due to fouling, slagging, and moisture content Transportation costs F.O.B. mine costs of low sulfur coal Potential ESP upgrade Potential coal pulverizer and coal and ash handling equipment upgrade	450–1000+[c]

Combustion control					
NO_x combustion control	Low to moderate controls commercially demonstrated (low excess air, overfire air, burners out of service, biased firing, low NO_x burners) Advanced low-NO_x burners and improved furnace geometries currently nearing commercial status	Retrofit options not currently available for cyclone or cell burners	Low excess air—0 to 15% Biased firing—5% Burners out of service—20% to 25% Overfire air—15% to 25% Low-NO_x burners—40% to 60%	Several options must be available due to wide variety of boiler designs in operation Retrofit costs and NO_x reductions attainable are highly site specific	100–300[d]
Furnace dry sorbent injection	Not commercially available in USA Full-scale demonstrations underway in Europe, Canada, and USA	Application may be limited by increased boiler solids loading	50% SO_2 removal likely; higher levels possible	Process not fully characterized Potential negative boiler impacts Particulate control upgrading required Potential adverse impact on ash handling and disposal systems	550–900[e]

Table 8-3. *(Continued)*

Technology	Commercial status	Applicability	SO_2 and NO_x reduction potential	Critical issues	Retrofit cost-effectiveness (\$/1,000 kg SO_2 removed)[a]
Combustion control *(continued)*					
Atmospheric fluidized bed combustion	Commercial for industrial applications 20 MW pilot plant in operation 160 MW demo scheduled for start-up in 1988 125 MW conversion—1986 start-up 110 MW new circulating bed boiler—1987 start-up	Wide fuel flexibility Approximately 150 boilers (20,000 MW) candidates for retrofit in USA	>90% SO_2 removal NO_2 levels approximately 0.2 lb/10^6 Btu[f]	Fuel feed system Turndown procedure Tube erosion and materials longevity Flyash recycle Limestone utilization Scale up to 500 MW or larger	550–1,650[g]
Postcombustion control					
Flue gas NO_x control	Commercial in Japan, West Germany, and Austria for SCR	Applicable to all fossil-fuel generation facilities; no experience with >1.2% S coal	80% NO_x reduction for SCR 30% to 60% NO_x reduction for urea or thermal DE NO_x	Catalyst life (SCR) SO_3 formation (SCR) NH_3 slip and by-products Heat rate penalty Reliability	1,500–4,000[h]

Lime and limestone flue gas desfulfurization (FGD)	Commercial—over 60,000 MW in operation or under construction in USA	Applicable to virtually all coals	>90% SO_2 removal	Retrofit costs typically vary from 1.1 to 1.4 times the cost of new units Impact on plant availability Retrofit may be constrained by space	500–1,500[i]
Spray Dry FGD	Commercial 18 systems operating or planned (7,500 MW)	Applicable to virtually all coals	Up to 90% SO_2 removal achievable on low-S coal Up to 90% removal indicated in short-term tests on high-S coal (3.5% S)	Baghouse or ESP improvements may be required for retrofit Process not suitable for gypsum production	850–3,100[j]
Dual alkali FGD (lime and limestone)	Commercial—3 systems operating in USA	Not applicable to low-sulfur coal (<1.5% S) Not applicable to gypsum production	>90% SO_2 removal	Limestone system difficult to control Same space and plant constraints as lime and limestone FGD	600–1,100[i]
Recovery of salable product FGD	MgO (1 plant) and Wellman-Lord (2 plants) commercial in USA	Applicable to virtually all coals Most applicable to sites where by-product disposal is constrained	>90% SO_2 removal	Markets for products must be available Processes are relatively complicated and expensive	850–1,600[i]

Table 8-3. *(Continued)*

Technology	Commercial status	Applicability	SO_2 and NO_x reduction potential	Critical issues	Retrofit cost-effectiveness (\$/1,000 kg SO_2 removed)[a]
Postcombustion control *(continued)*					
Advanced throwaway FGD	Horizontal—4 systems in operation or under construction in USA Chiyoda—2 systems in operation or under construction	All applicable over wide range of coals and operating conditions Additives may be advantageous in retrofits to offset design deficiencies	>90% SO_2 removal	Issues similar to those for lime and limestone FGD Little USA experience for most of these systems	450–1,150[i]
Postfurnace dry sorbent injection	Field tested at 20 MW on low-sulfur coal 100-MW unit operating 500-MW unit planned	Potentially applicable to all coals. However, current sorbents may restrict use to western plants due to cost Pressure-hydrated lime is under testing, which may broaden applicability and removal efficiency	70% to 80% SO_2 removal on low-S coal	Sorbent availability and cost Spent sorbent disposal—Potential leaching of soluble sodium compounds	1,500–2,900[k]

From EPRI (1985).

[a] All costs assume the following unless noted otherwise: End-of-year (EOY) 1982 dollars, 30-year levelized period (1983–2012), 65% capacity factor, 2×500 MW plant, Midwest location, 90% SO_2 removal.

[b] 27% sulfur removal from an Illinois basin coal.

[c] Lower value assumes no derating; upper value assumes major derating. SO_2 removal range 85% to 88%.

[d] Lower value assumes retrofit of a face-fired or horizontally opposed-fired boiler; upper value assumes retrofit of a tangentially fired boiler; costs in \$/1,000 kg of NO_2 removed.

[e] Lower value assumes an easy retrofit using 4% S coal; upper value assumes a difficult retrofit using 2% S coal; 50% SO_2 removal assumed.

[f] $0.2 \text{ lb } (10^6 \text{ Btu})^{-1} = 86 \text{ ng J}^{-1}$.

[g] Capital cost is an estimate based on the retrofit of a 95 MW pulverized coal boiler to AFBC (and upgraded to 125 MW) operating at a 40% capacity factor. Cost effectiveness ranges are based on no capacity credit (high value) and a 30 MW credit for replacing combustion turbine capacity (low value).

[h] Lower value assumes 80% NO_x reduction at 800 ppm inlet NO_x using 4% S coal; upper value assumes 60% NO_x reduction at 200 ppm inlet NO_x using 0.5% S coal. Costs are for new units; retrofit costs could be higher. Costs in \$/1,000 kg of NO_2 removed.

[i] Cost range basis same as footnote e except 90% SO_2 removal assumed.

[j] Lower value assumes an easy retrofit using 4% S coal and 80% SO_2 removal; upper value assumes a difficult retrofit on 0.5% S coal and 70% SO_2 removal.

[k] Lower value assumes a nahcolite system, reagent costs of \$55/1,000 kg, a 1% S coal, and 75% SO_2 removal; upper value assumes reagent (nahcolite) costs of \$165/1,000 kg, 1% S coal, and 75% SO_2 removal. Costs are for new units only; retrofit costs could be higher.

we expand on the prognosis for each technology based on research conducted through NAPAP or by the Electric Power Research Institute (EPRI, 1985; Maulbetsch et al., 1986; Offen et al., 1987; NAPAP, 1987). The costs in Table 8–3 are specific to electric utilities.

A. Fuel Changes (Precombustion Control)

Probably the most critical element in whether coal cleaning and coal switching would be a factor in meeting an SO_2 reduction strategy is the amount of SO_2 reduction required. Coal cleaning provides only a 20% to 30% SO_2 reduction; it is thus likely limited to a supplemental control technology. Where cleaning is ineffective, a moderate level of S removal (e.g., 25% reduction from existing emission levels) would likely be met predominantly by fuel switching because utilities can choose the lowest-cost units to switch. For such units, switching is generally cheaper than scrubbing. The degree of coal switching may be determined ultimately by politics, because coal switching is commonly seen as causing a loss of jobs in high-S coal mining states.

B. Combustion Process Changes

As seen in Table 8–3, there are several approaches to reduce NO_x emissions by controlling the combustion process. These generally involve either lowering the flame temperature or limiting the air content in the flame. Combustion control for NO_x depends on the development of a minimum-cost retrofit that can be installed on a wide variety of unit designs without requiring substantial alterations to existing equipment. The outlook for achieving this goal is optimistic, based on recent research. Full-scale demonstrations are being initiated on a variety of boiler types; commercial availability should follow. Even with the newest technology, however, not all boiler types can be retrofitted. This method of NO_x reduction could reduce emissions from boilers by as much as 50%.

Commercial application of furnace sorbent injection for SO_2 removal from coal requires resolution of the critical issues noted in Table 8–3. Ongoing and planned pilot-scale work should resolve these issues. Because of the differences between development programs in the United States and Europe (boilers, fuels, burners, and sorbent materials), the direct application of European results to the U.S. electric utility industry may be limited.

Atmospheric fluidized bed combustion (AFBC) is now a commercial process for industrial steam generation. Development for the electric utility industry is in progress and proving successful (Tavoulareas, 1987). Optimizing performance and demonstrating the reliability of the process are now necessary for its widespread application. Given the over 90% SO_2 removal and reduced NO_x emissions, AFBC is expected to impact both the replacement and addition of electric generating capacity in the next decade.

C. Flue Gas Changes (Postcombustion Control)

Widespread use of either selective catalytic reduction (SCR), urea injection, or thermal $DeNO_x$ in U.S. boilers, particularly those burning coal, will result only if (1) the critical issues in Table 8–3 are resolved and (2) the required level of NO_x reduction cannot be achieved by less expensive technology. The cost-effectiveness estimates in Table 8–3 are for new utility construction; retrofit costs of SCR could reach \$10,000/1,000 kg NO_2 removed. Potential cost reductions of SCR technology will depend primarily on ultimate catalyst life. Although much less expensive than SCR, the use of urea and thermal $DeNO_x$ centers on their lower level of NO_x removal—55%—compared to greater than 80% for SCR. Future regulations could require the more stringent NO_x control that SCR promises.

A variety of flue gas desulfurization (FGD) technologies have been commercial for many years. These include lime and limestone, spray dry, and dual alkali. Although these technologies are commercial, there are still areas for improvement, especially in operation and maintenance. These technologies remain an expensive option for SO_2 control. Except for lime or limestone scrubbing with gypsum production, recovery product FGD processes appear to have a limited future. The processes are complicated, revenue from sale of the product does not offset process cost, and product marketability can be a major problem. Some of the features of "advanced throwaway" FGD processes could be adopted widely and become standard practice for new utility coal-fired boilers. They are less attractive for retrofit applications. Currently, the postfurnace dry sorbent injection process has no advantages over more tested FGD technologies. With research, the SO_2 removal efficiency and economics of the process are expected to improve over the next several years.

D. Controls Specific to Smelters

In Canada and Mexico, the concern is emissions not from electric utility boilers but from nonferrous smelters. Techniques for controlling SO_2 emissions from smelters are well developed and use standard engineering concepts. Most smelters convert the SO_2 emissions to sulfuric acid. Accordingly, the only issue in Canada and Mexico is an economic one. The Mexican government recently constructed acid plants to control emissions at the two large northern smelters (Nacozari and Cananaea) (U.S. Department of State, 1987). These controls were estimated to cost \$50 million to \$65 million for Nacozari alone (Sun, 1985). Assuming 90% removal of $\simeq 0.6$ Tg yr^{-1} SO_2, the cost-effectiveness for such smelter controls is about \$150 to \$220/1,000 kg SO_2 removed in 1982 dollars.

The reason such controls have only recently been put in place in Canada and Mexico has to do with the past regulatory approach that focused primarily on the nearby health impacts of a point source. This allowed such technologies as intermittent control systems and tall stacks, which do protect the nearby public from health impacts but may not be effective in abating regional acidic deposition.

E. Coal Gasification and Liquefaction

An important component of energy research in the United States has been to develop methods for using coal cleanly. The research has focused on transforming coal to a liquid or gaseous fuel mixture while recovering the S from the coal. The technology is proving successful for generating electric power, as illustrated by the experience of the Cool Water Coal Gasification Program in California, a 120 MW demonstration of the Texaco syngas process (NAPAP, 1987). During its operation from mid-1984 through 1988, Cool Water attained an impressive record. Electric power has been generated from a synthetic gas mixture using a variety of both high-S and low-S coals (Currie and Springer, 1988). Both the SO_2 and NO_x emissions from Cool Water (~34 ng J^{-1}) are less than 15% of the new source performance standards (NSPS) for a coal-fired power plant burning high-S coal.* In comparison to the FGD methods of Table 8–3, the SO_2 emission rate of Cool Water is much less. Although the economics preclude retrofit applications, coal gasification holds the promise for significantly reducing SO_2 and NO_x emissions for the next generation of electric power stations.

F. Mobile Source Controls

The Office of Technology Assessment (OTA) of the U.S. Congress has studied the feasibility of more stringent nationwide motor vehicle emissions controls (OTA, 1988). The OTA conclusions on NO_x emission reductions are summarized in Table 8–4. The mobile source categories included in the table represent over 80% of the total NO_x emissions from highway vehicles. In terms of percent NO_x contribution in the United States, passenger cars and heavy-duty diesel trucks are comparable at about 30% each; light-duty gasoline trucks contribute about half as much at close to 15%; and heavy-duty gasoline trucks contribute about 5%.

The proposed NO_x emission rates in Table 8–4 are considered feasible by the authors of the OTA study, but the automotive manufacturers take strong exception to the technological feasibility of catalytic converters for heavy-duty gasoline trucks (OTA, 1988). The cost-effectiveness calculation for this vehicle type is therefore questionable. Further lowering of the NO_x emission rate for heavy-duty diesel trucks is not considered technologically feasible for the forseeable future. This is unfortunate because this class of vehicles, although not particularly large in numbers, emits close to 30% of the mobile source NO_x emissions.

Inspection and maintenance programs to help keep mobile source emission controls effective once the vehicle is on the road are also evaluated for cost-effectiveness. The range is from \$4,000 to \$6,500/1,000 kg NO_2, depending on whether a centralized or private garage arrangement is used. Studies of the effectiveness of the inspection and maintenance program in California (the state with the most stringent mobile source emission standards) show only about a 5%

*New source performance standards are 301 ng J^{-1} for SO_2 and 215–258 ng J^{-1} for NO_x.

Table 8-4. Mobile source emission control approaches.

Vehicle type	Proposed NO_x emission rate standard ($g\ km^{-1}$)	Technology	Critical issues/status	Cost-effectiveness ($/1,000 kg NO_2 removed)[a]
Passenger	0.25 (0.6)[b]	Catalytic converter; exhaust gas recirculation	Adopted for 1989 model year in CA	1,900
Light-duty truck	0.25 (0.7–1.0)[b]	Same technology as for passenger vehicles	Development work required to extend passenger car technology to heavier LDTs (>1,700 kg)	1,000
Heavy-duty truck (gasoline)	0.9–1.2 (3.0)[b]	Exhaust gas recirculation; develop catalytic converter	Requires engine redesign; poor catalyst life at high exhaust temperatures; not feasible before 1995 at earliest	550
Heavy-duty truck (diesel)	3.0 (3.0)[b]	Exhaust gas recirculation	Reduction of particulate emissions most critical; research into NO_x reduction technology in progress	NA

From OTA (1988).

[a] 1986 dollars.

[b] Current U.S. EPA emission standard from US-EPA (1985).

reduction in NO_x emissions, that is, not cost-effective. Because such programs would also help limit CO and hydrocarbon as well, however, a program's overall value is best judged by its effect on all three pollutants.

G. Summary of Emission Reduction Methods

Current knowledge suggests that operational methods exist or the technology is commercially available to reduce both SO_2 and NO_x emissions from existing North American industry by modest amounts (<40%); for much larger reductions at these facilities, there is much greater uncertainty of costs as well as fewer demonstrated approaches with proven applicability. This is especially true for large-scale industrial NO_x emission reductions. Continued research should provide better choices, including cleaner transportation options and innovative approaches to burn coal cleanly in new industries.

Cost effectiveness is a key factor that industry and government will use to decide which reduction method to employ in order to limit the growth or further reduce SO_2 and NO_x emissions. The cost-effectiveness numbers in Table 8–3 show that either fuel changes or combustion process changes are generally less costly than flue gas removal methods for both SO_2 and NO_x reduction from electric utility and industrial boilers. There is a trade-off, however, in the lower amount of SO_2 and NO_x reduction achievable with the less expensive techniques. At about $150 to $220/1,000 kg SO_2 removed, reduction of SO_2 emissions from smelters is slightly less expensive than the least expensive SO_2-reduction methods for coal-fired boilers. By a factor of five to one, reduction of SO_2 emissions from smelters is more cost-effective than a similar SO_2 reduction using FGD methods on coal-fired boilers. Examining the cost numbers for NO_x reduction methods (Tables 8–3 and 8–4) shows that the combustion-controls on boilers are less expensive than NO_x controls on mobile sources. However, flue gas NO_x controls for new coal-fired boilers are a factor of three or more more costly than catalytic reduction methods for mobile sources. The factor for retrofit of coal-fired boilers with SCR could go higher than ten to one.

V. Forecasting Future Emission Trends

There is considerable interest in knowing not only the emission rate of man-made acidic precursors but also the trend in those emission rates over the next few decades. Estimating emission rates is a very difficult and uncertain process, especially when one attempts to forecast trends over a 15- to 30-year period. Emissions forecasting consists of integrating individual trend forecasts of numerous different factors. First, a realistic economic growth rate is chosen. Next, assumptions are made regarding how each component of the economy (i.e., transportation, utilities, smelters, etc.) will grow, how industrial sectors within industries will change, and how each type of emission source is affected by the projected growth rate. Once the rate of expansion of a specific emission source is known, assumptions are made regarding fuel choice and any emission control equipment. The result is an estimate of emissions from a specific source type within a specific industrial sector. These individual forecasts are compiled by region or state in the United States and/or by source type (e.g., transportation, electric utility, industrial processes, smelters) and then integrated to yield the overall emissions forecast.

Several unique factors apply in forecasting electric utility emissions. These include the choice of energy for future power generation (e.g., nuclear, gas, oil, coal, hydro, geothermal, wind), transfer of electricity between the United States and Canada, elasticity of electricity use (i.e., response of customers to increased electricity costs), and any changes in reserve margins or plant capacity factors and operating efficiencies.

Emissions forecasts may be simplified but made less certain by using "permitted emissions" rather than "actual emissions." A forecast of permitted emissions implies a uniform capacity factor for a plant's operation. Although tracking published requirements, such a forecast has the shortcoming of not necessarily tracking reality because a plant may be emitting significantly more or less than its permitted emissions, depending on its operation in any given time period.

A. SO_2 and NO_x Emission Forecasts for the United States

Several forecasts of total SO_2 and NO_x emissions are listed in Tables 8–5 and 8–6. Hidy and others (1986) compiled a composite emissions forecast by combining actual emissions with data from Gschwandtner and others (1985) and a forecast from the United States–Canada Memorandum of Intent Document (MOI) (U.S.–Canada, 1982). The MOI forecast is also considered separately. A short-term EPA forecast provides a tally for total fuel combustion (U.S. EPA, 1987). These general forecasts are based on actual emissions coupled with assumptions of growth rate and no change in fuel mix or controls in the future, that is, simple extrapolation of historical emission rates.

All total emissions forecasts for SO_2 shown in Table 8–5 are within 15% of each other. This is remarkable consistency, considering the numerous variables involved. In addition, it is clear that no forecast shows a large increase in emissions through the year 2000. United States emissions of SO_2 from all sources were about 24 Tg yr^{-1} per year in 1980 and are forecast to increase slightly to a total of about 26 Tg yr^{-1} in 2000. On a national basis, an 8% to 13% increase in total SO_2 emissions is expected by the end of the century.

Few emissions forecasts for NO_x are available for several reasons. Interest in forecasting NO_x emissions was low until concern was voiced that NO_x may contribute to acidic deposition. For example, up until 1985, the various acidic rain proposals before the U.S. Congress focused solely on SO_2 emission limits and controls. Perhaps the major factor complicating the development of NO_x emission forecasts is that NO_x is emitted from a combination of area sources (e.g., automobiles and space heating) and point sources (e.g., electric utilities and factories). In addition, it is technically more difficult to estimate NO_x emissions

Table 8-5. Summary of SO_2 emissions forecasts from 1980 to 2010 with percent change from 1985 through 2000 and 2005.

Forecast group	SO_2 emissions ($Tg\ yr^{-1}$)							Percent change	
	1980	1985	1990	1995	2000	2005	2010	1985–2000	1985–2005
Hidy[a]									
West USA[b]	2.7	2.4	2.0	2.2	2.4	NA	NA	+0.0	NA
East USA[c]	21.8	21.6	21.5	22.0	23.6	NA	NA	+9.2	NA
Total	24.5	24.0	23.5	24.2	26.0	NA	NA	+8.3	NA
MOI[d,e]									
Utility only	15.8	15.8	15.9	16.0	16.1	NA	NA	+1.9	NA
Total	24.0	23.5	22.9	24.8	26.6	NA	NA	+13.2	NA
EPA[f]									
Fuel combustion	18.7	17.0	NA	NA	NA	NA	NA	NA	NA
Total	23.1	20.7	NA	NA	NA	NA	NA	NA	NA
EPRI utility only[g]									
Base case	15.5	14.9	14.3	13.6	12.3	11.7	10.3	−17.4	−21.5
Worst case	15.5	14.9	14.3	14.7	15.1	15.9	16.8	+1.3	+6.7
NCA utility only[h]									
West USA[b]	0.5	0.6	NA	NA	NA	NA	NA	NA	NA
Total USA	14.6	14.2	NA	NA	NA	NA	NA	NA	NA
WEST Associates utility only[i]									
West USA[b,j]	NA	NA	NA	NA	NA	NA	NA	+0.7	+2.0

[a] From Hidy et al. (1986).

[b] Eleven western states: AZ, CA, CO, ID, MT, NM, NV, OR, UT, WA, and WY.

[c] All other states.

[d] From U.S.-Canada (1982).

[e] Estimates for 1985 and 1995 are linear interpolations between reported values for 1980, 1990, and 2000.

[f] From EPA (1987).

[g] From McGowin et al. (1986).

[h] From NCA (1986).

[i] From WEST (1987).

[j] Relative change in permitted emissions only.

Table 8-6. Summary of NO_x emissions forecasts from 1980 to 2000 with percent changes from 1985 through 2000 and 2005.

Forecast group	NO_x emissions ($Tg\ yr^{-1}$)					Percent change	
	1980	1985	1990	1995	2000	1985–2000	1985–2005
Hidy[a]							
West USA[b]	3.3	3.2	3.2	3.6	4.3	+34.4	NA
East USA[c]	16.3	17.4	18.1	20.0	21.8	+25.3	NA
Total	19.6	20.6	21.3	23.6	26.1	+26.7	NA
MOI[d,e]							
Utility only	5.6	6.4	7.2	8.0	8.7	+35.9	NA
Total	19.2	19.3	19.5	21.7	24.0	+24.3	NA
EPA[f]							
Fuel combustion	10.1	10.2	NA	NA	NA	NA	NA
Total	20.2	20.0	NA	NA	NA	NA	NA
WEST Assoc. utility only[g]							
West USA[b,h]	NA	NA	NA	NA	NA	+11.2	+16.2

[a] From Hidy et al. (1986).

[b] Eleven western states: AZ, CA, CO, ID, MT, NM, NV, OR, UT, WA, and WY.

[c] All other states.

[d] From U.S.–Canada (1982).

[e] Estimates for 1985 and 1995 are linear interpolations between reported values for 1980, 1990, and 2000.

[f] From EPA (1987).

[g] From WEST (1987).

[h] Relative change in permitted emissions only.

than to estimate SO_2 emissions because measurements of NO_x emissions from various sources are rarely available and not easily calculated.

The NO_x emission forecasts shown in Table 8–6 are surprisingly consistent, within about 10%. Emissions of NO_x were about 20 Tg yr^{-1} in 1980 with an expected increase to about 25 Tg yr^{-1} in 2000, from all sources including electric utility emissions. Comparing the percent changes in Table 8–6 for NO_x with those in Table 8–5 for SO_2 shows some interesting differences between SO_2 and NO_x emissions. Whereas only a modest increase in SO_2 emissions is expected by the end of the century, a 25% increase is expected in NO_x emissions nationally. The percentage increase is especially significant in the west, where a 34% increase in total NO_x emissions is forecast. This is due primarily to population growth and resultant mobile source emissions.

B. Forecast of Emissions from Electric Utilities in the United States

Among the more detailed forecasts of utility emissions shown in Tables 8–5 and 8–6 are those of EPRI by McGowin and others (1986), the National Coal Association (NCA, 1986), and Western Energy Supply and Transmission Associates (WEST, 1987). Although the models used are quite different, each forecast used site-specific information provided by the utilities and the affected coal companies. The MOI utility forecast (U.S.–Canada, 1982) uses general information. The EPRI, NCA, and MOI forecasts are based on actual emissions, and the WEST Associates forecast is based on permitted emissions. Only the MOI and WEST forecasts include both SO_2 and NO_x.

Although electric utilities are often assumed to be experiencing large increases in emissions, no forecast shows any significant SO_2 increase. However, the range in forecasted utility SO_2 emissions through 2000 is from −17% to +2%. This wide range is one of the big controversies surrounding the acidic rain issue in the U.S. Congress (NCA, 1986). Both the EPRI base case and the MOI utility forecasts shown in Table 8–5 use historical trends and fuel properties as the starting point for the forecast. However, the MOI forecast uses an extrapolation of emissions based almost entirely on growth rates of electricity needs, not recognizing changes in emissions of SO_2 per unit energy. There is also some disagreement over the growth rate that is used, but that is not the most significant factor.

McGowin and others (1986) looked at the influence of specific factors in detail. The EPRI base case assumed that old power plants would be retired at 30 to 40 years of age (depending on if they were nuclear or fossil fueled) and that new generation would comply with NSPS (which are much more stringent than existing performance standards). It also assumed a representative annual electricity demand growth rate of 2.3%. This case shows significant emission decreases compared to the MOI utility forecast. The EPRI worst case used a 3.3% growth rate and assumed that *all* existing fossil fueled power plants would be kept on line an additional 20 years with no change in emission limits on those plants. This combination of worst-case assumptions, which are unlikely to occur, still shows only a 7% increase in utility emissions over the next 20 years.

The prevention of significant deterioration (PSD) provisions of the 1977 Clean Air Act (CAA, 1977) require new large stationary sources (essentially all fossil-fueled power plants) to install best available control technology (BACT). This means that as new generation is installed, it must include emission controls that are the best available, regardless of the ambient air quality impact. As seen in the previous discussion of the technology for SO_2 and NO_x emission reductions, new power plants will have emission limits much lower than current NSPS regulations. The WEST Associates forecast of western utility emissions incorporates BACT.

Table 8–5 also shows that in the western United States total SO_2 emissions will remain about the same for the next 15 years and electric utility permitted SO_2 emissions will increase about 0.7%. This net stability in total emissions is due to additional FGD systems being installed on power plants in the west and closure or control of smelters in the west, causing emissions reductions that offset the slight increase expected from utility emissions. Although permitted utility SO_2 emissions in the west are expected to increase only about 0.7% by 2000, total electricity generation will increase about 10.5%. In sharp contrast, Yuhnke and Oppenheimer (1984) estimated significant increases in SO_2 emissions from western power plants—between 30% and 70% by 2000. This estimate is unrealistic because it has shortcomings in three areas. It projects the construction of twice the coal-fired power plant capacity as is forecast by western electric utilities, assumes no application of BACT on the new power plants, and ignores the retirement of any older plants through the end of the century.

From Table 8–6, NO_x emissions from utilities are expected to increase as much as 36% by 2000. Western utility NO_x emissions are expected to increase 16% in the next 20 years, whereas utility SO_2 emissions (seen in Table 8–5) will increase only 2%. The explanation for this is twofold. First, some retrofit FGD systems are being installed on western power plants which will reduce SO_2 emissions. Only modest retrofit control systems for NO_x are currently available. Second, and more important, WEST Associates assumed that no major NO_x control systems will be commercially available in the next 20 years. Accordingly, even when BACT is applied, the emission rate for the newest plants is not much lower (about 25%) than some of the existing best-controlled plants.

C. Canadian and Mexican Emission Forecasts

The SO_2 emissions in eastern Canada are forecast to decrease by about 50% from 1980 to 1994, assuming that the planned additional regulatory measures in Canada are implemented (Lewis and Davis, 1986). Table 8–7 shows the emissions forecast by source type. Most reductions occur from reducing emissions from nonferrous smelters; electric utilities do not reduce emissions beyond 1990. The U.S. EPA has estimated that if no new regulatory changes are implemented in Canada, total SO_2 emissions will decrease only about 5% by the year 2000. The EPA estimated NO_x emissions for Canada to increase from 1.7 Tg yr^{-1} in 1980 to about 2.4 Tg yr^{-1} (U.S. EPA, 1984) by 2000. Yuhnke and Oppenheimer (1984) have estimated the smelter emissions from Mexico. They indicate that the

Table 8-7. Canadian SO_2 emissions and forecast changes.

	1980 ($Tg\ yr^{-1}$)	1990 ($Tg\ yr^{-1}$)	1994 ($Tg\ yr^{-1}$)
Nonferrous metal smelters	2.72	2.00	1.15
Electric utilities	0.73	0.45	0.45
Nonutility fuel use	0.60	0.40	0.30
Other	0.62	0.40	0.40
Total	4.67	3.25	2.30

From D. Lewis and W. Davis (1986); projections are for region east of Saskatchewan.

Mexican smelters emitted about 0.11 $Tg\ yr^{-1}$ in 1981 and forecast an increase to 0.57 $Tg\ yr^{-1}$ in 1988 unless additional controls are placed on the two largest smelters located in northern Mexico. A 1987 agreement between the United States and Mexico (U.S. Department of State, 1987) includes the construction of emission-reduction facilities at the smelters. With modern controls, the emissions from the smelters should be reduced to about 0.06 $Tg\ yr^{-1}$ (assuming 90% control) by 1989.

D. Emission Trends Summary

Forecasting trends in atmospheric emissions is difficult and uncertain. Many complex issues must be resolved to predict future emissions for acidic deposition precursors. Earlier in this chapter, we noted that point sources dominate North American emissions of SO_2 (about 90%), but point and area sources contribute about equally to NO_x emissions. Over the next 15 to 20 years, there is expected to be a modest increase (10%) in total SO_2 emissions, essentially no change, or perhaps a decrease in electric utility SO_2 emissions, but a significant increase (25%) in NO_x emissions. However, this increase in NO_x emissions is primarily related to nonpoint sources, especially in the western United States.

VI. Cost of Emission Reductions

The proliferation of acidic rain control proposals in recent years has spawned a large number of control cost estimates. Acidic rain control measures were introduced in the U.S. Congress in 1986 and 1987, as summarized briefly in Table 8–8.

Probably the most often quoted sources of cost estimates are EPA studies conducted by ICF, Inc., using the coal and electric utilities model (MIT, 1980; ICF, 1986a, 1986b, 1987a, 1987b); Congressional Budget Office (CBO) studies performed using the national coal model (CBO, 1986); Office of Technology Assessment (OTA) studies performed using the AIRCOST model (OTA, 1986); and Edison Electric Institute (EEI) analyses developed by Temple, Barker &

Table 8-8. Summary of acidic rain control proposals.

	1986		1987		
Proposal (sponsor)	S. 2813 (Proxmire) H.R. 4567 (Waxman)	S. 2203 (Stafford)	S. 300 (Stafford)	S. 321 (Mitchell)	S. 316 (Proxmire) H.R. 2666 (Sikorsky)
Target SO_2 reductions	9 Tg yr^{-1}	13 Tg yr^{-1}	13 Tg yr^{-1}	11 Tg yr^{-1}	9 Tg yr^{-1}
Compliance method for SO_2	Statewide average emission rate	Individual unit rate	Individual unit rate	Statewide cap/ statewide average rate/individual unit rate	Statewide average emission rate
SO_2 emission rate[a,b]	Phase I: 860 ng J^{-1} Phase II: 516 ng J^{-1}	NSPS at 35 years	387 ng J^{-1}	387 ng J^{-1}	Phase I: 860 ng J^{-1} Phase II: 516 ng J^{-1}
NO_x controls[b]	258 ng J^{-1} annual average	Best available control technology	Best available control technology	258 ng J^{-1} 3 hours average 3.6 Tg yr^{-1} reduction	258 ng J^{-1} annual average
Target date	Phase I: 1992–1993 Phase II: 1997	1991 SO_2 1995 NO_x	1991 SO_2 1995 NO_x	1996	Phase I: 1993 Phase II: 1997–1998
Tax subsidy scheme	H.R. 4567	No	No	No	H.R. 2666

[a] 430 ng J^{-1} = 1 lb (10^6 Btu)$^{-1}$.

[b] New source performance standards for coal combustion are 301 ng J^{-1} for SO_2 and 215–258 ng J^{-1} for NO_x.

Table 8-9. Annual cost estimates for acidic rain control bills in billions of 1985 dollars.

	1986		1987		
	S. 2813 H.R. 4567	S. 2203	S. 300	S. 321	S. 316 H.R. 2666
CBO[a]	\$2.5 (L)[b]	\$8.8 (L)[b]			
ICF[c]	2.7 (L)	22.6 (L)	\$22.6 (L)	\$9.4 (L)[b]	\$3.7 (L)[d]
OTA[e]	3.4–4.3 (L)[b]				
TBS[f,g]	9.2 (E) 5.0 (L)	10.8–15.4 (L)[h]	16.3–20.9 (L)	8.4–9.7 (E) 7.3–8.7 (L)	3.3–5.1 (L)

L = levelized (real dollars); E = early years.

[a] From CBO (1986).

[b] Does not include NO_x controls.

[c] From ICF (1986a, 1986b, 1987a, 1987b).

[d] Trading restricted to utility holding companies.

[e] From OTA (1986).

[f] From TBS (1986a, 1986b, 1987a, 1987b, 1987c).

[g] TBS costs are in 1986 dollars for H.R. 4567 and S. 2203. Other bills in 1987 dollars.

[h] Lower end of range for TBS analysis of S. 2203 represents interpreting NO_x controls as burner replacement rather than retrofitting SCR.

Sloane (TBS) (1986a, 1986b, 1987a, 1987b, 1987c). These studies, summarized in Table 8–9, show acidic rain control costs ranging from roughly \$2.5 billion to \$22.6 billion per year. The differences in estimates can largely be explained by (1) differences in proposals analyzed; (2) control options considered and cost of control alternatives; (3) institutional constraints; and (4) manner of expression of results. These are considered in detail below.

A. Alternative Proposals

Most of the difference between cost projections is usually explained by the fact that they are based on different control proposals. For example, the \$2.5 billion estimate cited earlier was a CBO analysis of H.R. 4567 (the 1986 Waxman bill), whereas the \$22.6 billion was an EPA-sponsored estimate of S. 2203 (the 1986 Stafford bill), a proposal with much more stringent controls.

Close examination of Table 8–8 shows obvious major differences in the targeted SO_2 reductions and the degree of NO_x control required. Another important difference is the method of compliance with SO_2 controls. Some bills place a cap on total statewide emissions, some place a limit on statewide average emission rates, and some place a limit on emissions for each electric utility unit. One bill, S. 321 (the 1987 Mitchell bill), uses all three methods. The bills also vary greatly in their compliance time frames, and only the House bills have attempted to include electricity taxes to subsidize the most expensive SO_2 reductions.

The source of differences in cost estimates is generally the scope of the proposed controls. In particular, the CBO and OTA analyses generally do not include the costs of NO_x control. This is often only a small problem for bills such as H.R. 4567, which have comparatively modest costs for NO_x control (usually estimated to be \$0.5 billion to \$1.0 billion per year). However, for bills such as S. 300 or S. 2203, which are generally interpreted as requiring SCR for NO_x control, this can be a significant factor. In fact, most of the difference between the CBO estimate of \$9 billion annually for S. 2203 and the EPA estimate of \$23 billion annually is the result of the exclusion of costs for NO_x control from the CBO analysis.

B. Cost and Variety of Options Considered

In all the models, SO_2 reductions can be achieved through either fuel switching or technological reduction alternatives. Technological reduction is usually limited to conventional flue gas desulfurization (FGD). There is a reasonable amount of agreement on the cost of FGD, because of the publication of EPA-sponsored analyses estimating retrofit factors for many electric generating units.

There is generally much more variation in the estimated cost of switching to lower-S coals. In the TBS and AIRCOST models, plants are usually limited to switching to 516 ng SO_2 J^{-1} quality coal (or blending such fuel). However, in some recent analyses of the more stringent control bills, TBS has also included switching to ultralow-S coals (344 ng J^{-1}).

In the CBO and ICF approaches, fuel premiums for lower-S fuels are projected by coal industry models. TBS and AIRCOST require that such premiums be estimated separately and entered as an input. Fuel premiums vary from bill to bill, depending on the usefulness of lower-S coals as a compliance strategy. Higher demand for lower-S coals will increase prices for these coals due to the depletion of easily and more cheaply minable reserves. Conversely, the reduced demand for higher-S fuels may depress those prices, contributing to the premium. This premium will also vary geographically, based on the relative proximity to higher- and lower-S coal fields. Proposals requiring a 7 to 9 Tg yr^{-1} reduction will generally have the highest premiums, with the range of projected premiums being around \$9 to \$16 Mg^{-1} of coal.

C. Institutional Constraints

Another difference among control cost studies is the degree to which attempts have been made to incorporate institutional barriers to optimal control strategies. Because of the use of a linear programming framework, the CBO and ICF models assume essentially nationwide optimization of both the coal and electric utility industries. This can cause an underestimate of control costs because such a degree of optimization is unlikely to be found in the real world. However, users of these models argue that more conservative assumptions made elsewhere in the modeling process can offset this problem, at least to some degree.

The CBO, ICF, and OTA models also assume that the electric generators in each state all operate as if there were only one or two utilities in the state. This may seriously distort the economics of compliance as the coordination envisioned may not match the reality of electric utility operations and regulations. However, TBS has generally assumed that each utility will meet control requirements individually, and ICF has also begun to include this type of constraint in some of its modeling efforts. The real institutional challenge will be for the states to develop methods and protocols to accomplish emissions trading for the utilities in their jurisdiction.

Another institutional barrier that may hinder optimal control strategies is the implementation of provisions requiring the use of locally mined coal. States producing large quantities of high-S coals may require the use of FGD, even if fuel switching to lower-S, out-of-state coals would be substantially cheaper. Only TBS has tried to include the possible effects of local coal provisions. The major drawback of trying to include these effects is that determining the degree to which fuel switching might be allowed can become a somewhat arbitrary exercise.

D. Expressing Results of Cost Studies

Results of control cost studies are usually expressed as total nationwide annual costs and as percentage rate increases for utility customers. Total compliance costs are usually expressed in dollars adjusted for inflation. Percent increases are usually shown for individual states or utilities. Sometimes, studies differ in the way these results are expressed.

One important difference is whether results are expressed on a "levelized" basis or on a traditional rate-making basis. The difference is important because, under current rate-making practices in the United States, revenue requirements for capital expenditures (such as FGD equipment) are much higher in the first several years after start-up than in later years. *Levelizing* is an engineering economics method that smooths out uneven annual costs. The drawback of using only levelized results to represent compliance costs is that significant early year rate impacts may be overlooked. The drawback of using only an early year traditional rate-making approach is that the results are valid only for the first few years, as the impact of controls in the later years would be lower.

Another difference in presentation is the method of calculating local rate increases. Often, only statewide average rate increases are shown. However, just as rate increases for a particular reduction proposal vary from state to state, rate increases within a state may vary markedly from one utility to another. For example, under many control proposals, the costs for SO_2 controls in Florida would fall largely on Gulf Power Company and Tampa Electric Company, even though these companies produce only a small part of the total electricity generated in the state. Average results for a state may, therefore, mask more severe effects in certain parts of the state. Recent House bills have included provisions that are intended to encourage the averaging of rate increases caused by acidic rain controls across all utilities in a state, regardless of which companies incurred the costs. The degree to which such a scheme would be successful remains in question.

Customers of lower-emitting utilities may be loath to pay for the pollution controls of higher-emitting utilities, many of whom may still enjoy lower rates than their lower-emitting neighbors.

In recent CBO studies, present values of compliance costs have been calculated, something not shown in other studies. Unfortunately, these have been widely misconstrued as total accumulated compliance costs. This is a serious misinterpretation because the present value is lower than total costs by a factor of between 1.5 and 2 in this case.

E. Summary of Costs of Emission Reduction

Table 8–9 shows the results of the analyses of recent bills. On a levelized basis, bills such as those recently sponsored by Proxmire and Waxman run between \$2.5 billion and \$5 billion annually. Costs in the first years after full compliance would run somewhat higher. Estimates of Stafford's bills, which include SCR for NO_x controls (apparently the current intention), generally run \$15 billion to \$20 billion annually. Early year costs would certainly run considerably higher because of the capital-intensive nature of the control requirements. Estimates of the cost of the Mitchell bill are similar to those estimates of the Stafford bills without the expensive NO_x controls, about \$8 billion to \$10 billion per year.

It can be seen, therefore, that most of the variation in the nationwide estimates of acidic rain control costs is explained by differences in the proposals analyzed, the scope of the analysis, and terms in which the results are presented. The remaining lesser differences are largely the result of the treatment of institutional barriers and differences in control cost assumptions, factors representing real uncertainties about the environment in which such massive emission reductions would have to take place. Research now under way through EPRI and NAPAP support will result in reductions of these uncertainties.

VII. Policy Options for Reducing Emissions

In both Canada and the United States, the primary regulatory principle is to limit emissions of pollutants to levels necessary to attain specified ambient air quality standards. These standards are established at levels that protect the public health and welfare with an adequate margin of safety. Once a standard is set, ambient air quality dispersion modeling is conducted in order to determine the maximum level of emissions that could occur without directly causing a violation of the ambient air quality standards. This emission level then becomes a regulation for the specific source.

Since 1977, the environmental regulatory principle has expanded to embody specific source emission limits for new sources based on technological feasibility of controls. These types of limits, the NSPS and BACT limits, are based on the concept that a source should control to the lowest emission level possible, regardless of the direct ambient air quality impact of the source. This regulatory philosophy slows the amount of air quality degradation that occurs from continued population growth and construction of new industries. The United States has formally codified such a philosophy for stationary sources in the PSD portions of the 1977 Clean Air Act Amendments (CAA, 1977).

A third regulatory philosophy, which has not yet been adopted by either the United States or Canada, is to establish source emission limits based on indirect, regional environmental impacts of the emissions. This approach is theoretically the one needed specifically to address acidic deposition. In this concept, the emissions from a source would be evaluated with respect to the indirect effects of the source at very large distances and through many stages of environmental transformations. However, the scientific knowledge necessary for such an assessment is not available and probably will not be available for quite some time (NAPAP, 1987). Accordingly, this approach is not being widely discussed in policy option evaluations, although parts of it do appear in various legislative and regulatory proposals.

For further acidic rain abatement, two fundamental emissions reduction approaches can be used. One is to require simple across-the-board reductions of either a percentage reduction or a tonnes-of-pollutants reduction. The second is to target specific source categories (e.g., power plants, mobile sources) for either a tonnage or percentage reduction in specific areas. The second approach is politically the most difficult to implement, but the most cost-effective. However, when one source category is responsible for much of the emissions, as is the case with smelters in Canada, this second approach is certainly feasible and is the more desirable from the standpoint of achieving the maximum benefit from the reductions. That approach is, in fact, what Canada is implementing (Lewis and Davis, 1986).

A. Timing of New Legislation and Regulation

There are essentially three options for the timing of new legislation and/or regulations to reduce emissions of acidic deposition precursors. One is to require emission reductions as soon as practicable, based on the assumption that enough is known about acidic deposition impacts to require the expenditure of funds and immediate action. The second option is to wait until the NAPAP research is complete in 1990 and then evaluate the results to identify the most effective and necessary reduction approaches and sources to be targeted. The third option is to establish new regulations and legislation now but build into that legislation midcourse correction opportunities as additional scientific evidence becomes available. This approach is probably the most cost-effective but is politically the most difficult. As a practical matter, it is very unlikely that a midcourse correction would ever be less stringent, regardless of the new scientific evidence.

B. Current and Planned North American Approaches

Canadian air quality regulations are aimed primarily at the direct health effects of emissions. Canada has established stringent ambient air quality guidelines, including a 1 hour SO_2 standard that is about 20% to 40% lower than the U.S.

standards. In the United States, however, the standards *must* be met by all states, whereas in Canada the standards are guidelines that *should* be met by the provinces. This means that Canadian provinces can legally adopt standards that are less stringent than the guidelines.

Quebec and Ontario are responsible for about three-fourths of the Canadian SO_2 emissions and have written emission regulations for specific sources to meet the guidelines (Lewis and Davis, 1986). In addition, a retrofit program is in place to reduce total SO_2 emissions in Canada by about 50% by 1994. According to Lewis and Davis (1986), Canada has no current plan to enact an NSPS, BACT, or PSD program or an indirect acidic deposition impacts regulation. In addition, because Canada does not have an indirect impact assessment program, there is no assurance that the emission limits being proposed will result in acidic rain abatement, although they will certainly result in reductions in airborne S concentrations and consequent deposition rates in areas now impacted by these sources.

The United States has a much more complicated set of regulations to address overall air quality. It combines the direct impact, ambient air quality concept with the NSPS, BACT, and PSD concepts, such that all new industrial sources of emissions must meet the most stringent technically feasible emission limits (CAA, 1977). These limits are often considerably more stringent than those necessary to maintain the ambient air quality standards. As indicated in previous sections, these provisions will hold SO_2 emission levels nearly constant over the next 20 years. However, the United States does not have an effective indirect impact regulatory system, even though the concepts for such a system are embodied in Sections 110, 115, and 126 of the 1977 Clean Air Act Amendments (CAA, 1977). Therefore, it is not certain that the existing set of regulations will alleviate any possible acidic deposition problems. Note also that the PSD, BACT, and NSPS provisions apply only to point sources. Currently, the CAA has less effective provisions to curb emissions related to increased use of automobiles as the population grows.

C. Future Policy Options

In light of the above conceptual alternatives and the data presented earlier in this chapter, the following legislative and regulatory approaches to the acidic deposition issue might be useful. First, there does not seem to be a significant upward trend in future SO_2 emissions in the United States. This is due to the fact that the existing BACT, NSPS, and PSD provisions for point sources will prevent SO_2 emissions from increasing by more than 10% over the next 20 years. To address a specific problem (e.g., Adirondack Lake acidification), it may be desirable to target specific, upwind sources (e.g., the 50 largest SO_2 emitters). This can be done through the Clean Air Act 110 and 126 sections (CAA, 1977) if the scientific evidence exists to show that controlling SO_2 will, in fact, alleviate or at least moderate the problem. Such evidence should be available at the end of NAPAP.

For NO_x, reductions may be warranted to limit future emission increases. However, new controls cannot be applied in a simple across-the-board fashion to point sources alone. Point sources contribute only about half of the NO_x emissions, and with the BACT, NSPS, and PSD provisions, increases of NO_x

emissions from point sources are not forecast to be large. However, the mobile source increase in emissions is forecast to be significant. On a dollars per 1,000 kg NO_x removed, additional regulation of mobile sources by available technologies is more cost-effective by factors of 1.5 to 3 than flue gas changes of NO_x emissions from new point sources (EPRI, 1985; OTA, 1988).

Despite their size, point sources are not the major regulatory problem for either SO_2 or NO_x. Because technology has continuously improved for point sources (the Cool Water plant is a good example of this) so that each new plant emits less than a previous plant on a normalized basis, the BACT provisions of the Clean Air Act ensure that there will not be a significant increase in emissions from individual point sources. However, existing Clean Air Act provisions to limit future mobile sources may be less effective. Therefore, new regulatory and legislative initiatives must also address the potential for mobile source emission increases in North America.

References

Clean Air Act (CAA). 1977. 42 U.S.C. 1857 et seq. as amended by PL 95-95.

Congressional Budget Office (CBO). 1986. *Curbing acid rain: Cost, budget, and coal market effects*. CBO, Washington, DC.

Currie, R. S., and B. K. Springer. 1988. Clean power from coal, paper 88-85.3. Proceedings of the 82d annual meeting, Air Pollution Control Association, Dallas, TX.

Drablos, D., and A. Tollan. 1980. *In Proc. Int. Conf., Sandefjord, Norway*. SNSF Project, Oslo, Norway.

Electric Power Research Institute (EPRI). 1985. *SO_2 and NO_x retrofit control technologies handbook*. EPRI CS-4277-SR. EPRI, Palo Alto, CA. 360 p.

Gatz, D. F., W. R. Barnard, and G. J. Stensland. 1985. Dust from unpaved roads as a source of cations in precipitation, paper 85-6B.6. Proceedings of the 78th annual meeting, Air Pollution Control Association, Pittsburgh, PA.

Gatz, D. F., W. R. Barnard, and G. J. Stensland. 1986. Water Air Soil Pollut 30:245–251.

Gschwandtner, G., K. C. Gschwandtner, and K. Eldridge. 1985. *Historic emissions of sulfur and nitrogen oxides in the United States from 1900 to 1980. Vol. I. Results*. EPA-600/7-85-009a. U.S. EPA, Research Triangle Park, NC.

Gschwandtner, G., K. Gschwandtner, K. Eldridge, C. Mann, and D. Mobley. 1986. J Air Pollut Control Assoc 36:139–149.

Harriss, R. C., and J. T. Michaels. 1982. *In Proceedings second symposium, composition of the nonurban troposphere*, 35–55. American Meteorological Society, Boston, MA.

Heisler, S. L., J. R. Young, F. W. Lurmann, J. F. Collins, and P. A. Hayden. 1985. Summary of a 1982 national emissions inventory for regional model development. Presented at the Second annual acid deposition emission inventory symposium, Charleston, SC.

Hidy, G. M. 1984. J Air Pollut Control Assoc 34:518–531.

Hidy, G. M., D. A. Hansen, R. C. Henry, K. Ganesan, and J. Collins. 1984. J Air Pollut Control Assoc 34:333–357.

Hidy, G. M., J. R. Young, R. W. Brocksen, D. W. Cole, M. M. El-Amamy, and A. L. Page. 1986. *Acid deposition and the West: A scientific assessment*, Doc. P-D 572-503. ERT, Inc., Camarillo, CA. 250 p.

Hutchinson, T., and M. Havas. 1980. *Effects of acid precipitation on terrestrial ecosystems*. Plenum Press, New York.

ICF, Inc. 1986a. *An economic analysis of H.R. 4567*. Washington, DC.

ICF, Inc. 1986b. Summary findings of all scrubber/all NO_x control analysis (memo to EPA). Washington, DC.

ICF, Inc. 1987a. Summary findings of the 12 million ton SO_2 reduction case (memo to EPA). Washington, DC.

ICF, Inc. 1987b. Preliminary analysis of the Proxmire Bill assuming emissions trading restricted within utility holding companies (memo to EPA). Washington, DC.

Lewis, D., and W. Davis. 1986. *Joint report of the special envoys on acid rain* (presented to the President of the United States and the Prime Minister of Canada). U.S. Government Printing Office, Washington, D.C.

McGowin, C. R., D. J. Leadenham, J. B. Parkes, M. J. Miller, and S. C. Fan. 1986. *Sensitivity analysis of electric utility SO_2 emissions in the U.S.* EPRI, Palo Alto, CA. 35 p.

Maulbetsch, J. S., M. W. McElroy, and D. Eskinazi. 1986. J Air Pollut Control Assoc 36:1294–1298.

MIT. Energy Laboratory. 1980. *The ICF, Inc., coal and electric utilities model: An analysis and evaluation*. Report No. MIT-EL-81-015. Cambridge, MA.

Munger, J. W., and S. J. Eisenreich. 1983. Environ Sci Technol 17:32A–42A.

National Academy of Sciences (NAS). 1983. *Acid deposition: Atmospheric processes in eastern North America*. Report of the Committee on Atmospheric Transport and Chemical Transformation in Acid Precipitation. National Academy Press, Washington, DC.

National Academy of Sciences (NAS). 1986. *Acid deposition: Long-term trends*. Report of the Committee on Monitoring and Assessment of Trends in Acid Deposition. National Academy Press, Washington, DC.

National Acid Precipitation Assessment Program (NAPAP). 1987 *Interim assessment report*. NAPAP, Office of the Director of Research, Washington, DC.

National Coal Association (NCA). 1986. *The downward trend in sulfur dioxide emissions at coal-fired electric utilities and 1985 update*. NCA, Washington, D.C. 35 p.

Oden, S. 1968. Swedish Nat Sci Res Council. Ecology Committee Bull 1:68.

Offen, G. R., D. Eskinazi, M. W. McElroy, and J. S. Maulbetsch. 1987. J Air Pollut Control Assoc 37:864–871.

Office of Technology Assessment (OTA). 1986. Analysis of a 1986 acid rain control proposal in response to Congressman Henry Waxman. OTA, Washington, DC.

Office of Technology Assessment (OTA). 1988. *The feasibility and costs of more stringent mobile source emission controls*. Sierra Research Inc., Sacramento, CA. 154 p.

Oppenheimer, M., C. B. Epstein, and R. E. Yuhnke. 1985. Science 229:859.

Roth, P., C. Blanchard, J. Harte, H. Michaels, and M. T. El-Ashry. 1985. *The American west's acid rain test*. Res. Rep. 1. World Resources Institute, Washington, DC.

Schlesinger, W. H., and W. T. Peterjohn. 1988. Soil Sci Soc Am J 52:54–58.

Science. 1986 Letters: Acid deposition in the western United States 233:10–14.

Stensland, G. J., W. R. Barnard, and D. F. Gatz. 1985. The flux of Ca, Mg, K, and Na into the atmosphere for acid rain neutralization. Presented at the Air pollution control association acid rain symposium, Boulder, CO.

Sun, M. 1985. Science 229:949.

Tavoulareas, S. 1987. Atmospheric fluidized combustion technology: 1987 update. Presented at the Fourth annual Pittsburgh coal conference, Pittsburgh, PA.

Temple, Barker & Sloane, Inc. 1986a. *Economic evaluation of H.R. 4567*. TBS, Washington, DC.

Temple, Barker & Sloane, Inc. 1986b. *Economic evaluation of S. 2203*. TBS, Washington, DC.

Temple, Barker & Sloane, Inc. 1987a. *Economic evaluation of S. 321*. TBS, Washington, DC.

Temple, Barker & Sloane, Inc. 1987b. *Economic evaluation of S. 316*. TBS, Washington, DC.

Temple, Barker & Sloane, Inc. 1987c. *Economic evaluation of S. 300*. TBS, Washington, DC.

United Nations Conference on the Human Environment (UN). 1972. Air pollution across national boundaries. The impact on the environment of sulfur in air and precipitation. Sweden's Case Study, Stockholm.

U.S.–Canada Memorandum of Intent on Transboundary Air Pollution. 1982. Work Group 3B, pp. 55–56. U.S. EPA, Washington, DC.

U.S.–Canada Memorandum of Intent on Transboundary Air Pollution. 1983. *Executive summaries—work group reports*. U.S. EPA, Washington, DC.

U.S. Department of State. 1987. Annex to 1987 Agreement of Cooperation between the United States of America and the United Mexican States Regarding Transboundary Air Pollution Caused by Copper Smelters along Their Common Border. U.S. Department of State. Washington, DC.

U.S. Environmental Protection Agency (U.S. EPA). 1984. *The acidic deposition phenomenon and its effects: Critical assessment review papers*. Vol. I. EPA-600/8-83-016 AF. U.S. EPA, Washington, DC.

U.S. Environmental Protection Agency (U.S. EPA). 1985. *Mobile source emissions standard summary*. U.S. EPA, Washington, DC.

U.S. Environmental Protection Agency (U.S. EPA). 1986. *Development of the 1980 NAPAP emission inventory*. EPA-600/7-86/057A. U.S. EPA, Research Triangle Park, NC.

U.S. Environmental Protection Agency (U.S. EPA). 1987. *National air quality and emissions trends report, 1985*. EPA-450/4-87-001. U.S. EPA, Washington, DC.

Western Energy Supply and Transmission Associates (WEST). 1987. *In* Testimony presented before the U.S. Senate Subcommittee on Protection of the Environment, June 17, 1987.

Young, J. R., E. C. Ellis, and G. M. Hidy. 1988. J Environ Qual 17:1–26.

Yuhnke, R. E., and M. Oppenheimer. 1984. Safeguarding acid-sensitive waters in the intermountain west. Paper presented at Conference on acid rain, Gothic, CO.

Index

GPSR Compliance
The European Union's (EU) General Product Safety Regulation (GPSR) is a set of rules that requires consumer products to be safe and our obligations to ensure this.

If you have any concerns about our products, you can contact us on

ProductSafety@springernature.com

In case Publisher is established outside the EU, the EU authorized representative is:

Springer Nature Customer Service Center GmbH
Europaplatz 3
69115 Heidelberg, Germany

www.ingramcontent.com/pod-product-compliance
Lightning Source LLC
LaVergne TN
LVHW052243200926
848481LV00002B/253
9781461287810